Studies in the History of Philosophy of Mind

Volume 28

The aim of the series *Studies in the History of Philosophy of Mind* is to foster historical research into the nature of thinking and the workings of the mind. The volumes address topics of intellectual history that would nowadays fall into different disciplines like philosophy of mind, philosophical psychology, artificial intelligence, cognitive science, etc. The monographs and collections of articles in the series are historically reliable as well as congenial to the contemporary reader. They provide original insights into central contemporary problems by looking at them in historical contexts, addressing issues like consciousness, representation and intentionality, mind and body, the self and the emotions. In this way, the books open up new perspectives for research on these topics.

This book series is indexed in SCOPUS.

For more information, please contact the publishing editor at christi.lue @ springer.com

More information about this series at http://www.springer.com/series/6539

Daniel Heider

Aristotelian Subjectivism: Francisco Suárez's Philosophy of Perception

Springer

Daniel Heider
Faculty of Theology, Department of Philosophy
and Religious Studies
University of South Bohemia in České
Budějovice
Kněžská 8, Czech Republic

ISSN 1573-5834 ISSN 2542-9922 (electronic)
Studies in the History of Philosophy of Mind
ISBN 978-3-030-67343-7 ISBN 978-3-030-67341-3 (eBook)
https://doi.org/10.1007/978-3-030-67341-3

This Springer imprint is published by the registered company Springer Nature Switzerland AG
The registered company address is: Gewerbestrasse 11, 6330 Cham, Switzerland

The book is dedicated to my teachers, Stanislav Sousedík and Karel Floss, who each embody a distinct part in the object-subject polarity of Suárez's philosophy of perception.

Acknowledgements

This book could not have been finished without the generous help of many persons over the last many years. First, I would like to thank Dr. Lukáš Lička for carefully reading through the whole manuscript. His highly constructive remarks and insightful comments were extremely helpful and improved the text in many ways. My big thanks go also to Dr. Claus Asbjørn Andersen for his conceptual recommendations. My thanks are due to two anonymous reviewers (of whom one is not entirely anonymous to me, I have to admit) for their valuable glosses, which helped me to clarify my formulations and motivated me to rearrange some parts of the book. I am also grateful to my two former postdocs, Dr. David González Ginocchio and Dr. Emanuele Lacca, for giving me useful feedback on a number of papers I delivered at many places in Europe and overseas, which served as the preparatory basis for this monograph. I cannot forget to express my sincere thanks to Prof. Therese Cory for her wonderful hospitality during my sabbatical at the University of Notre Dame in the winter semester of 2018/19. During this sabbatical, I was able to work undisturbed on a major part of the book at the excellent Hesburgh Library, which I consider to be one of the best university libraries for philosophy in the world. I am obliged to Doc. Rudolf Svoboda, Dean at the Faculty of Theology, University of South Bohemia, for granting me a free semester which dispensed me of the administrative burden connected with my position as Head of the Department of Philosophy and Religious Studies. My appreciations are due to Mgr. Světla Hanke Jarošová for her (as always) careful proofreading and to Ing. Eva Křížková for (as always) relentlessly supplying me with literature. Last but not least, I must thank my beloved wife Tereza for her endless patience and persistent support.

This book is a result of research funded by the Czech Science Foundation as the project GA ČR 14-37038G "Between Renaissance and Baroque: Philosophy and Knowledge in the Czech Lands within the Wider European Context."

November 2020 Kamenný Újezd

Contents

Abbreviations

CDA	Francisco Suárez. (1978–1991). *Commentaria una cum quaestionibus in libros Aristotelis de anima*, 3 Vols. S. Castellote (Ed.). Madrid.
DA	Aristotle (2000). *On the Soul. Parva Naturalia. On Breath*. Translated by W. S. Hett. Cambridge, MA/London: Harvard University Press.
DM	Francisco Suárez. (1861). *Disputationes metaphysicae*. Vols. 1–2. In *Opera omnia*, Vols. 25–26. Paris: L. Vivès.
GC	Francisco Suárez. (1575). *Thesaurus doctrinae circa libros Aristotelis De Generatione et corruptione*.
Ord.	John Duns Scotus. (1950–2013). *Ordinatio*. In *Opera omnia*, Vols. 1–14. Civitas Vaticana: Typis Vaticanis.
STh.	Thomas Aquinas. (1888–1906). *Summa theologiae*. In *Opera omnia*, Vols. 4–12. Rome: ed. Leonina.
TDA	Francisco Suárez. (1856). *Tractatus De anima*. In *Opera omnia*, Vol. 3. Paris: L. Vivès.

Chapter 1
Introduction

1.1 Philosophy of Perception and Suárez's *Commentary on Aristotle's On the Soul*

The goal of this book is to reconstruct the complex philosophical theory of sense perception of Francisco Suárez (1548–1617), which is presented in his *Commentaria una* cum *quaestionibus in libros Aristotelis De anima*.[1] There is no such comprehensive and complex study available as yet in international scholarship. This is all the more surprising if we consider that Suárez's philosophy in general constitutes one of the crucial milestones in the transition from scholastic thought to early modern philosophy and that Suárez is rightly regarded as one of the most important representatives of the remarkable cultural-ideological phenomenon called Second scholasticism, which is confirmed by the large amount of secondary literature that has been published on his philosophy in the last decades and above all in the last several years.[2]

[1]The Commentary was written in the first half of the 1570s when Suárez taught philosophy in Segovia, his first teaching period after his studies at the University of Salamanca (1566–70). The text was edited in 1619–21 and published posthumously by Baltasar Álvares in Lyon (1621). In the autumn of his life Suárez himself started to edit and revise the juvenile text to compose a volume that was supposed to serve as a fundamental psychological study for his theology, in the same way as *Disputationes metaphysicae* (1597) as an essential metaphysical text stood in the service of theology. However, he only managed to revise the first 12 chapters of the first book before he died. Symbolically, he stopped writing in the middle of his work at the arguments for the immortality of the human soul. For Álvares's note about the genesis of the text inserted in his edition, see Suárez (1856c), 'Intermissi operis ratio, eiusque supplementum ex eodem auctore', 69–70. For a synoptical overview of the genesis of the text, see Castellote in Suárez (1978, xxxix–xvi). For Álvares himself, see Santiago de Carvalho (2019).

[2]For a rich overview of the recent literature and conferences organized on Suárez's philosophy only in the Iberian Peninsula in recent years, see Lázaro Pulido and Zorroza (2018). For an almost exhaustive list of the secondary literature on Suárez's philosophy in general, see Penner (2020).

D. Heider, *Aristotelian Subjectivism: Francisco Suárez's Philosophy of Perception*, Studies in the History of Philosophy of Mind 28,
https://doi.org/10.1007/978-3-030-67341-3_1

One of the perennial problems of a philosophical theory of sense perception and cognition in general is the issue of intentionality.[3] How can we explain that our vision, often regarded by philosophers not excluding Suárez as *the* dominant sensory modality, is *about* this tree or that it is *related to* this tree or that our visual experience *tends* toward this tree that stands in front of us? What perceptual processes and what changes in the eyes have to take place for this intentional act directed to this tree to occur? It is not only the issue of the genesis of these acts but also the structural item that causes problems here. In what kind of relation does our mental state stand to the visual representation of this tree? Is it a relationship of similarity, or only a causal relation? Moreover, our mental events are often related to nonexistent fictitious objects or objects that are apprehended differently from the way they exist. The question is how to explain the intentionality of these operations. Should we say that they refer or must refer directly to some intramental objects with a special intentional being, the being that an extramental object acquires when it is known? Considering the principle of explanatory simplicity, should we not posit these immanent objects in all veridical acts including those related to existent objects? Admittedly, if these questions were answered affirmatively, we would get a representationalism or indirect realism according to which material objects are perceived indirectly by means of perceiving intramental representations, which was a view popular in early modern philosophy. If not, the doctrine of direct or naïve realism according to which external sensible objects are apprehended directly without this mediation, the theory dominant (but not exclusive) in medieval and post-medieval Aristotelian-scholastic philosophy, would be the outcome.

The problem of the intentionality of perceptual acts is also linked to the query about perceptual activity and passivity.[4] Is perception a basically passive process that mirrors sensible objects by receiving their stimuli and likenesses? Or is it rather an active process searching for an object, much like a headlight searching for something in the darkness?[5] If the second were the case, it would be clear that perceptual intentionality would not be reducible to the reception of the stimuli that are emitted from sensible objects. Reception would be only a part of the whole story. The production of perceptual acts would also require an activity or, more precisely, an attention or intention on the side of the percipient. Without attending no act could be produced and the reception of stimuli would be 'dead'. If "mixture theory', that is, the theory that includes both aspects were the case, the question is which phases can be distinguished in the overall genesis of perceptual operation. As is well known, significant problems concern these stimuli. Are these to be conceived as beings that are specifically distinct from the represented qualities, or should it rather be said that these are of the same kind as the represented items? If the first were the case, should

[3]For an introduction to the medieval problems of intentionality constituting the background of Suárez's theory, see Perler (2004, 1–30).

[4]For a representative volume on the issue of activity and passivity in sense perception, see Silva and Yrjönsuuri (2014).

[5]For these metaphors, see Abrams (1953).

we posit such stimuli conceived by scholastics as representational items called intentional species actually in all the external sense modalities, that is, including the senses of touch and taste, which as 'contact senses' seem to operate in immediate contiguity with their sensibles?

An important part of the problem of intentionality is constituted by the topic of consciousness and self-consciousness.[6] The philosophy of perception is expected to answer not only the enquiry into how percipients are related to the extramental world (the objective aspect) but also how they become aware of their own mental states (the subjective aspect). On the one hand, a philosopher can give a purely material and naturalized account describing only the physical and physiological changes caused by a visual scene in the eye, the nerves and the brain. This account, based largely on the third-person perspective, however, has its own pitfall in ignoring the subjective aspect of consciousness, which is a paramount phenomenon of the first-person perspective. On the other hand, we can render a completely phenomenological and, in a way, idealistic explanation of how perception proceeds and is established from the self. From the point of view of scholastic epistemological realism, this exposition, however, will have to face the difficulty that is related to the transition from the nonconscious world to conscious perception. In other words, this theory will bypass this transition since it does not start from the world. This is also why a sound theory of perception will be required to find a balance between these two distinct requirements.[7] Yet, if such a theory were granted, how would we become aware of our vision of a tree? Does our seeing of the tree turn out to be the object of a higher-order perceptual act of reflection? *Or* do we experience this act of seeing non-reflectively, that is, simultaneously or by a same-order act with our vision of a tree?[8]

As accepted by all Aristotelians, perception constitutes the primary source of our knowledge about the external world. If perception did not provide us with reliable information, all post-sensory forms of cognition would turn out to be fallible and unreliable. This is the reason why epistemological questions about the veridicality of our perception are important for epistemology in general. But how certain and reliable are our senses?[9] Can we speak about truth and falsity in perceptual acts, or do these belong only to the second mental operation of the intellect, which is the only power that can generate judgments since it is the only faculty that can compose and

[6]For an overview of the issue of consciousness and self-consciousness at the levels of both perception and intellection in medieval philosophy, see Cory (2021). For a systematic and historical introduction to this topic based on the fundamental distinction between the Aristotelian and the Augustinian model, see Perler and Schierbaum (2014, 11–68).

[7]Concerning this formulation of the philosophical problem of sensation in Suárez, see South (2001a). As regards the identical formulation of the issue in contemporary philosophy of perception, see Fish (2010, 1–3).

[8]For a detailed description of these two kinds of self-awareness in connection with Suárez, see Rode (2015, 13–6).

[9]Concerning the issue of scepticism in medieval philosophy, see Perler (2006) and Lagerlund (2010).

divide subject and predicate? And what are the conditions of veridical perception? Can they be stated? Given both the object-oriented Aristotelian epistemology and the brute fact of illusory perceptions, an Aristotelian is expected to develop an account of the conditions under which perception is veridical, and of those under which it is not. Moving the issue of perceptual scepticism to the level of thought experiments, which due to their theological setting abound in scholasticism, the question can be raised whether God or a wicked demon can intervene in our perception in such a way as to produce in us a visual percept of absent or nonexistent objects. Provided that no logical or metaphysical contradiction is involved in this state of affairs and this scenario is *de absoluta potentia Dei* possible, what ontological assumptions are implied? If this were possible, would we be able to discern veridical perceptions from false ones? If so, what would the criteria of such discernment be?

These questions are intimately connected with the extensive topic of the penetrability of perception by rationality. We not only receive data from the external environment through the external senses but also process this manifold information. Considering the *explanandum* in the form of a phenomenological criterion of perceptual unity (our perception is unified in many ways), what are the cognitive mechanisms by virtue of which we are aware not only of atomic qualities, such as the proper sensibles of colour, sound, odours etc., proper to the particular external senses, but also of a sensible *object* that has all these qualities together with other aspects and modalities, due to which we perceive it in the practical terms of its convenience or inconvenience, in its presence or absence? How do these aspects accrue to the fundamental sensation of those present atomic qualities? And to what degree is this epistemic processing realized by the intellect and its judgmental, inferential and discursive capacities, and to what extent is it part of the working of the internal senses and common to nonhuman animals that are also endowed with these senses? Moreover, if these acts of processing, such as judging, storing, recalling, interpreting, inferring, (re)identifying etc., were parts of internal sense apprehension, should we attribute them to really distinct senses or should we assign them only to a single interior sense? And, in general, are there any significant differences in the cognitive functions of both the external and the internal senses between human and nonhuman animals?

These and other issues revolving around the central problem of intentionality are pressing issues not only for contemporary philosophers of mind[10] but were discussed also by Suárez in his *Commentary on De anima* (henceforth: *CDA*). Like his *Metaphysical disputations* (*DM*) *CDA* is not a text closely following Aristotle's model. It is much more a *systematic reconstruction* of Aristotelian theory of soul, which incorporates a large number of philosophical and theological sources from ancient philosophy, patristics, medieval and renaissance scholasticism including Arabic philosophy, and last but not least the Galenic and Renaissance medical tradition.[11] The highly systematic and pedagogical character of the *Commentary*

[10]For contemporary authors, see Fish (2010) and Siewert (2017), among many others.

[11]For this assessment of Suárez's manner of commenting, see Simmons (1999).

significantly facilitates the precise demarcation of the textual segment corresponding to the main theme of this book. From the overall number of 14 disputations[12] the topic of sense perception both at the level of the external and the internal senses is *ex professo* treated in disputations 5–8 (*CDA* disp. 5–8). In *CDA* disp. 5 titled 'De potentiis cognoscitivis in communi' Suárez treats cognitive powers in general, that is, he approaches them indiscriminately with respect to whether they are external senses, internal senses, or the intellect. They are all conceived only in so far as they are cognitive faculties. If not explicitly restricted to this or that kind of apprehension, all conclusions from disp. 5 are applicable to both intellectual and sensory cognition. In fact, these conclusions concern the issues of the nature of the cognitive act, its principles and terminus, the questions regarding apprehension, judgment and discourse, and topics related to attention and its division. The following *CDA* disp. 6–7 are explicitly devoted to the external senses. While *CDA* disp. 6 called 'De sensibus in communi' analyses sense perception from a general point of view, that is, without specification related to whether an external sense is this or that particular sensory modality, such as sight, hearing etc., disp. 7 titled 'De sensibus exterioribus in particulari' is the most focused. Besides the comparison of the particular senses, this disputation aims at the specific features of the particular external senses, which largely bear upon the question about the nature and kinds of the proper sensibles. However, it is especially *CDA* disp. 6 that significantly complements *CDA* disp. 5. It is the analysis of the notion of the agent sense (*sensus agens*) that appends the debate about the principles of the perceptual act, that is, the issue of the activity and passivity of perceptual powers and the question about the nature and origin of the sensible species. *CDA* disp. 6, q. 4 elaborates the issue of perceptual self-awareness at the level of both the external and the internal senses. This elaboration complements Suárez's discussion of perceptual judgment in *CDA* disp. 5, q. 6. The *quaestiones* related to the sceptical issues on the plane of both the natural and the supernatural possibility of error (*CDA* disp. 6, qq. 3 and 5) are laid out as the systematic consequences of *CDA* disp. 5. Much of what is said in *CDA* disp. 5 and 6 is exploited also in *CDA* disp. 7. Although this disputation is largely dedicated to Suárez's scientific or, more precisely, anatomophysiological views on perception, a

[12]While the early text, which in its various manuscripts was the source of Castellote's edition, is divided into disputations, questions and paragraphs (numbers), Álvares's *editio princeps*, called *Tractatus de anima* (henceforth: *TDA*), is structured into books, chapters and numbers. In Álvares's edition the overall treatment of sensory perception, together with the treatise on cognitive powers *in communi*, can be found in the third book (Suárez, 1856c, 613–712). Despite Álvares's statement 'Non erat tamen cur omnia plene prout ab eo accepimus non hic, tibi, Lectore benevole, immutata daremus [...]' (ibid., 567), this edition, as many scholars are not hesitant to note, does not correspond to the authentic manuscripts. This is also why I will make use of Álvares's edition especially in the parts related to the first 12 chapters of the first book that were revised by Suárez at the end of his life, i.e., in the sections 2.1 and 2.2. I will consider the other parts only where Castellote's critical edition seems to be either obviously false or where Suárez's text of the critical edition stands in need of clarification. The third book of Álvares's edition has to be considered also due to its frequent references to *DM* (especially) in the titles of the particular paragraphs. For the 'mutilated character' of Álvares's edition, see Castellote in Suárez (1978, xli).

substantial emphasis is put on philosophical issues. This disputation gives us evidence of the interconnection between philosophy and science, especially medicine in Suárez's account. It also shows that the Jesuit always deals with science in respect to philosophical problems. The theory of the sensible species and its propagation *in medio*, which is taken by Suárez as the necessary principle in *all* sensory modalities, represents one of the crucial philosophical topics in this disputation. Finally, in the brief two questions of *CDA* disp. 8 'De sensibus interioribus' Suárez approaches the issue of post-sensory modalities. Two topics dominate the discussion here. First, the issue of the number of the post-sensorial powers; second, the question about the cognitive functional scope of internal sense apprehension. Both topics are closely connected with what Suárez says about the external senses in the previous disputations and systematically anticipate and include what he says in *CDA* disp. 9 that is devoted to the subject matter of the intellect. In sum, and especially when compared with two other main early Jesuit commentaries on *On the Soul* from Francisco de Toledo (1532–1596) and the Coimbran Manuel de Góis (1543–1597), which follow more closely Aristotle's text,[13] these four disputations give witness to how systematic and pedagogically mature the mind of the young Suárez was. It is all the more admirable if we recall that his philosophical psychology, as that of any philosopher who in a broad line follows Aristotle, is substantially determined by and issues from the *fundamenta* of his natural philosophy (physics) and metaphysics. Even though only in his mid-twenties, in the first half of 1570s he had already elaborated some of the metaphysical assumptions of his cognitive theory in some of his commentaries on Aristotle's other texts.[14]

1.2 State of Research

For many decades, most research in Suárez's philosophy was devoted largely to his metaphysics, philosophy of law and political philosophy. This is understandable, inasmuch as his views related to these disciplines are discussed in most of his philosophical volumes of the *Opera Omnia* and given their impact on early modern

[13]Francisco de Toledo, *Commentaria una cum quaestionibus in tres libros Aristotelis de Anima* (Hildesheim, Georg Olms Verlag, 1985; reprint of Cologne edition, 1615–6) and Collegium Conimbricensis, *Commentaria in tres libros de Anima* (Hildesheim: Georg Olms Verlag, 2006; reprint of Cologne edition, 1609).

[14]Castellote in Suárez (1978, xxxviii–xxxix) says the following: '[. . .] la producción filosófica de Suárez realizada en los años de su docencia en Segovia (1571–1575). Durante este período compuso Suárez sus tratados filosóficos, comentarios casi todos a las obras del Estagirita, y que han quedado inéditos, si exceptuamos su curso De Anima, publicado póstumamente por el P. Baltasar Alvares en Lyon 1621 [. . .] La existencia dé estos tratados filosóficos ha sido atestiguada por el mismo Suárez, quien en su tratado De Anima cita en varias ocasiones los siguientes: *Peri Hermeneias*, *Libri Posteriorum*, *Libri Physicorum*, *Libri Praedicamentorum*, *De Caelo*, *De generatione et corruptione*, y, posiblemente, una *Metafísica*. Además, por crítica externa, sabemos que ésta era la práctica de la docencia filosófica en las cátedras universitarias de la época'.

philosophy as such. It was chiefly due to Salvador Castellote's critical edition of *CDA* in particular and to the medievistic turn to the philosophy of mind in general that the situation has somewhat changed in the recent decades. Indeed, as it can be easily noticed, Suárez's philosophical psychology has become a widely discussed topic worldwide. This applies also to his philosophy of perception and cognition which has enjoyed different interpretations.

In this section, I aim to feature several representative studies on Suárez's cognitive theory by sorting them into distinct clusters according to the way of their exegesis. I dare say that his theory of cognition has been read in the following five distinct ways: (1) as a theory *sui generis* against the background of Aristotle's (peripatetic) and Aquinas's (Thomist) positions; (2) as a position criticized and (ideologically) rejected by orthodox Thomism; (3) as a doctrine defended against these attacks by Suarezians (frequently coinciding with (1)); (4) as a philosophy construed largely in light of Augustine's philosophy and medieval Augustinianism; and (5) as a conception anticipating the Cartesian turn to subjectivity (often overlapping with (2)).

(1) The first reading is undoubtedly the most widespread and historically most justified toward Suárez. In the English language scholarship, the most representative contributions exemplifying this reading have been written by James South. In his 'Suárez and the Problem of the External Sensation'[15] South assesses Aquinas as an advocate of a passivist account of perception for whom sensation is said to be equivalent to nothing more than the reception of 'spiritualized' sensible species emitted from the sensible objects. Unlike the interior senses and the intellect, the external senses do not produce internal representations called 'expressed species' (*species expressa*). If they did, the direct realism of external sensation would be jeopardized. As South claims, the equivalence between the production of a sensorial act and the mere reception of species makes an act of perception in this theory into 'a black box'.[16] Such a theory leaves aside the necessary explanation of the perceiver's contribution. Only the 'activist approach', embraced by Suárez, which leads to a substantial revision of Aquinas's theory of the sensible species, can result in a successful explanation of perceptual intentionality, South argues.

In his 'Suárez on Imagination'[17] South discusses two topics which constitute Suárez's main agenda concerning the internal senses. While the first one concerns the issue of the number of the internal senses, the second one has to do with the query about the relation of the interior senses to the external senses and the intellect. Again, Aquinas's theory of four internal senses constitutes the main doctrinal background of his presentation of Suárez's reductionist view of a single internal sense power called 'phantasy'. Some attention is devoted also to the consequent reduction of the functional scope of this (single) interior sense. Suárez's view of the intellect and the privileged epistemological position of the external senses are introduced as two

[15]South (2001a).

[16]Ibid., 224.

[17]South (2001b).

reasons of this reduction. The single interior sense is by Suárez conceived as a mere transmitter of sensory experience coming from the external senses to the intellect. The Jesuit's doctrine of a single power is contrasted with Aquinas's view motivated by the necessity to 'bridge' a gap between the senses that apprehend only singulars and the intellect which cognizes only universals.[18] The paper's second topic concerns the origin of the *species phantastica*. Suárez is said to recur to the device of the agent sense productive of this species, which he identifies with the soul. The soul produces the sensible species on the occasion of the external sense's actual cognition that serves as an exemplary cause for the soul's 'depiction'. In an analogy to the production of the intelligible species, an act of the external sense is not the efficient cause but only an *exemplum* or *occasio* for the soul's production of the species. Employing the lexicon of *emanation*, the *species phantastica* is said by South to flow from the soul through the power of the phantasy in the same way as the intelligible species flows from the soul through the agency of the agent intellect on the occasion of exemplary phantasms. As South concludes, this emanation gives us evidence of 'Suárez's latent and dispositional innatism'.[19]

In his later 'Aspects of Intentionality in two 16th Century Aristotelians',[20] South supplements his older analysis of Suárez's view of the external sensation with a comparison with the doctrine of perception in Jacobo Zabarella (1533–1589). Zabarella is indicated as an author who identifies the same problems in Aquinas's theory as Suárez does. Zabarella sees the Angelic Doctor as a philosopher who undermines the immanent and the vital character of cognition by attributing too much of the causal efficiency to the sensible species. Both Zabarella and Suárez make sure that the intentionality of perception cannot be explained by a mere reception of the 'spiritual' species because the sensible species are only the material and imperfect *traces* of sensible objects. Perceptual intentionality must be explained (also) through a reference to the factor of selective attention. Additionally, South formulates a hypothesis related to a possible historical influence: Suárez's activist view of perception was not influenced by Augustine, but rather by some authors of the Aristotelian tradition, especially Averroes.[21]

The Thomistic point of reference represented by Cajetan's stance can be found in Cees Leijenhorst's study 'Cajetan and Suarez on Agent Sense: Metaphysics and

[18]For the centrality of this interior sense in Aquinas, see recent Lisska (2016, 194–218).

[19]South (2001b, 158).

[20]South (2017).

[21]Ibid., 739: 'Suárez, obviously, has a much more extensive acquaintance with Augustine, but does not refer to him in the context of his discussion of sensory experience. But if not Augustine, what is the source of his critique of the Aristotelian position? What follows is partly speculative on my part, but I do have some evidence'. What follows after this passage is a quotation from Averroes's *Long Commentary on the De Anima*, in which the Arab philosopher raises the question about the agent sense as a potential 'elevator' of sensibles to the level of intentional species. In Sect. 3.2 and elsewhere I show that this conception was sharply criticized by Suárez and that he refers also to Augustine (not frequently, though) in contexts that are relevant for his activist view of perception.

Epistemology in Late Aristotelian Thought'.[22] In this chapter Leijenhorst develops the topic of the sensible species against the backdrop of the premise of the entire debate regarding the origin of these species that amounts to the axiom of the impossibility of ascendant causality. How can a sensible object generate the sensibles species, characterized as 'the form without matter', if the sensible species is more perfect than the sensible forms that are represented by it? Clearly, sensible species do not corporeally affect an organ in the way in which the reception of colour would result in colouring the eye. In this context, Leijenhorst presents an illustrative comparison of the Jesuit's theory of the agent sense with that of Cardinal Cajetan (1469–1534) who is frequently conceived by Suárez as a doctrinal arch-opponent. While for Suárez the agent sense is not necessary in the production of the sensible species of the external senses, Cajetan does endorse this 'ontological lift' that elevates the material sensibles to the level of the immaterial sensible species.

The same methodological approach is employed also in his 'Suárez on Self-Awareness'. In this study Leijenhorst compares Suárez's theory of sensory self-awareness with that of Aquinas.[23] As many others, he argues that the best way to understand Suárez's view is to conceive it as an attack on Aquinas's view. Aquinas is presented as denying self-awareness to the external senses.[24] This is why the Angelic Doctor stipulates the common sense as the power responsible for noting the acts of the external senses. Distinguishing between knowledge *quid est*, that is, scientific knowledge of what an act is, and knowledge *an est*, that is, knowledge *that* there is an act, Suárez insists that the external senses are aware of their acts since their simple apprehension is a vital affection (*immutatio vitalis*).[25] They are able to exercise not only same-order self-awareness but also a non-complex (imperfect) judgment. In this context Leijenhorst underlines the significance of the sensory memory in the argument for this self-awareness. There must be the '*an est*–experience' of a sensorial act since we remember not only an object that we saw in the past but also the act through which we saw it.[26]

(2) With increasing awareness of the doctrinal distinction between the philosophy of Aquinas and Suárez in the first decades of the twentieth century, the criticism of

---

[22] Leijenhorst (2007).

[23] Leijenhorst (2012); for sensory self-knowledge, see esp. 138–44.

[24] It must be said, however, that this holds only for some of his texts and only for some periods of his carrier. For this, see Cory (2021). See also Sect. 3.7 below.

[25] For this for Suárez important notion and its employment in late medieval philosophy, see Kaczmarek (1990); Rode (2015, 337–42); for Suárez (ibid., 377–87).

[26] Leijenhorst (2012, 143–4). In his recent study (Perler, 2014a), beside a useful classification of the various types of self-awareness in Suárez, Dominik Perler takes the same methodological point of departure. Additionally, he raises a question about the certainty of this self-awareness. He points out that unlike the possibility of error concerning the various kinds of sensibles, especially common and incidental sensibles, Suárez never raises a doubt about the certainty related to the cognition of these acts. This shows that the cognition of these acts is on a safe ground. When we attentively perceive an object *X*, it is not possible for us not to be aware of our sensorial act through which we see this object.

the Jesuit theory of cognition from the position of orthodox Thomism increasingly emerged. This approach is still espoused today as well. The incoherence related to the alleged representationalism, immanentism, voluntarism and scepticism that plague Suárez's philosophy of perception and his theory of cognition in general has become the target of many critiques written by Thomist-oriented authors. One of the recent Anglophone criticisms can be found in Anthony Lisska's book *Aquinas's Theory of Perception. An Analytic Reconstruction.*[27] In this book Lisska derives his evaluation largely from the earlier censure that had been formulated by John Deely.[28] If the sensible species ceases to be the formal cause and comes to be a 'seedy' virtual representation that generates only efficient causality, and if the external senses are the essentially active powers that produce their own termini (expressed species), the result can be nothing else than epistemological representationalism. Suárez leaves the preserve of direct realism centred in an object in favour of representationalism centred around the subject. In this Suárez unhappily anticipates early modern subjectivism and representationalism.

A similar evaluation can be found in Leopoldo Prieto López's articles.[29] Prieto López speaks about 'Suárez's problematization of the intentionality of cognition'. This 'problematization' originates in Suárez's admission of the possibility of the state of affairs, in which God (like the Cartesian *mauvais génie*) causes in the external senses an intuitive cognition of nonexistent objects. There is no contradiction in this hypothetical scenario since each secondary efficient cause can be substituted by the primary (divine) cause. Since the sensible species depends on the sensibles only in the order of efficient causality, its efficiency can well be supplanted by God. For Prieto Lopéz this testifies to an unfortunate influence of late medieval philosophers, such as Peter Auriol (ca. 1280–1322) and William Ockham (ca. 1287–1347), on Suárez's cognitive theory as such.[30]

(3) One of the most extensive critical responses to these, albeit much older, Thomistic accusations of representationalism, immanentism etc.,[31] and one of the most thorough defences of Suárez's direct realism can be found in the book *La gnoseología del Doctor Eximio y la acusación nominalista* written by Suarezian José María Alejandro.[32] Alejandro rejects the abovementioned imputations by pointing to various aspects of Suárez's theory which are not consistent with these charges. These are Alejandro's claims: the perceptual act, which pertains to the category of quality,

---

[27]Lisska (2016).

[28]Deely (1994, 130, 135 and elsewhere).

[29]See especially Prieto López (2016).

[30]Ibid., 42–4.

[31]For a negative assessment of Suárez's theory from a Thomistic perspective of an earlier date, according to which the Jesuit is to be seen as a 'grand corrupteur de Saint Thomas', to be blamed for his 'erreur subjectiviste' and for having been responsible for early modern doctrinal errors, such as occasionalism, see Farges (1921, 183–7).

[32]Alejandro (1948).

is transcendentally related to extramental beings;[33] the act-quality is *vitally* assimilated to extramental things;[34] a *pictorial* theory of the impressed and expressed species is not a doctrine endorsed by Suárez;[35] intentional species are 'ignota'; expressed species are not 'that in which as if in the mirror' (*id in quo*) an extramental sensible object is apprehended but 'that by which' (*id quo*) it is grasped etc. All in all, for Alejandro Suárez is a proponent of the 'inmediatismo cognoscitiva real absoluto'.[36] Although in vision and in the operations of the other external senses we are always at the same time aware that we see, Suárez's theory of the expressed species should be classified as substantially *object-oriented*.[37]

(4) Unlike studies devoted to the topic of the agent sense and the activist view of perception, which largely remain within the preserve of a broadly conceived Aristotelian tradition, in his 'Introduccion a la teoria del conocimiento de Francisco Suarez'[38] Santiago Fernández Burillo goes beyond this tradition toward the broadly conceived lore of Augustinianism. Although he focuses on intellection, his conclusions (in line with Suárez's own approach) can be extended to perception too. A crucial trait of Suárez's cognitive and affective theories is the emphasis on the immanency and the vitality of the soul's acts. Suárez's cognitive theory and his philosophical psychology is said to be the philosophy of a 'Molinist theologian'.[39] For a Molinist theologian, however, the most important power is free will. It is free will that postulates the crucial capacity of self-determination. No external influence, whether coming from God, stars, body etc., can compromise Molina's libertarian account of freedom. As Burillo argues, this libertarian account of freedom constitutes a model for *all* immanent and vital operations in Suárez's philosophical psychology. Consequently, this emphasis leads to a substantial relativization of the Aristotelian-Thomistic axiom 'Whatever is moved is moved by another'. Rejecting this axiom gives rise to several important consequences, of which the most relevant one for us is the emphasis on the activity of the cognizant and appetent. This emphasis has led several scholars to a historical re-evaluation of Suárez's cognitive theory: Especially due to the influence of Scotus and Olivi with his primacy of the power of will and his will's activity conceived as the model for all cognitive activities, Suárez ought to be classified as a representative of medieval Augustinianism, rather than of Aristotelianism.[40]

---

[33]Ibid., 213.

[34]The crucial notion of *assimilatio vitalis* in connection with cognition is a leitmotif of Alejandro's book.

[35]Ibid., 194.

[36]Ibid., 207. '¿Se podrá hablar de un *inmediatismo* sensitivo del Doctor Eximio? Creemos que sí.', 237.

[37]Ibid., 222.

[38]Burillo (1992).

[39]Ibid., 212.

[40]For this evaluation, see also Mendía (1948); Müller (1968, 169–72); Abercrombie (1938, 81–4); Spruit (1995, 300): 'Thus, Suárez's philosophy of mind resumed essential aspects of the Augustinian and Olivean lore'.

Clearly, this Augustinianism is strongly mixed with Aristotelianism. While Benito Mendía regards Suárez's synthesis of the Aristotelian and the (Neo)Platonic–Augustinian elements as constituting a coherent unity, others take it to be an amalgam of heterogeneous doctrines, Heinrich Rösseler formulates the following critical evaluation of Suárez's cognitive theory:

> [. . .] in his teaching Augustinian-Scotistic thoughts are grafted onto Aristotelian-Thomistic foundations[41] [. . .] However, since Suárez's system is constructed of building blocks that have not yet been sufficiently stripped of their mutually contradictory natures, the doctrine we present reveals in its entirety a lack of inner unity and integrity (*Geschlossenheit*).[42]

While Augustinian features are represented by the activity of the sense powers and the essential impassibility of the soul through the efficiency of the material world, the *virtual* representation of the intentional species and the production of the expressed species in all the cognitive powers including the external senses, the Thomistic foundations are mirrored in his theory of the unicity of the soul, its parts and the doctrine of the real distinction of the soul from the powers.[43]

(5) Several recent studies which have been devoted especially to the topic of (perceptual) self-awareness show that in Suárez's theory lurks a modern Cartesian subject. They emphasize Suárez's claim that already on the level of external sensation we are aware of our perceptual processes through which we perceive the extramental objects. We perceive them not through a higher-order act but simultaneously with the sensation of extramental objects. Christian Rode takes Suárez's theory of inner experience as being intimately connected with his theory of the soul, attention and consciousness, as substantially *transcending* the traditional Aristotelian psychology with its cognitive division of labour.[44] Behind internal experience a modern Cartesian subject operates, conceived as a main distinct agent.[45] Like other scholars, such as Joseph Ludwig,[46] Rode analyses the metaphysical underpinning of

---

[41]Rösseler (1922, 196).

[42]Ibid., 198.

[43]Ibid., 192–5. Clearly, this evaluation cannot be taken without a reservation. It does not always hold that when two say the same that it really is the same. As I will show below in Sect. 2.6, the last theory, namely the teaching of the real distinction between the soul and the powers conceived as *res*, cannot be attributed to Aristotle or Aquinas.

[44]Rode (2015, 384–5).

[45]For this historiographical assessment, see also Rozemond (2012, 167): 'But at the same time it is hard to resist the impression that some sort of Cartesian subject lies at the centre of Suárez's complex soul'.; ibid., 172: 'Buried, as it were, in Suárez's complex Aristotelian Scholastic soul there is a single, unitary entity, whose involvement in its activities relies on attention, suggestive of Cartesian unitary conscious subject, an entity that is 'the same soul' that senses, imagines, and understands'.

[46]Although old (published in 1929), this monograph still belongs to the best publications that have been written on the topic of the relationship of Suárez's philosophical psychology and his metaphysics (natural philosophy). It refers to a large number of useful texts from Suárez's *CDA* and especially *DM* that have been neglected by later scholars. For a recent mapping of Suárez's theory of the sympathy of powers rooted in a single coordinating soul and various epistemological applications of this theory in *CDA* and *DM*, see also García Cuadrado (2005), Knuuttila (2014)

Suárez's theory of substantial attention which he considers to be complementary to his theory of the sympathy of powers which are really distinct from the soul and which are all 'rooted' in the same soul. As one of the historical sources of this theory, Rode and other scholars mention Francisco Vallés (1524–1592) who, as we will see, can be regarded as a major authority for Suárez in physiological questions related to topics of the sense organs of the external senses.[47]

This last interpretative approach seems to me to provide a fitting explanatory key to many issues of Suárez's philosophy of perception. Regardless of his object-centred Aristotelianism, the subjective 'colouring' of Suárez's peripatetic philosophy is a leitmotif of the whole part of *CDA* related to his cognitive theory. The emphasis on the soul's immediate activity in its operations accompanied, among others, by the subject's concomitant introspective consciousness can be regarded as a typical trait of Suárez's theory. As will become clear in Sect. 6.3, this subjective modification leads me to label his theory of perception 'Aristotelian subjectivism'.

1.3 Goals and Methodology

Although in recent years interest in Suárez's epistemology and cognitive theory has increased, no comprehensive treatment of his philosophy of external and internal perception in book format is available yet. The goal of this book is to amend this by presenting Suárez's *comprehensive* doctrine of the external and internal senses with a special (but not exclusive) focus on the human sensory powers, this being in line with the Jesuit's own approach.[48] Besides providing analyses of the particular issues related to perception, the primary goal of the volume is, on the basis of a thorough text analysis, to bring a *systematic* and *contextual* light on Suárez's account of perception, which is lacking in contemporary research as it remains largely at the level of case studies. My approach, however, will not be restricted to highlighting

---

and Tropia (2014). For this theory and its significance for Suárez's theory of free will, see the recent Anfray (2017).

[47]Rode (2015, 394–7). For a relationship between Vallés and Suárez, see Amaral (1987–8). Of recent studies on Suárez and the medical tradition, see García Cuadrado (2017a). It must be said that while Suárez's theory of cognition in general and perception in particular have become a relatively frequent topic of research in recent years, Suárez's views about the particular sensory modalities and his scientific views have been much less elaborated. Leaving aside a brief outline in Knuuttila (2015, 205–7) and a survey in Castellote (1962, 143–155), there are no in-depth studies on topics related to the individual external sense modalities, which are treated in *CDA* disp. 7.

[48]Although Suárez lays emphasis on *human* sense perception and on the *rational* soul, there are several more extensive treatments in *CDA* that are *ex professo* devoted to a comparative analysis of perception in human and nonhuman animals. These treatments concern issues of perceptual judgment and discourse (*CDA* disp. 5, q. 6), interior sense operations in perfect and imperfect animals (*CDA* disp. 8, q. 2), and questions related to comparison of human and nonhuman external senses (esp. *CDA* disp. 7, qq. 15–6). All of these will be considered below too.

the connections between the acts of the external senses, of the interior senses and of the intellect. It will also include an introductory part on the metaphysical premises that systematically determine Suárez's perceptual theory. These metaphysical assumptions have to do with the definition of the soul, its relation to the body, its parts and powers. Besides this systematic and contextual perspective this monograph also aspires to be innovative in introducing issues that have been neglected or have been treated only marginally by scholars. These topics are associated especially with the particular sensory modalities, the division of attention and with the problem of corporeal pain in its relation to the sense of touch. Focusing on all these issues will allow us to see the subjective twist of Suárez's philosophy of perception more clearly. Unlike an uncharitable reading charging Suárez with internal inconsistency,[49] I will mostly apply a charitable reading.[50] I will argue that Suárez strives to find a dynamic balance between the abovementioned two theoretical demands which are supposed to be met by a successful philosophy of perception, namely between the epistemological (objective) aspect and the phenomenological (subjective) aspect. Considering the dichotomy of cognitive activism and passivism, it may be said that while the receptivist account of perception can readily accommodate the epistemological standard, it has to face difficulties related to the phenomenological criterion. On the contrary, while the activist view neatly fulfils the phenomenological canon, it finds itself in trouble when it comes to meeting the epistemological standard.

What do I mean by a *contextualizing* account? Although Suárez's exposition in *CDA* is a systematic adaption of Aristotle's *On the Soul* and as such is clearly structured, it is a historical fact that while Suárez was writing *CDA* he had already finished some of his commentaries on other Aristotle's texts, especially on his *Physics* and a part of his *Organon*. Commenting on the Stagirite's texts was a part of the standard curriculum at Spanish universities in his time.[51] Unfortunately, with the exception of the commentaries on the *De anima* and the *De generatione et corruptione*, these texts have never been published and, unfortunately, seem to be lost.[52] As shown by a large number of references throughout *CDA*, at the time of its composition Suárez had already formulated the basics of his metaphysics and natural philosophy. These commentaries and their conclusions are most often quoted in

[49]It must be admitted, however, that there are many issues that evince 'dialectical tensions' in Suárez, especially if they are seen from the perspective of post-Suárezian philosophy. For these tensions in the Jesuit's metaphysics, see Novotný (forthcoming 2021).

[50]One of the few exceptions will be the issue of corporeal pain and its relation to the object of touch (for this see Sect. 4.5.2).

[51]For the philosophical triennium at Jesuit universities, with the first year focused on logic (the *Organum*), the second on natural philosophy (*Physics*, *On the Heavens*, *On Generation and Corruption* and *Meteorology*) and the third year spent on *Metaphysics* and *On the Soul*, see Grendler (2019).

[52]Scorraille (1911, vol. 2, 416) admits his search failed.

disp. 1–3, in which Suárez presents his theory of the soul, its parts and powers. However, some references can also be found in the disputations devoted to cognitive theory.[53] It is not difficult to determine why Álvares did not include them in his editorial plan. The reason could have been nothing else than the later publication of the comprehensive *DM*. In fact, we may be sure that *DM* covers all the issues of those juvenile texts. Although scholars have rightly pointed out some doctrinal shifts between *CDA* and *DM*,[54] some premises, theories and arguments connected especially with Suárez's theory of causality are nonetheless shared by the two texts, which are separated by a quarter of a century.[55] Accordingly, one of the traits of my procedure will be the *intertextual* incorporation of the relevant parts of *DM* and, in a lesser degree, also of some theological texts, not only in the exposition of the metaphysical assumptions of Suárez's perceptual theory in *CDA*, but also in other parts of the book.

In my interpretation of Suárez's theory, I will not restrict myself to the delineated segment of disp. 5–8, in which he *ex professo* deals with the issue of cognition and perception, and disp. 1–3. In *passim* I will be referring to the *whole* corpus of *CDA* since this commentary makes up a remarkable systematic whole. Accordingly, I will contextualize the Jesuit's views on perception also *intratextually*, namely by rearranging several parts of *CDA*.[56] Suárez's theory of perception is not an isolated segment in his *scientia de anima*. It cannot be separated from the context of the other disputations. Due to an important parallelism between sensory and intellectual processes it is above all impossible to omit *CDA* disp. 9, focusing on intellectual knowledge. It is not only a structural similarity related to the nature of the cognitive act, its principles and endpoint (*terminus*) that make Suárez's treatment of sensory and intellectual cognition parallel. Intellectual knowledge is important also for a contrasting reason. The intellect and its acts constitute a crucial point of comparison for issues such as (perceptual) self-awareness and judgment, cognition of incidental sensibles, memory, cognition of universals etc. Last, but not least, Suárez's treatise of the intellect cannot be set aside from an analysis of his view of perception also for another reason, which is related to his reductionist approach inspired by 'Ockham's Razor'. As he denies the real distinction between the agent and the passive sense, so he rejects the real distinction between the agent and the potential intellect. The same

---

[53]These references are related especially to his metaphysical theories of efficient causality and the categories.

[54]Des Chene (2000, 187) and Schmaltz (2008, 147), to name a few.

[55]It is due to Suárez's genius that as a very young scholar he had already worked out the foundations of his later metaphysical system. Ángel Poncela González (2015, 275–6) characterizes this juvenile 'anticipation' as follows: 'Para este trienio filosófico (1571–1574) compuso, a modo de apuntes, una serie de comentarios a todo el corpus aristotélico. En estos apuntes, en su mayor parte perdidos, puede localizarse el sustrato de lo que, veintitrés años más tarde, serían las Disputaciones Metafísicas (1597)'.

[56]In Chap. 4 I will premiss parts on the number and comparison of the external senses, introduced by Suárez in the last two questions of *CDA* disp. 7, before all the topics analysed in the preceding questions of the disputation, among others.

reductionism can be observed also in Suárez's treatment of the appetitive powers. In a direct connection with his denial of the plurality of the internal senses, Suárez dismisses Aquinas's real distinction between the concupiscible and the irascible appetite.[57] The affinity of his perceptual and appetitive theory is clear also from the issue of whether bodily pain constitutes the tactile power's proper sensible object, or whether it is rather an emotion. At the same time, as I have said in the previous section, Suárez's conception of the rational appetite (analysed in *CDA* disp. 12) makes a prototypical model of vital and immanent operating that needs to be considered too.[58]

Not differently from most studies devoted to interpreting the Jesuit's philosophy, in historiographical contextualization (in line with the first abovementioned reading) I will consider Aristotle and Aquinas as the main referential points and the immediate background and contrast for Suárez's views of perception. Although his text is framed as a commentary on Aristotle's *On the Soul*, it can be at the same time regarded as an ample gloss on Aquinas's 'Treatise on Man' (*Summa theologiae* I, qq. 75–89). This contrasting will occasionally go hand in hand with references to Augustine and medieval Augustinianism. Although the primary goal of this book is *not* historiographical in the sense of analysing a plethora of historical influences and antecedents of Suárez's theory of perception, I will claim that Suárez's views can often be best understood against the background of other medieval and especially late medieval authors, such as Peter John Olivi (ca. 1248–1298), John Duns Scotus (1266–1308), Peter Auriol (ca. 1280–1322).[59] Especially in the treatises of the particular sensory modalities I will refer to the theories of the representatives of the medical tradition, such as Claudius Galen (ca. 130–ca. 210), Francisco Vallés (1524–1592) and Andreas Vesalius (1514–1564). Above all in the context of the theory of the sensible species and its propagation *in medio* I will consider the theories developed by proponents of the medieval science of optics (*perspectiva*) as well. Especially in the historical contextualization of the problems of the internal senses and the agent sense I will delve into the Arabic tradition represented by Avicenna and Averroes, and in the case of the latter I will take into account the tradition of medieval and Renaissance Latin Averroism represented by John of Jandun (ca. 1285–1328). From a contrastive point of view, I will be frequently referring to the doctrinal perspectives of Cardinal Cajetan who, rather than Aquinas,

[57]For his denial of the real distinction between these two powers, see King (2002) and Heider (2016a).

[58]Additionally, I will not leave aside the last disp. 14 'De anima separata'. In this disputation, Suárez provides the reader with information about the ontological status of the human soul (its essence and subsistence) in an embodied composite. However, only occasionally will I also consider topics related to the vegetative part of the human soul (*CDA* disp. 4) and the locomotive power of the sentient part of the soul (*CDA* disp. 13).

[59]In connection with Olivi, who is not explicitly quoted by Suárez but whose influence seems to be the undercurrent in many places of *CDA*, I will refer especially to Toivanen (2013b), a recent comprehensive account of the Franciscan's theory of perception.

is often taken as *the* main rival by Suárez. Although my goal is not to locate Suárez's theory in the context of early Jesuit philosophy, let alone in the context of Second scholasticism and early modern philosophy represented by Descartes, Locke and others,[60] I cannot entirely refrain from comparing Suárez with some of his Jesuit contemporaries, such as Francisco Toledo (1532–1596) and the Coimbran Manuel de Góis (1543–1597), who both wrote commentaries on the *De anima*. Nor can Suárez's scholastic followers be completely left aside. Every so often I will refer to Pedro Hurtado de Mendoza (1578–1641) and Francisco de Oviedo (1602–1651). Of non-Jesuit authors I will occasionally employ theories formulated by the Thomist John Poinsot (1589–1644) and Scotists Bartholomeo Mastri da Meldola (1602–1673) and Bonaventura Belluto (1600–1676), whose insights into Suárez's philosophy can be regarded as exceptional. However, considering the book's central topic and size, the theories of these medieval and post-medieval authors can be discussed only in a cursory way.

Considering all these and other primary sources I hope that this volume will, in its modest degree, contribute to contemporary historiography of philosophy by filling in the gap that exists in the mapping of the continuities and discontinuities between medieval (first) and early modern (second) scholasticism. This book thus subscribes to the 'Continuity Thesis', a highly indigent approach in contemporary historiography of medieval and post-medieval scholasticism, according to which there is actually no radical discontinuity between the intellectual developments in the Middle Ages, the Renaissance and the early modern period.[61]

[60]Each of these contextualizations would deserve a special book. Without a doubt these comparisons are a challenge for the future.

[61]For this 'Continuity Thesis', see Pasnau (2011a).

Chapter 2
Metaphysics of the Soul, Its Parts and Powers

A philosophy of perception presupposes many metaphysical assumptions. Leaving aside now those that come from the side of objects, such as the ontological status of the proper sensibles and their representational forms, the general premises on the perceiver's side, namely the soul, its parts and the powers, are analysed in the first three disputations of *CDA* called 'De substantia animae in communi', 'De substantia trium animarum in particulari' and 'De potentiis animae in communi'. In all of them Suárez starts from the fundamentals of Aristotle's natural philosophy to which he adds his own twist. Before dealing with these topics, in the *Proemium*,[1] among other topics, he situates the *scientia de anima*, which means for him largely *scientia de anima rationalis*, in the context of other sciences. What is relevant for our topic is his claim that the treatise of the rational soul is absolutely (*simpliciter*) a part of natural philosophy. Since a soul in general, including the rational soul, is essentially characterized through its ordering to the body, investigating it must belong to the field of natural philosophy. At the same time, however, Suárez is sure that in a certain respect (*secundum quid*), that is, in respect to its incorporeal, subsistent and immortal character, treating of the soul pertains also to metaphysics.[2] Given this and considering the abovementioned metaphysical framework of the treatise of causality and the categories in *DM*, his conclusions and theories concerning the topics of the soul, its parts and powers can be indicated as the *metaphysical* premises of Suárez's philosophy of perception.

These premises can be divided into four thematic clusters:

---

[1]This *Preface* (*CDA* t. 1, 4–53) is a comment on Aristotle's *On the Soul* I.1.

[2]See *CDA*, *Proemium*, t. 1, n. 16, 26. This stretch between metaphysics and natural philosophy in Suárez and in other Jesuit authors comes as a reply to this basic question: How can the soul be both a form of the body and (due to its incorporeal, subsistent and immortal character) an independent entity at the same time? For this dilemma in Suárez, see Aho (2009, 59). For this question regarding the philosophical discipline to which the inquiry concerning the rational soul belongs, which was a hot issue in the sixteenth century, see Bakker (2012, esp. 11–2).

D. Heider, *Aristotelian Subjectivism: Francisco Suárez's Philosophy of Perception*, Studies in the History of Philosophy of Mind 28,
https://doi.org/10.1007/978-3-030-67341-3_2

1. The first cluster applies to Suárez's reading of Aristotle's two basic definitions of soul (including the rational soul) from his *On the Soul*. In the first definition the soul is characterized as '[. . .] the first actuality of a natural body potentially possessing life; and such will be any body which possesses organs'.[3] The second definition conceives the soul as '[. . .] the origin of the characteristics we have mentioned, and is defined by them, that is by the faculties of nutrition, sensation, thought and movement'[4] or '[. . .] that whereby we live and perceive and think in the primary sense'.[5] While the first formula, which unlike the two formulations of the second is applicable to all kinds of soul, features the soul in terms of formal causality, the two formulations of the second definition describe it through efficient causality. The metaphysical interplay between these two definitions, which in Suárez, as we shall amply see, is substantially determined by a reification and 'dualization' of both hylomorphic principles, constitutes an important part of the metaphysical background of Suárez's philosophy of perception on the side of the percipient.
2. Due to the validity of the scholastic axiom 'actions are of supposits' (*actiones sunt suppositorum*) and with respect to the fact that for Suárez the soul becomes an agent *sui generis*, it is impossible to omit his theory of the partial subsistence of the rational soul. If perception is a doing, it must somehow come from a subsistent entity, whether it is a hylomorphic composite or a soul. Typically, in the last *CDA* disp. 14 the rational soul in its embodied status is qualified by Suárez as a 'semi-person' (*semipersona*).
3. An important part of the theoretical point of departure of Suárez's view of perception is constituted by his theory of the unicity of a substantial form (the soul) connected with its indivisibility and its holenmerical character, that is, the claim that the soul exists wholly in every part of the body. In the reasoning for these conclusions, Suárez introduces his theory of the sympathy or harmony of the powers 'rooted' in the same soul, which is operative especially in Suárez's theory of the origin of the species of the internal sense.
4. The last set of the metaphysical agenda is connected with Suárez's theory of the powers, namely with their reified character, their real distinction from the soul and from the material composite and their origin or resulting from the soul. I will argue that the theory of the 'reified' hylomorphic principles is directly correlated with precisely this view of the reified powers.[6]

I will not delve into a detailed elaboration of these metaphysical issues as it would lead me too far from my main concern. First, I will take into account only what Suárez says in *CDA* and the later *TDA*.[7] Second, I will refer to *DM* and his other

[3] Aristotle, *On the Soul* II.1, 412 a29–412 b1, 69.

[4] *Ibid*., II.2, 413 b12–3, 77.

[5] *Ibid*., II.2, 414 a12–3, 79.

[6] For a detailed elaboration of some of these issues in the context of other early Jesuits' theories as often adumbrating Cartesian philosophy, see Des Chene (1996, 17–251).

[7] For *TDA* and its use in this book, see the notes 1 and 12.

treatises only to the extent to which its formulations help clarify what he says in *CDA*.

2.1 The Soul as the First Actuality of a Natural Body Potentially Possessing Life

In line with Aristotle, Suárez starts with commonsensical observations. Living beings are distinguished from non-living beings through self-movement in a broad sense including augmentation and self-sustenance. If an animal can move and nourish itself, it is alive, if not, it is dead. This is why we call still waters 'dead waters' and moving waters 'living waters'. It is also evident that living beings are not simple. They are composed of parts. Some parts are passive and moved, while others are active and moving. The ancients called the active parts 'the principle of life'.[8] A perennial philosophical problem is what this principle of life amounts to. A reader of *CDA*'s introductory passages soon finds out how clear and structured Suárez's reply to this question is.[9]

He starts by addressing the question of the *genus* of this vital principle, and then he asks about its specific difference.[10] Following the first part of Aristotle's first definition of the soul (the soul as the first actuality) Suárez declares that the principle of life or the moving part, sc. the soul, can be neither an accident, such as the harmony (*temperamentum*) of the first qualities, nor a body, nor (prime) matter, but the *substantial form of a living body*.[11] Although a certain harmony of the first qualities of the cold, the hot, the dry and the wet on the side of the organic parts is necessary and constitutes the material cause for the reception and conservation of the substantial form,[12] the principle of life cannot be characterized through this temperamental account of primary qualities associated, among others, with the names of Galen and Epicur.[13] Referring (without precisely pinpointing a *locus*) to Augustine's *De immortalitate animae* Suárez replies to this position by simply claiming that there are the rational soul's operations, such as those of love and understanding, which obviously exceed the virtues of these qualities and thus are irreducible to them.

---

[8] *CDA* disp. 1, q. 1, n. 3, t. 1, 58–62.

[9] In defining the soul Suárez follows the first two chapters of Aristotle's *On the Soul* II.

[10] *TDA* book 1, ch. 1, n. 1, 4.

[11] *CDA* disp. 1, q. 1, n. 5, t. 1, 66. In *TDA* book 1, ch. 1, n. 2, 467 Suárez explicitly refers to *DM* disp. 15, s. 1 where he argues for the existence of substantial forms. I will lay out his argument based on the distinction between accidental and substantial forms below in 2.4 when treating the issue of the unicity of the soul.

[12] *TDA* book 1, ch. 1, n. 10, 470.

[13] *CDA* disp. 1, q. 1, n. 4, t. 1, 64; *TDA* book 1, ch. 1, n. 6, 469. The Galenic temperamental account of the soul was widespread in the medical tradition in the Renaissance. For this conception, see Giglioni (2019). For the importance of Galen's materialist psychology for early modern Aristotelianism, see Buchenau and Lo Presti (2017, "Introduction," 3).

Moreover, the Jesuit notes that the soul equipped with the virtue of temperance often resists passions that come from a complexity of the primary qualities. This resisting soul cannot be identified with such a (resisted) temperamental mixture.[14] And since the tempering of these qualities amounts to the qualities themselves together in a proportion or to a simple quality resulting from them, the principle of life cannot be an accident either. Life cannot be caused by an accidental principle since for a living being it is substantial to be alive.[15]

Assuming that the principle of life is a substantial item, in his progressive elimination of the other candidates Suárez rejects all the remaining substantial candidates except of substantial form. First of all, he rejects matter. Matter cannot be the principle of life since it is not a principle that distinguishes the living from the non-living, but quite the contrary it is the principle *common* to them. Consequently, the vivifying principle cannot be a body either.[16] In fact, a body can be understood either as 'matter with accidents' or as 'a compound of matter and form'. Neither possibility is plausible. The principle of life cannot be the former because neither matter nor accidents are suitable candidates. The principle cannot be a hylomorphic compound, either, since no substance can be the principle of life of a distinct body or a distinct substance.[17] In line with this last refutation Suárez also denies two 'dualistic' versions of hylomorphism. First, it is the theory of universal hylomorphism according to which the human soul is not only a substantial form, but it is an entity composed of spiritual matter and form.[18] Second, he dismisses a view stemming from Averroes according to which the soul (intellect) is similar to an Aristotelian intelligence or motor of the spheres.[19] Suárez's denial of these views evinces the close interconnection between Aristotle's two definitions of soul. As we shall see more in 2.2, Suárez argues that the local movement of the spheres cannot be regarded as a vital act in the proper sense of the word since the movers of these spheres are not intrinsic to the moved planets. They are extrinsic to them in the way a sailor steering his ship is extrinsic to his ship. The principle of the circular movement of the spheres can be characterized only as an assisting form (*forma assistens*). This separated form or really distinct substance moves the spheres through a transitive operation. It produces its effect in a distinct subject, not in itself. This efficiency is not immanent because its effect does not remain in the agent and this movement does

[14]*TDA* book 1, ch. 1, n. 9, 470. For the early Jesuits' (especially that of Suárez and Francisco de Toledo) critique of Galen's conception, see Des Chene (2000, 67–76).

[15]*CDA* disp. 1, q. 1. n. 7, t. 1, 66. Cf. Aristotle, *On the Soul* II.4, 415 b13–5, 87.

[16]For a similar rejection of matter and body as the first principle of life in Aquinas, see *STh*. I, q. 75, a. 1, c., 194–5.

[17]*CDA* disp. 1, q. 1, n. 8, t. 1, 68; *TDA* book 1, ch. 1, n. 13, 471.

[18]For the same rejection in Aquinas, see *STh*. I, q. 75, a. 5, c., 202. In this Suárez also differs from especially thirteenth-century Franciscan authors such as Bonaventure and Olivi who considered spiritual matter to be the material constituent of the soul. For Olivi, see Toivanen (2013a, *passim*).

[19]For Averroes, see his *Commentarium magnum in Aristotelis De anima libros*, lib. 3, text 5, 387–413.

not perfect the agent itself.[20] A *forma assistens* can be called 'a soul' only *metaphorically*.

Having eliminated all the potential rival candidates for the position of the principle of life, Suárez concludes that the principle of life can be only a second kind of form, namely an informing substantial form (*forma informans*).[21] Both in *CDA* and in *TDA* this conclusion is corroborated with an argument based on the criterion of the per se unity of an ensouled composite. It is impossible for a soul to be related to its body as a sailor is related to a ship.[22] Why? Besides the abovementioned transitive operation between the two substances, the reason is that such a relation would entail the composition of a merely accidental unity. Only the soul *qua* an informing form can constitute a true substantial unity per se with matter.[23] However, if compared to Aquinas, Suárez's definition of per se unity acquires a new meaning. This new concept of essential unity represents a fundamental metaphysical premise of his philosophy of perception. Despite the aforesaid denial of the dualistic expositions of hylomorphism, in which Suárez basically follows Aquinas, this new concept is after all more dualistic than that of Aquinas.[24] It is related to the *reified* concept of prime matter and substantial form.[25] Contrary to Aquinas for whom per se unity was characterized through a single act of being given by the substantial form to the composite and prime matter,[26] for Suárez this unity derives from the union of two distinct *incomplete beings* or *incomplete substances*. Mereologically speaking, if the parts were not entities, they could not together constitute a *tertium quid*, namely a hylomorphic compound. Importantly, this characterization of per se unity based on Suárez's concept of substantial form and prime matter can be dated back to the early *CDA*. In *DM* disp. 15, s. 5 Suárez characterizes the substantial form succinctly as follows: 'Form is a certain simple and incomplete

[20]For the same argument from immanent operating in Aquinas, see *STh*. I, q. 76, a. 1, c., 209. For a critique of the theory according to which the soul is a *forma assistens* (a view that was criticized not only by Suárez but by all other early Jesuits), see Des Chene (2000, 76–81).

[21]*TDA* book 1, ch. 1, n. 14, 471.

[22]Suárez notes that Aristotle left the question of whether the soul as an actuality bears the same relation to the body as a steersman to his ship, or not, open in the beginning of the second book of his *On the Soul*. See Aristotle, *On the Soul* II.1, 413a7–9, 73.

[23]*CDA* disp. 1, q. 1, n. 8, 68. Properly speaking, this *informing form* is only a necessary but not a sufficient condition of this unity. Since God can cause the intimate co-existence of both without the formation of a substantial unity, it is necessary to add a third ontological item in this substantial make-up, namely the substantial mode of unity. For this, see Perler (2020a).

[24]*DM* disp. 4, s. 3, n. 8, 128: "Igitur haec unitas per se in hoc consistit, quod resultat *ex rebus* constituentibus completum ens in aliquo genere, intercedente unione eorum inter se, ad tale ens constituendum accommodata, quae in compositione unius naturae est per modum actus et potentiae substantialis." [my italics]

[25]For the reified conception of the hylomorphic principles (a *locus communis* in the contemporary research on Suárezian causality), see Heider (2004, 2008), Hattab (2009, 2011, 2017), Åkerlund (2015), and many other authors.

[26]For a recent distinct (non-Suarezian) analysis of the unity and being in relation to the theory of prime matter as *pure* potential being without any form and actuality, see Wu (2017).

substance which, as the act of matter, constitutes with it the essence of a composite substance'.[27] Analogically, prime matter is conceived as an incomplete substance that together with a substantial form constitutes the complete essence of a material substance. It makes no sense to conceive prime matter as a pure 'ens in potentia' and say that it constitutes a composite with a substantial form. If generation and corruption are to be true generation and true corruption and not merely metaphysical movements of creation and annihilation, prime matter as the common 'continuant' must persist in substantial change. Without an enduring subject corruption would become a transition from being *tout court* to non-being *tout court*, and generation a change from non-being *tout court* to being *tout court*.[28] Material substances can possess unity per se only if they are composed of two distinct incomplete substances which both are individuated by their own entity.[29] Since it holds that there is no entity without identity, these principles must have not only an entity, but a unity as well.

In line with the rejection of Aquinas's real distinction between essence and existence in created beings (another metaphysical common element between *CDA* and *DM*[30]) Suárez declares that substantial form cannot be the principle that gives being to prime matter.[31] According to his theory of the real identity between essence and existence, if prime matter has its own essence (*esse essentiae*), it must also have its own existence (*esse existentiae*). Through divine intervention prime matter can even be conserved without a substantial form.[32] Prime matter is not a mere 'quo' that cannot exist without a substantial form. In correlation to his 'reification' of the substantial form, prime matter as such is endowed with an entitative act.[33] It does not receive its existence from the substantial form[34] but directly from the efficient cause, that is, from God.[35] Importantly, prime matter can become an immediate

[27]*DM* disp. 15, s. 5, n. 1, t. 1, 517. [I take this translations of *DM* 15 from Suarez (2000, 77).]

[28]For this, see *DM* disp. 13, s. 4, n. 17, 414.

[29]For Suárez's principle of individuation, see *DM disp.* 5, s. 6, n. 15, t. 1, 186. On this view it holds that no entity *X* can be individuated by a distinct entity *Y*. To give two examples, the entity of a substantial form cannot be individuated by the entity of prime matter and the entity of an accident cannot be individuated by the entity of its subject (substance). For the various issues of Suárez's metaphysics of individual unity, see Gracia's "Introduction" in Suárez (1982, 1–27).

[30]See *CDA* disp. 2, q. 4, n. 17, t. 1, 284–6.

[31]In *STh* 1, q. 66 a. 1, c., 154 Aquinas says: "Ipsum autem quod est actus, est forma. Dicere igitur materiam praecedere sine forma, est dicere ens actu sine actu: quod implicat contradictionem."

[32]*DM* disp. 15, s. 9, n. 3ff, t. 1, 533ff. Aquinas (*Quodlibet* III, q. 1 a. 1, c.) again thinks otherwise: "Esse autem actu repugnat rationi materiae, quae secundum propriam rationem est ens in potentia. Relinquitur ergo quod non possit esse in actu nisi in quantum participat actum. Actus autem participatus a materia nihil est aliud quam forma; unde idem est dictu, materiam esse in actu, et materiam habere formam. Dicere ergo quod materia sit in actu sine forma, est dicere contradictoria esse simul; unde a Deo fieri non potest."

[33]*DM* disp. 13, s. 5, n. 10, t. 1, 416.

[34]*DM* disp. 15, s. 8, n. 7, t. 1, 527: " [. . .] forma non est propria causa materiae, dans illi formaliter proprium esse, quo materia existit."

[35]*DM* disp. 13, s. 4, n. 16, t. 1, 413.

subject of the accident of continuous quantity and through it a remote subject of qualities.[36] The same theory of the reified hylomorphic principles can *in nuce* be found also in *CDA*.[37] It is not incorrect to speak about Suárez's 'reification' of the hylomorphic principles as early as in his *CDA*.

After this metaphysical digression to the reified character of soul and prime matter let me come back to *CDA* in which Suárez proceeds by highlighting the 'act-potency correlation' of both principles conceived in a reified form! For Suárez the soul as the substantial form is *essentially* ordered to its subject. Due to the possibility of a divine miraculous separation of incomplete substances including (sic!) sentient souls and inanimate substantial forms that are traditionally called material forms (leaving aside the natural separability of the rational soul from the body) the definition of this ordering cannot be captured in terms of the actual substantial form's informing but only in terms of the *aptitudinal* form's informing. The soul's essence consists in the internal capacity (*aptitudo*) to inform the body, not in its actual informing.[38] As *part* and as an *incomplete* substance the soul must be essentially ordered to the whole and to its correlative part. *Mutatis mutandis*, matter can also be defined only as an entity which is *apt* to receive a substantial form. As Suárez makes it clear, this *aptitudo* or the adjustment of the two reified principles to each other (*coaptatio*) does not involve an accident of relation. A categorial relation cannot slip into a substantial composition. If it did, the categorial order would be violated. Referring to *DM* 47, in *TDA* Suárez applies to this ordering a conceptual commonplace in Thomistic metaphysics, namely the notion of transcendental relation (*relatio transcendentalis*).[39] The soul is related to the body through such a transcendental relation, that is, the soul keeps being related to the body even when the body has been corrupted. Suárez nevertheless underlines that the soul must be conceived as an *absolute* entity which takes its essence through the *aptitudo* to the body. Formally speaking, this body is not its terminus. If it were, the soul would be formally dependent on the body since a relation can exist only if a terminus exists. If the soul, which is intrinsically adjusted

[36] *DM* disp. 14, s. 3, nn. 10ff, t. 1, 474: "Secunda sententia est [that embraced by Suárez], materiam primam ex vi suae entitatis esse sufficientem causam materialem formarum accidentalium sibi proportionatarum; primo quidem et immediate quantitatis, et, ea mediante, caeterarum." In *STh.* I, q. 76, a. 6, c., 229, on the contrary, Aquinas says that "Esse autem in actu habet per formam substantialem, quae facit esse simpliciter, ut iam dictum est. Unde impossibile est quod quaecumque dispositiones accidentales praeexistant in materia ante formam substantialem; et per consequens neque ante animam."

[37] For a suggestion about prime matter being the subject of accidents, see *CDA* disp. 3, q. 3, n. 4, t. 2, 120.

[38] *CDA* disp. 1, q. 3, n. 4, t. 1, 90.

[39] *TDA* book 1, ch. 2, n. 1, 471–2. See also *DM* disp. 47, s. 3, n. 10ff, t. 2, 797–9. As any other scholastics, for Suárez transcendental relation is a relation that is necessary or essential to a thing that it cannot be without it. A removal of this relation would means a change or destruction of its essence. Interestingly, in *CDA* Suárez rejects this terminology. For this, see *CDA* disp. 1, q. 3, n. 11, t. 1, 102–4.

to the body, can endure without its body, the body should be conceived as its *intrinsic end*, rather than its formal terminus.[40]

Having assigned the generic feature of the principle of life as a substantial informing form that has the aptitude to inform its body, what is the principle's specific difference? Suárez makes it clear that the specific difference can be found in the second part of Aristotle's first definition, namely in the phrase 'of a natural (organic) body potentially possessing life'.[41] Contrary to the substantial forms of inanimate beings, a soul is essentially ordered to an *organically structured body*. The soul bestows a much more perfect being on matter than inanimate substantial forms constituting with it a more perfect composite whose operations are more perfect and more varied. Proportionally, such a more perfect being requires a more perfect body. It necessitates a body that is appropriately structured by heterogeneous (organic) parts endowed with the manifold material dispositions that constitute what is called a proper bodily *temperamentum* or *temperies* of the first qualities.[42] How does Suárez conceive such organicity? And which 'life' is meant according to him in the second part ('a body potentially possessing life') of the second Aristotelian definition?

Replying to the first question gives Suárez an opportunity to distinguish between two kinds of organization of matter.[43] While the first structuring consists in accidental organization, the second is characterized by a substantial setup. According to the first organization an organic body is equivalent to matter informed by the accidents of continuous quantity and qualities. The qualitative variety of the parts is caused by the qualitative variety of accidents. Since these accidents form a proximate disposition for the introduction or eduction of a substantial form, their causality amounts to proximate material causality. As said, Suarezian prime matter is already endowed with an entitative act that makes it robust enough to sustain the accidents of quantity and quality. The second kind of organization, sc. substantial organization, is based on the various ways (*modi*) in which the soul informs matter. In spite of the fact that the rational soul is indivisible in its entity and exists whole in all the parts of its body,[44] it is multiple 'in virtue', that is, it virtually contains its lower grades. Therefore, it is not contradictory for the indivisible soul to be divisible through the various modes of its informing the body.[45] While the effect of accidental organization is matter plus accidents, the result of the second (substantial) organization is a matter-form composite.[46] While an 'organic body' in the first sense is

[40] *CDA* disp. 1, q. 3, n. 11, t. 1, 104.

[41] *Ibid.*, q. 4, n. 4, t. 1, 108.

[42] *Ibid.*, q. 3, n. 6, t. 1, 92–4.

[43] See *TDA* book 1, ch. 2, n. 7, 473–4 and *CDA* disp. 1, q. 3, n. 8, t. 1, 96–100. For an elaborate analysis of these kinds of organization of matter, see Des Chene (2000, 94–102).

[44] For this, see 2.5.

[45] *TDA* book 1, ch. 2, n. 11, 474–5.

[46] Suárez rejects the Scotistic theory of *forma corporeitatis*. See *TDA* book 1, ch. 2, n. 2, 472 and *DM* disp. 15, s. 10, nn. 7–15, t. 1, 537–540. For Suárez's and Fonseca's critique of the Scotistic corporeal form, see Des Chene (2000, 88–93). For Suárez's criticism of this corporeal form, see also Perler (2020a).

structured through accidents caused by a generating efficient cause, an 'organic body' in the second sense is structured by the soul.

Suárez is well aware of problems affecting both these interpretations related to Aristotle's dictum 'a natural (organic) body that potentially possesses life'. While accidental organization compromises the substantial character of the soul-body composite because no accident can be prior to a substance, substantial organization boils down to an already fully constituted living thing. According to this second organization this organic body will be identical to a living being. However, this is posterior to a soul since it is a hylomorphic composite which already includes a soul. Clearly, an organic body cannot be defined through it. It cannot be defined through something posterior (*per quid posterius*).[47] In his replies to these two objections Suárez employs a distinction between various orders of causality. He refutes the first objection related to the accidental kind of organization by claiming that although it holds that a substance is prior to accidents in its perfection and final causality due to the soul's *entelechy*, some accidents can be prior to a substantial form in the order of material causality. In distinct causal orders priority and posteriority in respect to the same entity may vary. As it has been said, Suárez's view of prime matter that possesses its entitative act is even better suited for this kind of accidental structuring than prime matter in Aquinas is. In his rejoinder to the second caveat, Suárez again distinguishes between two orders of causality. This time he refers to the order of final causality and that of efficient causality. Even though in the order of execution a composite is posterior to the substantial form or a soul, this is not so in the order of final causality. In the order of final causality, a material composite is always prior to a form since a form is essentially ordered to the constitution of a material composite. Since a soul's relation to composite and matter is the same in the order of final causality an organic body can be conceived as a matter informed by a soul.[48]

Despite his *prima facie* neutral evaluation of both views of the organization of matter indicating that both can be regarded as probable and compatible,[49] in *TDA* Suárez after all hints his preference for accidental organization which is more in line with his theory. He declares there that accidental organization better expresses the proper meaning of Aristotle's first definition of soul. He substantiates this preference by his correlation of the notions of accidental and substantial organization of matter with two kinds of life in the definitional idiom 'potentially possessing life'.[50] Following Aristotle and Aquinas,[51] Suárez distinguishes between substantial and accidental life. The first kind of life is contrasted with death. Substantial life abstracts from particular manifestations of life and corresponds to Aristotle's dictum '[...] for

[47] *CDA* disp. 1, q. 3, n. 9, t. 1, 100.

[48] *Ibid.*, n. 9, 100–2.

[49] In *CDA* Suárez even indicates that Aristotle considered 'organic body' to be organized by substantial organization. For this, see *CDA* disp. 1, q. 4, n. 6, t. 1, 110–1.

[50] *TDA* book 1, ch. 2, n. 17, 476–7.

[51] See Thomas de Aquino, *STh*. I, q. 18, a. 2, c., 226–7.

living creatures existence is life'.[52] On the contrary, accidental life marks a *particular* vital operation or a particular manifestation or exercise of substantial life. Such accidental lives are introduced by Aristotle in his *Nicomachean Ethics* where he speaks about the contemplative and practical lives (acts) leading to the blessed life. Suárez holds that if life in Aristotle's definition is taken in the substantial sense, then only an organic body that is proximately disposed through accidental organization can potentially have substantial life.[53] As stated, the inclusion of accidents in the definition is not detrimental to the substantial constitution of a hylomorphic composite since the accidents are prior to the soul and the composite only in the order of material causality. If life is understood in the sense of accidental life, the organic structure and the organization of matter will be substantial.[54] Crucially, Suárez makes it clear that if Aristotle's first definition were embraced in this sense of accidental life, it would not be much different from the second definition according to which 'the soul is that whereby we live and perceive and think in the primary sense'. These definitions would come to be similar since the order of the soul to accidental life and the order to operating are actually one and the same.[55] Accordingly, this conclusion is more compatible with Suárez's reification and dualization of the hylomorphic principles representing a significant metaphysical twist of Aristotle's and Aquinas's theories.

2.2 The Soul as the Intrinsic Principle of Vital Operations

It has been indicated that Aristotle's two definitions of the soul expressing the soul's two different causalities are systematically interconnected. It cannot be said that Suárez replaces the soul's formal causality with the soul's efficient causality.[56] As seen in the previous section, the soul which is the principle of operating must be the same as that which is the principle of being.[57] The efficient principle of vital acts cannot be a form that would not be the *first* actuality of the organic body or the *informing* substantial form. Despite some tendencies which will be highlighted below, a motor moving a body from the outside or a sailor moving his ship are not the right models for the soul according to Suárez.

---

[52]Aristotle, *On the Soul* II.4, 415 b13, 87.

[53]*TDA* book 1, ch. 2, n. 30, 481: '[...] corpus organicum [...] includere tantum accidentales dispositiones, quia per illas sufficienter praeparatur, et in potentia proxima ad vitam substantialem componendam vel recipiendam, constituitur'.

[54]*TDA* book 1, ch. 2, n. 33, 482.

[55]*Ibid*., ch. 3, nn. 8–11, 487–8.

[56]For a tendency to this Replacement Claim, see Pasnau (2004, 38–9); Pasnau (2011a, 101–18); Hattab (2009, 64); Perler (2015, 136–9); López Prieto (2006); Knight (1957); Ewbank (1987). For an important caution that Suárez does not set aside the formal causality of the substantial form in favour of the efficient cause, see Richardson (2015).

[57]*CDA* disp. 1, q. 4, n. 3, t. 1, 106–8; *TDA* book 1, ch. 12, n. 16, 559.

Before coming to the issue of how Suárez explains Aristotle's second definition with its central notion of 'actus vitalis', let me briefly note how he evaluates the two definitions and how he views their relationship. He regards both definitions as *optimae* or *bonae*, which means that they are neither broad nor narrow.[58] The extension of both definitions holds for all animated beings. However, concerning their intension Suárez says that, unlike the first, the second definition cannot be called 'quidditative' but only 'descriptive'.[59] The soul's giving substantial being to the composite and being the formal cause of matter is prior to the soul *qua* the principle of operations. Since 'action follows being' (*agere sequitur esse*) the first must be considered as a proper accident (*proprium*) of the second.[60] It is not right to say that the soul gives being because it operates, but rather vice versa that it operates because it animates the body.[61] While the soul immediately and through itself gives being to the composite, for operating (as we will see in 2.6) it requires a really distinct instrument, sc. a power.[62] Unlike its essential ordering to matter and the composite, the soul is not primarily and essentially related to its operations but to the constitution of the integral essence.[63] This is also the reason why operations cannot be, properly speaking, conceived as the soul's final causes. The demonstration of the first definition from the second, which is better known to us, cannot be conceived as an *a priori* demonstration. Contrary to some early Jesuits,[64] this demonstration can proceed only as going from the effect to the cause, that is, *a posteriori*. 'Actus vitales' are the effects with which we are most familiar. They are much more familiar to us (*prout nobis*) than their principles, such as the soul, powers and habits. This is also why the second descriptive definition is more known to us than the first formula. Only from the (phenomenological) cognition of acts we can infer their principles and, consequently, only from the second definition we can infer the first definition. With respect to our knowledge, it is more evident that the soul is the principle of operating than that it is the formal cause.[65]

How does Suárez precisely characterize the notion of 'actus vitalis', which he regards as an effect of the internal efficient principle, sc. the soul? Contrary to scholastics of the first half of the seventeenth century, Suárez does not give a

[58] *CDA* disp. 1, q. 4, n. 4, t. 1, 108. For this explanation of Suárez's characterization of the first definition as 'bona', see Kronen and Reedy (2012, 86, note 236).

[59] *CDA* disp. 1, q. 4, n. 9, 114.

[60] *Ibid.*, n. 14, 122.

[61] *TDA* book 1, ch. 3, n. 18, 489–90.

[62] *Ibid.*, n. 19, 490.

[63] *Ibid.*, n. 20, 490; see also *CDA* disp. 1, q. 4, n. 15, 122.

[64] For a theory according to which the first definition is proved *a priori* from the second definition as from its final cause, endorsed by Francisco de Toledo, and for a distinction between Toledo and Suárez related to their views of the relationship between the two definitions of the soul, see Des Chene (2000, 105–7).

[65] *TDA* book 1, ch. 3, nn. 15–6, 488–9; *CDA* disp. 1, q. 4, n. 17, 124–6.

systematic account of what it is to be a 'vital act'.[66] However, an important lesson can be drawn from Suárez's formulations that occur in the various places of the first, second and fourth question of the second disputation of *CDA* and of the third and the fourth chapter of the first book of *TDA*, in which he asks whether the substantial form, which is the principle of the vegetative, sensitive and intellectual operations, is a true (informing) soul, or not. In the case of *CDA* disp. 2, q. 4 the second disjunct of the dilemma is not whether the principle is an *inanimate* substantial form, but whether it is an *external* moving principle or 'extrinsic motor'. In these two contexts Suárez comes up with several qualifications related to the notion of 'actus vitalis' that are relevant for his theory of perception.

He makes it clear that the least problematic case is represented by the sensory soul. The form that is the principle of sentience must be a true soul since a sensorial act can be understood only as coming from the soul as from its internal principle. Referring to the first-person phenomenological perspective, he says that we experience that our perceptual and appetitive acts come from within, from us and that they are caused by a principle which exists in us. In this context we can also find what can be regarded as a first indication of Suárez's perceptual activism in the whole *CDA*: *Sentire* is not a mere passive reception caused by a distinct entity such as a sensible species or an external object but a vital activity coming from the perceiver. A perceptual act is *vitaliter operari*.[67] If an external entity were this principle or cause, let alone the only cause, a perceptual act would become nothing else than a transitive act. Accordingly, it would cease to be vital. Cognitive and appetitive acts are examples of perfect vital operations since, as it will be shown below, they—as immanent operations—are produced and received in the *same* power.[68] In this respect they differ from the operations of non-living beings, such as fire. Although it might seem that fire brings about some sort of vital acts through its growing, drawing and converting, as if 'digesting', a matter such as a piece of wood, paper or other combustible, it is clear that it cannot be ranked among living beings. Suárez rejects such panpsychism by saying that this 'digesting' proceeds merely *per iuxtapositionem*, that is, through a series of partial transitive processes in which one part acts on the adjacent part. Fire's spreading takes its course between the distinct partial *agens* and *patiens*, say, of a log.[69] However, he is also clear that immanency in an unqualified sense cannot be considered to be the sufficient condition of vitality. In his negative demarcation of the notion of vital act, Suárez

[66]For the systematic account of the notion of vital act, its kinds and the question about the possibility of the divine creation of this act in a subject which does not intrinsically produce it, see Mastrius and Belluto, *Philosophiae ad mentem Scoti cursus integer*, vol 3: *In libros de anima*, 1727, disp. 2, q. 2 'Quid sit actus vitalis', 46–9; q. 3 'An actus vitalis pendeat ab anima essentialiter, ut sine illius concursu nequeat produci', 49–50.

[67]*CDA* disp. 2, q. 2, n. 11, t. 1, 160. Elsewhere (for instance, *CDA* disp. 5, q. 5, n. 8, t. 2, 378) Suárez says that 'Cognoscere nihil est aliud quam vitaliter operari, formando quodammodo rem ipsam cognitam'.

[68]*TDA* book 1, ch. 4, n. 5, 493.

[69]*Ibid.*, n. 17, 497.

differentiates a vital act from the operations caused by per accidens efficient causality. What kind of the efficient causality does he mean by this causality? He illustrates it with the example of the efficiency of the substantial form of water. As said in *DM* disp. 15, s. 1 and elsewhere, when water has been heated and the heating has stopped, the water starts to return from this accidental state of being hot to its connatural state, sc. to being cold. In line with the ancient physics of the first qualities constituting the elements he says that the primary quality of cold is for water its connatural property. The reduction to the original quality is and can be caused only by the efficient causality of the water's substantial form (not, e.g., by its environment or any other external agent). This efficacy is not an example of a per se vital operation since the water's reduction to its connatural coldness happens to be only per accidens, that is, after it has been heated and received its accident of the hot.[70] The water's accidental 'deflection' from its natural state makes this process of cooling only per accidens efficient causation. Clearly, no such 'deflection' from its connatural state can be found in the soul's vital operating. In order to vitally operate the soul does not need to be accidentally 'deflected' from its original state. Vital operating is a connatural state for the soul.

As we shall see more in 3.5 on the example of cognitive acts and their end terms, in *DM* Suárez claims that vital acts conceived as qualities in the kind of disposition caused through an action, understood as an ontological mode of dependence upon a cause, can be produced transitively in Peter through an external principle, namely through a divine intervention, without his soul's active causal concurrence.[71] If these termini or vital operations are produced by a distinct action conceived as a separable mode, this action must be a variable item.[72] However, in *TDA* he, at least implicitly, comes with a distinction which was conceptualized by later authors as the distinction

---

[70]*TDA* book 1, ch. 4, n. 12, 495; cf. also *DM* disp. 15, s. 1, nn. 8–13, t. 1, 500–2.

[71]If this conclusion were Suárez's final word (we will see that it is not), he would deviate from the theory of Aquinas and Thomists headed by Cajetan who endorsed the necessary and essential dependence of *opera vitae* on their powers. Accordingly, Suárez would espouse Scotus's view according to which the vital acts, conceived as absolute entities (qualities) rather than actions, could be transitively inserted to Peter's mind by God. For the former conclusion, see, among many *loci*, Aquinas, *STh.* I, q. 51, art. 3, c., 18: '[...] ad id quod est proprium viventium [...] cuius est potentia, eius est actio; unde nihil potest habere opus vitae, quod not habet vitam, quae est potentiale principium talis actionis'; for Cajetan, see *In STh.* I, q. 26, art. 3, IV, 304: 'Impossibile quoque est operationem vitalem meam non ab intrinseco meo esse, cum in hoc ratio meae vitae consistat: esset enim vitalis et non vitalis. Et propterea, cum beatitudo sit operatio vitalis beatorum, implicat divinam essentiam esse illam'. For Scotus's view, see, e.g., *Quodl.* q. 13, n. 32 (Vivès 25: 581): '[...] illa qualitas, quae vel est operatio vel includitur in operatione, non habet relationem ad subiectum magis essentialem quam aliae qualitates; et ideo, si ponatur alias non esse essentialiter relatas ad subiectum, de quo non est modo quaestio, nec ista poneretur essentialiter relata ad subiectum'. For Scotus's view of the creatibility and insertability of vital operations into finite minds, see Pini (2011, 61–3) and Cross (2014, 119).

[72]*DM* disp. 48, s. 2, n. 12, 877: 'Potest ergo Deus efficere hos actus [perfect vital acts such as sensations and intellections], quatenus sunt quaedam qualitates, sine concursu activo potentiarum, sive illos efficiat informantes potentias, sive non [...] ergo dependentia harum qualitatum a potentiis non est de essentia et entitate harum qualitatum; ergo est aliquis modus ex natura rei

between the acts *qua* qualities (*in ratione qualitatis*) and the acts envisaged as operations of seeing, intellection, etc. (*in ratione operationis*).[73] This distinction and other textual indications show, so my claim, that not only sentient operations but all cognitive and appetitive acts *qua* vital operations (in the second sense) must for Suárez come from the internal principle. None of these operations can be substituted by the divine primary causality without ceasing to be vital. It is impossible for Peter to perceive and understand through an external principle, even though it were divine. God cannot implant cognitive or appetitive acts in Peter without the proper concurrent efficiency of Peter's soul.[74] Moreover, even in the beatific vision (*visio beatifica*), which is a manifestation of life in its fullness, God cannot become the total cause of Peter's beatific vision. This vision cannot be fully implanted in Peter without Peter's concurrent vital activity.[75] If *per impossibile* this were the case, Peter's vital and immanent operation would turn to be transitive and would lose its essential intrinsicness. In such case, a cause originating from one supposit (God) would be received in a distinct supposit.

Against this background it is not surprising to find Suárez rejecting Averroist monopsychism in *CDA*. According to Averroes a separated supraindividual intellect communicates its own acts to particular human minds.[76] For Suárez the intellective principle must be only the informing form. This testifies the basic cohesion of both Aristotle's definitions. I experience that it is *me* who understands, sees, wills, etc. In all these acts I know not only an external object or its quality, such as a tree or a dog, but also an act through which I see this tree, this dog. I also somehow experience and

---

distinctus ab ipsis'. This passage became 'the locus communis' and was quoted by a great number of post-Suarezian scholastics in the seventeenth century.

[73]For this useful distinction, see also Raffaele Aversa, *Philosophia metaphysicam physicamque complectens in duos tomos distributa*, vol. 2, q. 48: ‚De Actibus vitalibus Animae', s. 1: ‚An hi actus sint essentialiter vitales, & esentialiter ab anima pendeant', 671.

[74]This does not mean that God cannot causally intervene and insert a sensible species to Peter's senses. There is no logical contradiction in this state of affairs because this item is non-vital and as such it is transitively caused by an external object. For this divine intervention and the issue of scepticism, see 3.9.

[75]For Suárez's claim about a contradictory occurrence of a vital act of the beatified intellect that comes to be without the concurrence and efficiency of the internal principle of life, see Suárez (1856a), *Opera omnia*, vol. 1, *De deo uno et trino*, book 2: *De attributis Dei negativis*, ch. 10, n. 7, 79–80: 'Addo denique valde probabile esse, implicare contradictionem, fieri actum, qui de se et natura sua est actus vitalis sine concursu et efficientia principii vitalis. Quod altero e duobus modis affirmari potest. Primo afferendo talem actum essentialiter postulare dependentiam a tali principio et ideo sine illa fieri non posse, quod tamen difficile suaderi aut probari potest, de quo videri possunt dicta in Metaphysicae disp. 48, sect. 2, n. 9, et seqq. Secundo dicendo, quamvis hujusmodi actus possit fortasse fieri et conservaria Deo in entita sua, absque activo concursu causae secundae, quia Deus eminenter continet omnem efficientiam: fieri tamen non posse, ut talis actus exerceat proprium munus actus vitalis, id est constituat cognoscentem, amantem, vel aliquid simile sine concursu ejus, in quo talis effectus exercetur'.

[76]*TDA* book 1, ch. 12, n. 11–3, 556–8; see also *CDA* disp. 2, q. 4, n. 4, esp. 256–60. For the critical approach of the early Iberian Jesuits in general to secular Aristotelianism (Averroism and Alexandrism) with its bastion in Italy, see Lohr (1976) and Heider (2016b, 1–11).

am aware of these operations and, importantly, I know that these operations are *mine*. These acts are not free-floating absolute entities; they inhere in my mind or (in the case of senses) in the powers that are subjected both in the soul and in the body. I do not get to know these acts primarily from a third person perspective. I always apprehend them in the first person perspective as being mine. If the vital principle were a separated intellect related to my body accidentally as a sailor is related to his ship or if they were inserted to my mind from outside even by God, it could not be *me* who understands, sees, wills, etc. It would have to be someone else who would exercise them and who would be active through myself being the medium. I would thus be operated through someone or something else, as a puppet is operated by an actor. A mere reception of these acts-qualities, however conceived as the absolute entities would not justify their essential selfhood and ownness which we experience.[77]

Regarding what was later called imperfect vital acts, such as vegetative operations,[78] a following objection can be adduced, though. Cognitive and appetitive operations differ from vegetative acts, such as nutrition and growth since the latter are, in a way, *transitive* operations. They proceed through a series of alterations, such as the heating or cooling of one part of the body from its other moving parts. The heart moves and makes the blood circulate in the body; gastric juices dissolve food in the stomach.[79] Moreover, God can substitute for a vegetative soul's efficiency and cause these processes.[80] The growth of a tree could be caused by the primary cause who would apply nutrition to it.[81] However, if this can be the case, these operations do not differ substantially from the spheres' circular motion caused by the intelligences. Consequently, they cannot be evaluated as vital.[82] In his reply, Suárez makes it clear that also vegetative acts are vital operations. He notes that it is not necessary that all vital acts are perfectly immanent, though. Although all immanent acts are vital, the contrary is not true. While immanent activities in the strict sense can be found in cognitive and appetitive processes, vegetative operations can be called vital, albeit not entirely immanent. However, at the same time their transitivity is limited.

[77]*CDA* disp. 2, q. 4, n. 4, t. 1, 258. For this, see also *TDA* book 1, ch. 12, n. 14, 558: 'Ergo actus intelligendi elicitus ab intelligentia separata intra me existente, nunquam potest ita communicari meae cogitativae, ut per illummet actum ego fiam intelligens actu: sicut neque per intellectum ejus possum ego constitui formaliter intellectivus'.

[78]It was only in post-Suarezian philosophy in which a systematic account distinguishing the imperfect (*actus vitalis imperfectus*) and the perfect vital acts (*actus vitalis perfectus*) became established. While the perfect vital acts, such as cognitive acts after the production of their endpoints continue in their vital operation by intentional stretching into and unifying with an object known, the imperfect vital acts, such as generation, nutrition or local motion after the production of their own termini cease to be vitally operating. For this distinction, see Mastrius and Belluto, *Philosophiae ad mentem Scoti cursus integer*, vol 3: *In libros De anima*, disp. 2, q. 3, n. 41, 49.

[79]*TDA* book 1, ch. 4, nn. 7–8, 493–4. For the vegetative soul and its operations, see *CDA* disp. 4, qq. 1–9, t. 2, 152–281.

[80]For this, see also Des Chene (2000, 59–60).

[81]*TDA* book 1, ch. 4, n. 9, 494.

[82]*Ibid.*, n. 16, 496–7.

In *TDA* Suárez distinguishes between two kinds of immanent acts. While one is immanent in a narrow sense, the other kind is called immanent in a broad sense. An immanent act in the narrow sense is an act that remains both in the same supposit and the same *power* through which it was produced. From this point of this definition vegetative acts are not immanent. In the second sense, however, an immanent act can be conceived as an act that proceeds transitively from one part to another but remains in the same supposit.[83] Conceding again the possibility that the effects of vegetative operations may be divinely caused (see the quote from *DM* disp. 48, s. 2), Suárez's reply is clear: A material identity of termini is not sufficient for unity of action. God can cause these vegetative acts, indeed. If the *modi efficiendi* are diverse, one true vegetative or cognitive act coming from inside the soul and the other coming from outside, sc. from God cannot be considered identical even though their effects seem to be such. Crucially, the 'vitality' of vital acts *cannot* be identical with the absolute entity of an endpoint. The formal character of vitality as such lies in a terminus's (act's) essential and inextricable *relation to* and *denomination from the vital principle*, sc. the soul. No doubt, the same holds also for cognitive and appetitive acts, that is, for perfect vital acts. Although God could create an absolute quality amounting to a cognitive operation in Peter, if this quality were not caused by the soul it could not be, properly speaking, called a *vital* operation since for any vital operation it holds that it must be necessarily caused by the soul via a pertinent power. As created by God it would be no more than an item *in ratione qualitatis*.[84]

Importantly, Suárez does not conceive this denomination from the soul as the remote causal influence through the capacities but as the soul's *proximate* causal influence in vital acts. In *DM* disp. 18, s. 5 a vital act is succinctly characterized as an operation that bears:

> [. . .] an *intimate* connection with its principal formal principle, in order to be able to be effected in a way consonant with its nature and also in order to be able to affect its own principle or suppositum in the vital mode.[85]

The key difference between the operations of living and non-living beings lies in the former's *intimate* connection with a soul. Accidental life or vital operations can be caused not only through the immediate influence of a power. Regarding proximate

[83] *Ibid.*, n. 13, 496.

[84] *Ibid.*, n. 16, 496–7: 'Unde concedo non repugnare, eamdem rem quae per actionem vitalem fit, posse fieri ab altero agente absque vitali actione, si diverso fiat. Nam fortasse etiam ipsas entitates, seu qualitates actuum vitalium rigorose permanentium potest Deus facere sine concursu potentiae, et tunc actiones *non* essent vitales, licet res, seu qualitates productae essent eaedem [. . .] vitalitas actionum non dicit aliquid absolutum in ipso termino, sed *habitudinem* ad tale principium *cum denominatione* ab illo [. . .]' [my italics]. This question about the separability and the (potential) vitality of these separated acts became a hot issue between the Thomist-oriented and the Scotus-oriented scholastics in the first half of the seventeenth century (see abovementioned Mastrius and Belluto, Aversa, and many other Baroque scholastics).

[85] My italics. *DM* disp. 18, s. 5, n. 3, t. 1, 628: '[. . .] hanc intimam connexionem cum suo principali principio formali, ut modo consentaneo ad naturam suam fieri possit, et suum principium vel suppositum vitali modo afficere valeat'. (The translation is taken from Suárez (1994, 123)).

causal influence, a power alone (with a sensible species) is not enough to produce a vital act. Employing technical nomenclature related to kinds of causes, Suárez says that it is the soul as the *proximate* principal principle *quo* of acting (with the accidental principle *quo* of acting, sc. the power) that necessarily concurs in the production of vital operations.[86] An intellect separated from the soul through a divine intervention, a bizarre but possible state of affairs for Suárez due to its real character and its real distinction from the soul, cannot be conceived as *in actu* understanding.[87]

2.3 The Rational Soul as 'Semi-Person'

It has been said that the soul as the first actuality of the organic body is at the same time the principle whereby a living being generates its vital operations. It has been indicated that all the souls and even all inanimate forms are not only 'quos' but also reified 'quods'. All these forms can be miraculously conserved in their being independently of their matter since their essence is defined only through the aptitudinal informing. Deploying a parallel with the Eucharist (an important methodological regulative for Suárez's philosophy as such),[88] if the quantity of bread in a host can be conserved independently of the substance of bread, the same must *a fortiori* hold for a substantial form.[89] Having said that for Suárez the rational soul is incorporeal, subsistent and immortal[90] and material souls (of animals and plants) and inanimate substantial forms are neither subsistent nor immortal, it is important to consider an additional metaphysical feature that appends to the rational soul. When embodied, the rational soul possesses not only a reified entity characterized by the notions of partial and incomplete being (substance) but also its own partial and incomplete *subsistence*. This subsistence makes the soul a crucial diachronic 'continuant' stretched between its fleshy status and the resurrected state. As said in *DM*

[86]For a classification of the causes into principal and instrumental 'quo/quod', see *DM* disp. 17, s. 2, n. 7, t. 1, 586. See also 3.4.

[87]For this, see *CDA* disp. 3, q. 1, n. 22, t. 2, 80–2: 'Sed quaeres [rursus], utrum intellectus separatus possit intelligere, sicut calor calefacere. Respondetur: Non [posse], quia intellectio est actio vitalis, et ideo intrinsece pendet a primo principio vitae, sine quo esse non potest'. See also 2.6.

[88]For the methodological significance of the transubstantiation notion of the Eucharist in Suárez's view of continuous quantity conceived as an accident that is really distinct from substance, see Heider (2013).

[89]See *DM* disp. 15, s. 9, n. 1, t. 1, 532.

[90]For the immortality of the rational soul as being one of the important determinants of the early Jesuits' psychology in general and for Suárez's incipient dualism conceived as a reaction to Pomponazzi's view of the inextricability of matter from form, see Heider (2011b) and South (2012).

34 entitled 'De prima substantia, seu supposito eiusque a natura distinction',[91] both material forms and rational souls have their own partial entity. A numerically one and the same prime matter can exist under the substantial form *X* at t1 and under the substantial form *Y* at t2. However, regarding forms it is only the rational soul that has its partial subsistence in the composite. We will argue that precisely this feature makes the rational soul, especially when compared to the lower souls, even more ontologically and epistemically 'robust'. Considering the scholastic axiom 'actions are of supposits',[92] the attribution of a partial subsistence or 'semi-personhood' to the embodied rational soul cannot be set aside as irrelevant to Suárez's conception of perception.[93] This theory is all the more important if we take into account that in this context Suárez sometimes willy-nilly slips into a Neoplatonic assertion according to which the human body is an *instrument* used by the soul and, and in line with his emphasis on its efficient causality, considers the rational soul to be an agent *sui generis* also in sensory operations.[94] Additionally, this view gains more of a methodological relevance if collated with Cajetan's doctrine which, as in many other issues, represents the important doctrinal opponency of Suárez's view.

The theory of the rational soul conceived as semi-person (*semi-persona*) introduced in *CDA* is grounded in Suárez's metaphysical theory of the divisibility and communicability of subsistence.[95] Before we present his conclusions related to this theory, two preliminary brief remarks are in place. The first concerns the definition

---

[91] Unfortunately, when compared to other disputations, this disputation which demonstrates how the mystery of the Incarnation of Christ was operative in Suárez's metaphysics has not received much attention from scholars. For an exception, see Söchting (2019).

[92] In the seventh section of *DM* 34 entitled 'Whether subsistence has any causality, and in which sense actions are said to be of the supposits', when speaking about efficient causality, Suárez formulates two main conclusions. While in the first conclusion he says that the created subsistence does not have a proper and immediate influence in the actions of a supposit (*DM* disp. 34, s. 7, n. 6, 413) and only the form does, in the second upshot he notes that subsistence is far from being related to an agent's activity only concomitantly or accidentally but bears upon it antecedently and necessarily (per se) (*DM* disp. 34, s. 7, n. 10, 415). Importantly, this holds not only for complete subsistent entities but also for the rational soul. Despite its incomplete and part-like nature, it is possible to attribute action also to the rational soul since it is 'subsistens singulariter ac per se' (*DM* disp. 34, s. 7, n. 11, t. 2, 415).

[93] James South (2018, 64) states that the attribution of semi-personhood to the rational soul indicates Suárez's important departure from Aquinas: 'Once he makes this move the *other traditional topics* [and we may be sure that also his cognitive theory including the theory of perception] he addresses take on new dimension'. [my italics].

[94] *CDA* disp. 14, q. 1, n. 2, t. 3, 446: 'At in homine, licet quod praecipue subsistit sit homo, tamen haec subsistentia praecipue est ratione animae, quae unitur corpori, non ut substentetur ab illo, sed potius ut utatur illo *ut instrumento coniuncto*'. [my italics]. This characterization comes close to Augustine's definition of man as a soul using a body, indeed. See Augustine (1978, ch. XXVII, 289): 'Homo igitur, ut homini apparet, anima rationalis est mortali atque terreno utens corpore'.

[95] *CDA's* disputations on the immortality of the human soul and its separated existence going beyond the model of Aristotle's *On the Soul* represent a crucial link to the metaphysics of subsistence. It must be said that for early Jesuits introducing these topics into commentaries on the *De anima* was a commonplace.

of the notions of subsistence, supposit and person; the second applies to the broad issue of the 'additum' of created subsistence to an individual nature.

In the beginning of *DM* 34 Suárez identifies supposit with what Aristotle in the *Categories* calls primary substance (this man Peter, this horse Bucephalus, etc.). A primary substance is an individual substance which, unlike accidents, exists per se. Unlike secondary substances, it cannot be communicated to another subject. It can neither inhere in nor be predicated of a subject.[96] The notion of 'person' differs from this concept of supposit with respect to the rational nature. As Boethius's famous definition has it: 'Person is an individual substance of a rational nature'.[97]

In his extensive analysis of what a *suppositum* adds to an individual nature Suárez compares a complete individual nature of a created substance to a supposit. What is compared is Peter's individual humanity that is composed of his soul and his body, called also the form of the whole (*forma totius*). In the first negative step, Suárez denies Aristotle's view, according to which there is no *ex natura rei* distinction: '[. . .] the essence and the individual thing are [. . .] the same'.[98] According to this view, the supposit of Peter is nothing above the composite of Peter's soul and Peter's body. It is superfluous to say that a supposit brings an additional element in the metaphysical make-up of a first substance. As can be expected, Suárez the theologian is highly critical of this position. Although he agrees that natural reason cannot prove its falsity, it is the Christian mystery of the Incarnation that clearly shows that a supposit *adds* something to an individual nature and that both these items are *in re* distinct. From the Incarnation we learn that the singular humanity of Christ was assumed and hypostatically united in the single person to the divine Word. It was not a created supposit that was assumed by the substantial *Logos* but only Christ's individual humanity.

After a long segment of *DM* disp. 34, 2–4, in which Suárez delves into a very complex and sophisticated critique of a whole sequence of 'additum-theories',[99] he formulates his own position: Personality is added to an individual human nature to

---

[96]Aristotle, *Categories* ch. 2, 1b2–4, 15.

[97]The Greek word 'ὑπόστασις' is synonymous to 'suppositum' or 'person.' The Greek 'ὑπό' signifies the Latin 'sub', which is part of the term '*sub*stance' and '*sub*sistence'. Cf. *DM* disp. 34, s. 1, n. 14, t. 2, 352. For Boethius's definition, see Boethius (1847, 1343).

[98]Aristotle, *Metaphysics* VII.11, 1037b1–2, 371.

[99]Besides the theory claiming that an individual nature and supposit differ only conceptually, he rejects Scotus's view according to which a supposit adds only a *negation* to the individual nature, namely the negation of dependence (both actual and aptitudinal) on another supposit (*DM* disp. 34, s. 2, n. 8, 355). He also denies the theory saying that the supposit of a material composite is determined by accidents or by a principle of individuation added to the specific nature (*DM* disp. 34, s. 3, t. 2, 359–67). He also declines the view that the supposit is the substantial *esse*. He takes this position to be consistent with the theory of the real distinction between the essence and existence of created beings, which he sharply denies (*DM* disp. 34, s. 4, nn. 1–16, 367–72). Lastly, he criticizes the stance associated with Cajetan according to which created subsistence is *ex natura rei* distinct from the essence and existence of a substantial nature and being prior to existence it constitutes a real composition with its essence (*ibidem*, n. 17, 372). This rich part of *DM* 34 is undoubtedly awaiting elaboration.

provide it with a last complement in its existence, or (in other words) it is posited to complete the nature's existence in respect to its subsistence. Personality is not a terminus or mode of an individual nature according to its essential being but according to its existential being. Existence cannot be subsistence since it denotes only an entity's being outside its causes. As such existence is indifferent to the modalities of per se and per accidens existence.[100] Employing the ontology of modes, which is largely operative not only in *DM* but (as we shall see) also in *CDA*, Suárez states that what a created supposit adds to the nature is not distinct from it as a *res* but only as a *modus*.[101] Subsistence differs from an existent individual nature modally. Such modal distinction fully suffices to constitute a metaphysical framework in which the mystery of the Incarnation can be conceived as a possible state of affairs.[102] Showing its possibility is one of the main tasks for Suárez the metaphysican. A created personality constitutes a true metaphysical complex of the nature completed by subsistence conceived as a mode.[103]

It seems that if subsistence as the last component of an individual nature makes the nature subsistent, it makes it also *entirely* incommunicable in other possible respects. As absolutely incommunicable created subsistence thus seems to be fully indivisible. If it were divisible, one part of subsistence would have to be united to another part. No doubt, this would be at odds with the subsistence's absolute incommunicability. Those who advocate this absolute unshareability are consequent in denying that the essential parts of the individual nature, viz. matter and substantial form, can be conceived as subsistent entities. Suárez is aware of this view and says that this view is actually that of Cajetan for whom subsistence is something distinct from the whole composite nature. For him subsistence is applicable only to a hylomorphic compound. Given that the rational soul can be separated from the body, an important consequence of this theory will be this: On the separation from the body the rational soul will acquire an entirely *new* subsistence which it did not have when it had been joined to the body. It will acquire a new subsistence that it did not possess as being part of the soul-body compound. Therefore, this partial subsistence or semi-personhood will belong to it only after its separation from the body. Due to Cajetan's emphasis on the unity of the soul-body composite, being in line with the spirit of Aquinas's reasoning, the rational soul and the body in its embodied state will subsist only through a single indivisible and spiritual subsistence. Since subsistence primarily follows form (in the case of man the rational soul), human subsistence will have to be spiritual. Accordingly, as it can be expected, it will somehow ontologically elevate the human body with all the powers that also inhere in it. Consequently, this will 'elevate' also all the sensory cognitive and appetitive powers and approximate them to the level of intellectuality and the level of 'sentient

[100] *DM* disp. 34, s. 4, n. 23 and n. 28, t. 2, 374, 375.

[101] For Suárez's theory of modes, see Pasnau (2011a, 253–258) and Menn (1997); see also 6.2.

[102] *Ibid.*, n. 33, 377.

[103] *Ibid.*, n. 38, 379.

thinking', the view that is, as we shall see especially in Chap. 5, consistently denied by Suárez.

Suárez considers this Cajetanic view to be absolutely false.[104] It makes no sense to say that a human body subsists through a spiritual subsistence. The human body cannot be understood as a spiritual entity or as in any way 'spiritualized'. It does not make sense to assert that the being of the soul is the same as the being of a man either.[105] For this reason Suárez states that no subsistence of a material nature (except of spiritual entities) is indivisible.[106] Since the subsistence of a man is not spiritual, it cannot be indivisible. In general, the subsistence of material things is composite in two ways. First, it is composed of integral parts, such as hands, feet and heart; second, it is composed of essential parts.[107] Regarding the second (for us here relevant) composition, the subsistence of a human composite contains the partial (incomplete) subsistences of the body and the soul.[108] There is no contradiction in the co-existence of the rational soul's mode of partial subsistence and its substantial mode of unification with the body. The soul's subsistence, that is, its independence of the body, is fully compatible with its being the substantial form of the body.[109] Since it holds that only subsistence can be a proper terminus of creation and the rational soul is created by God, the soul must belong to the subsistent items. At the same time, the rational soul is created as essentially (aptitudinally) informing the body. If there is no tension between these two basic characteristics in the very beginning of the soul's existence, there can be no incompatibility between these at a later point. Moreover, as Suárez argues, Cajetan himself recognizes that after the resurrection the rational soul in a glorified body will be a subsistent entity. However, if this is possible for a glorified body, the same should be possible also for a mortal body. Contrary to Cajetan, upon separation the rational soul receives *no* new subsistence. It retains the old one that it possessed in the embodied state. Death is not an all-or-nothing affair for Suárez.[110] What is changed is only the soul's mode of

---

[104] *DM* disp. 34, s. 5, n. 4, t. 2, 381: '[. . .] difficilius haec doctrina defendi potest, et ideo absolute falsam illam censemus, et in omni opinione insufficientem ad hanc rem explicandam'.

[105] *DM* disp. 34, s. 5, n. 28, 387. Cf. also *CDA* disp. 2, q. 4, n. 17, t. 1, 284–6. Of course, this theory is structurally similar to the theory according to which the *esse* of the rational soul is the same as the *esse* of the whole material composite. As we saw above, Suárez criticized this theory already in *CDA* as well.

[106] *DM* disp. 34, s. 5, n. 7, t. 2, 381.

[107] *Ibid.*, n. 9, 382.

[108] *Ibid.*, n. 25–7, 386–7.

[109] The harmonization of these two antinomic *desiderata* is the content of the long *CDA* disp. 2, q. 4, t. 1, 249–318.

[110] For Robert Pasnau (2002a, 387–8) only this (one can say 'Suarezian') theory claiming that the soul is part of the person can guarantee the necessary continuity between the corrupted and the resurrected body: '[Only this theory] should have been Aquinas's view, given the rest of what he wants to claim.' Following Pasnau, it might be said that Suárez's theory of the divisibility of subsistence fills in the explanatory gap in Aquinas's account.

existence, that is, its union with the body. After separation it loses the substantial mode of union but not its partial subsistence.[111]

In this partial subsistence the rational soul substantially differs from material forms. Although Suárez agrees that the integral subsistence of a material substance is composed of the partial modes of matter and form, it is only matter (and not material form) that possesses partial subsistence in such a composite. This might seem to be an obscure claim given Suárez's assertion that the subsistence of a material supposit is a whole composed of the two partial modes of form and matter. If matter possesses the mode of partial subsistence, the form also seems to have to be endowed with it after all. But this is precisely what Suárez denies. Material forms cannot possess this partial subsistence since they are not independent of body. They need a body as a subject in which they inhere. But subsistence goes against inherence. Although a partial mode of a material form somehow concurs in the composition of the integral subsistence of the composite together with the partial subsistence of matter,[112] it is not partially subsistent. As he remarks, it is not necessary that everything composing the whole participates in the denomination of the whole. Suárez concludes with this 'asymmetrical' statement according to which the mode of a material form is not a mode of its partial subsistence but only:

> [...] some complement of integral subsistence, which does not make such a form to be subsistent in itself, but [the material form] together with the subsistence of matter completes the integral terminus of the whole supposit.[113]

2.4 Unicity of the Soul and the Formal Inclusion

In the last questions of *CDA* disp. 2 Suárez analyses several important attributes of this reified and partially subsistent embodied rational soul which is causally proximately active in vital operations. They are as follows: (i) the rational soul is the only soul in the human supposit (the Unicity View); (ii) it is indivisible; and (iii) it exists as a whole in the whole body and also holenmerically, that is, as a whole in all of its parts.[114] Unlike *DM*, in *CDA* the latter two features are also attributed to the souls of

---

[111] *DM* disp. 34, s. 5, n. 31, 388–9.

[112] In this regard Suárez says that matter is more perfect than form since it can naturally be independent of a particular material form. Cf. *ibidem*, n. 38, 391: '[...] quanquam enim forma sit absolute perfectior quam materia, in hoc tamen est secundum aliquid minus perfecta, quod plus pendet haec determinata forma ab hac determinata materia, quam e converso; non enim potest eadem forma migrare a materia in materiam, sed, eo ipso quod ab hac materia separatur, perit; materia autem migrat (ut ita dicam) a forma in formam, vel potius eadem manet subdiversis formis in ea succedentibus'.

[113] *Ibid.*, n. 51, 396: '[...] complementum quoddam integrae subsistentiae, quod non constituit hujusmodi formam secundum se subsistentem, sed cum subsistentia materiae complet integrum terminum totius suppositi complete subsistentis'.

[114] For the term 'holenmerism' (holenmerianism), see Rozemond (2015). The term was coined by Henry More (1614–1687), an English Cambridge Platonist, for a view advocated much earlier by

perfect animals, such as mammals. In this section, I will focus only on Suárez's arguments for (i), that is, for his theory of the single soul and for a supplemental theory related to the manner the lower parts of the soul are contained in the higher soul.

While the vegetative and sentient souls can be separated from the rational soul,[115] the rational soul cannot exist without its inferior 'parts'.[116] A lower soul is related to a higher soul as a generic grade is related to a specific grade. The Unicity View faces a serious doctrinal opponency in the theory of the plurality of (substantial) forms (the Plurality View), which was popular in medieval scholasticism.[117] According to this theory, it holds that there are several (two or three) really distinct souls or substantial forms in a single human supposit.[118]

Suárez leaves no doubt about the truth of the Unicity View: One and the same soul, being the only substantial form in a supposit, is the principle of intellectual, sentient and vegetative acts.[119] This conclusion is confirmed by a series of quotations from Scripture, the Councils, Church Fathers and theologians headed by Aquinas.[120] As can be seen from his theory of universals, Suárez often rejects projections of merely conceptual distinctions to extramental reality.[121] If no real distinction exists between a generic grade, say, the vegetative one and the specific difference of, say, an olive tree, and if no real distinction obtains between the specific nature of an olive tree and its individual difference that would make an olive tree this olive tree, there can be no real distinction between the vegetative, sensitive and the intellective grades in Peter either. In *DM*, unlike *CDA*, Suárez adds that there is no *ex*

Plotinus, Augustine and Aristotelian scholastics, among others, according to which the mind is in the body as a whole in a whole and whole in the parts. The human soul as a spiritual substance is unextended and does not have parts that can be spread out through a body.

[115]Although according to Aristotle there are five kinds of powers (nutrition, sensation, appetite, locomotion and intellection) and four grades of living beings (vegetative, sensitive, locomotive and intellective), there are only three kinds of soul. Sensation, appetite and local motion are never entirely separated. *Locomotio* is the common effect of sense apprehension and sensory appetite. All the powers and grades can be reduced to three principles, namely vegetative, sentient and intellectual. See *CDA* disp. 2, q. 6, n. 14, t. 1, 352–4; *TDA* book 1, ch. 7, nn. 8ff, 508–10. For this Suárez's reduction, see also Des Chene (2000, 159–160).

[116]*CDA* disp. 2, q. 2, n. 6, t. 1, 154.

[117]For the plurality of forms debate between unitarians and pluralists in late medieval scholasticism and Descartes, including many references to the literature, see Pasnau (2011a, 574–605).

[118]For Suárez's long critique of the plurality of forms theory advocated by Scotus, Ockham and other medieval authors, see *DM* disp. 15, q. 10, t. 1, 536–57.

[119]*CDA* disp. 2, q. 5, n. 4, t. 1, 322; see also *DM* disp. 15, s. 10, n. 61, t. 1, 554. The Unicity Thesis was included also in a brief decree including the list of seventeen propositions that teachers were supposed to teach that was sent to all Jesuit colleges in November 1565 by the Superior General of the Society of Jesus, Franciscus Borgia (1510–1572). For this, see Sander (2017, 57–61).

[120]*CDA* disp. 2, q. 5, n. 4, t. 1, 322–6. For Aquinas, see *STh*. I, q. 76, art. 3. c., 220–1.

[121]For this, see Heider (2015a). Cf. also *DM* disp. 15, s. 10, n. 4, t. 1, 536–8.

natura rei distinction between the metaphysical grades either.[122] No Scotistic formal distinction can be assumed to exist in a supposit.[123]

Suárez introduces three arguments against the Plurality of View. The first one is based on his definition of the real distinction. The real distinction is a distinction of which the extremes are mutually separable. Really distinct extremes are such that they can exist without each other.[124] If Peter's sentient soul were *realiter* distinct from his rational soul, the two souls would have to be separable at least by divine power. However, as separated Peter's sentient soul would have to cease to be of a generic nature and would have to become constituted through a specific difference in its own species. No generic individual can exist extramentally without being determined to this or that particular species. There can be no animal *ex natura rei* that is not of this or that specific nature. There is neither a generic animal nor a kind of animal that would not be determined to this or that individual of this or that kind. Peter's *per impossibile* separated sentient soul would have to constitute a species that would differ from the soul of a horse. Obviously, this statement entails a number of absurd consequences. First, due to his sentient soul Peter would turn into an irrational animal. Peter would be a human and a non-human at the same time. A part of his nature would be a non-rational animal. Second, Peter's rational soul as an act would come to be related to an already completely constituted supposit. The rational soul would be informing a non-rational soul and Peter would become a 'spiritual non-rational animal' or 'spiritual beast'. Third, if there were *in re* three souls in Peter, there would have to be three different generations and three different corruptions in him. This manifold generation and corruption would make the matters even worse considering that the rational soul is created *ex nihilo* and vegetative and sentient souls are merely educed from the potentiality of prime matter.[125]

However, it is the second *a posteriori* argument that is most widely used by Suárez throughout the whole of *CDA* and in this reasoning for the existence of the single soul. This argument, no doubt the strongest one for Suárez, is equivalent to what has been above called the theory of the 'consonance', 'sympathy', 'cohesion', 'bondage' (*colligantia*) or 'harmony' between the acts of the distinct powers.[126]

---

[122]Metaphysical grades are the non-physical parts of an extramental thing (substance) that as *fundamenta in re* correpond to the logical parts or abstract concepts, such as 'animality' and 'rationality', that are introduced in the Tree of Porphyry.

[123]For Suárez's rejection of Scotus's formal distinction conceived as a true and actual distinction independent of the human intellect, see *DM* disp. 7, s. 1, n. 13, t. 1, 254 and many other *loci* in *DM* disp. 5 and 6. Just this rejection, among others, has led me in Heider (2009) to evaluate Suárez's metaphysics as 'nominalized Scotism'.

[124]For this definition of the real distinction, cf. *DM* disp. 7, s. 2, n. 9, t. 1, 264.

[125]*CDA* disp. 2, q. 5, n. 5, t. 1, 326–8. For this argument see also *DM* disp. 15, s. 10, nn. 20–1, t. 1, 541–2.

[126]This argument assumes the real distinction between the soul and its power which is the topic that will be treated in 2.6.

Applying abductive reasoning or inference to the best explanation,[127] a kind of argument often employed by him, Suárez introduces the phenomenon of the acts' mutual interferences. As every philosopher knows, too much of intellectual attention can significantly impede sentient and vegetative acts. After a long reading of a difficult scholastic text or after writing a philosophical essay our sentient operations are significantly diminished and our sensomotorical orientation is affected. At the vegetative level headache and gastric neurosis can often be the consequences of long-term intensive study. These and other negative interferences between the operations of the faculties show that one and the same soul must co-ordinate all these acts. Since it has a finite attentional reservoir (for this see 3.6), an intensive attention or, technically speaking, a strong substantial causal influence of the soul through a power to one kind of object necessarily weakens attention and capacity in a distinct power. This holds also for operations within the external senses' modalities. An attentive visual operation often affects our auditive acts. We do not listen attentively to someone who is calling us if our visual attention is strongly captured by an object. Vice versa, we do not see a person standing in front of us when we are deeply engaged in a debate or listening to music.[128]

In the third (this time *a priori*) reasoning, in line with Aquinas,[129] Suárez argues that if two forms are subordinated so that one presupposes the other, the higher form that gives the ultimate perfection can also give the grade of an inferior soul with less perfection since the two are related to each other as generic and specific (differential) metaphysical parts. This is *a fortiori* true if one considers that the main function of the lower grades is to *serve* the higher ones. The vegetative soul in animals is in service of sentient operations, sentient operations minister to the rational soul. If this is the case, all the grades have to come from one and the same soul.[130]

Suárez's stance on the issue of the unicity or plurality of forms is clear. The lower grades are somehow included in the higher soul. However, what is not clear is *how* they are contained in it. How does the rational soul include the grades 'being vegetative' and 'being sentient'? And how does the sentient soul of a horse contain the grade 'being vegetative'?

---

[127] Abductive reasoning or inference to the best explanation is a form of logical inference that starts with an observation and seeks to find the simplest and most likely explanation. One can understand abductive reasoning as an explanatory reasoning in justifying hypotheses. For this kind of reasoning in Suárez in *DM* disp. 15, q. 1, see Shields (2012).

[128] See *CDA* disp. 2, q. 5, n. 5, t. 1, 328–30; *DM* disp. 15, s. 10, n. 22, t. 1, 542 (in this paragraph this argument is called 'optima ratio'). In connection with this theory of hamorny of powers Suárez often refers to Francisco Vallés's *Controversiarum medicarum et philosophicarum libri decem* (1556). For this, see chapter VI 'De sympathia & idiopathia & protopathia & anthipathia & consensu' of the fourth book of his *Controversiarum medicarum et philosophicarum libri decem*, 1582 (194–200). This theory might have even been brought to Suárez by Vallés. For this connection, see Rode (2015, 391–4).

[129] Thomas de Aquino, *STh*. I, q. 76, art. 4, c., 223–4.

[130] *CDA* disp. 2, q. 5, n. 5, t. 1, 330.

On the (as Suárez says) common view it holds that the higher soul contains the lower soul eminently (*eminenter*).[131] What does this 'eminently' mean? In scholastic jargon the term 'eminently contained' means 'contained in a higher or better way', or in the case of attributes such as wisdom and justice being included in God it can mean 'contained in a supreme or infinite way'.[132] Suárez first presents the arguments for this common theory. On this view, the rational soul cannot be formally vegetative and sentient since these two kinds of soul specifically differ from the rational soul and since the latter is immaterial while the former two are material. This kind of inclusion can also be found in Aristotle who employed analogy with geometrical figures:

> For both in figures and in things which possess soul, the earlier type always exists potentially in that which follows; e.g., the triangle is implied by the quadrilateral and the nutritive faculty by the sensitive.[133]

Although the quadrilateral contains the triangular, it can contain it only potentially, not formally. That is why a triangular occurs in a quadrilateral in an eminent way. A quadrilateral, formally speaking, cannot be a triangular. Analogically, a rational soul can contain the vegetative and sensitive souls only eminently. Suárez infers that if these grades do not exist formally in a higher grade and exist in it only eminently, they of course cannot be formally distinct from the rational grade.[134]

Suárez considers this view to be troublesome (*difficillima*).[135] He agrees that with respect to efficient cause it is correct to speak of the eminent or virtual containment of the lower grades of the soul in the higher soul.[136] However, at the same time he makes clear that this inclusion in the order of efficient causality is *not* sufficient for the soul's *formally* giving being to a composite. Sunlight eminently includes heat since it causes it. But this inclusion is not enough for the sun's *formally* giving heat to its subject such as air. Not only the soul's efficient causality but also *formal* causality needs to be considered. As we have seen in 2.2, it is the order of being from which the order of efficiency arises. The rational soul therefore must *formally* give vegetative and sentient being to its supposit. Contrary to the thrust of his debate on universals in *DM*, in *CDA* Suárez seems to after all embrace the Scotistic formal distinction.[137] If the sentient and rational grades were not formally distinct it would

131 Since he compares the containment of the vegetative and the sentient soul in the rational soul to the containment of a tetragonal shape in a pentagonal shape, Aquinas can be regarded as an advocate of eminent containment. See Thomas de Aquino, *STh*. I, q. 76, art. 3, c., 221.

132 For this definition, see Wuellner (2011, 39).

133 Aristotle, *DA* II.3, 414b28–32, 83.

134 *CDA* disp. 2, q. 6, n. 2–3, t. 2, 332–34.

135 *Ibid*., n. 4, 334.

136 The notion of virtual containment is commonly related to the effect's existence in the cause. The effects are said to exist virtually in the cause. Nevertheless, it is also applied to the virtual existence of lower levels of life in a higher form.

137 In the whole of *CDA* I have not found a single attempt to define this distinction, though. One may conjecture that Suárez understands it in the Scotistic way, i.e., as a distinction that exists *in re*

not be possible to conceive a univocal agreement between the soul of a man and the soul of a horse. The univocity of the 'ratio' of 'being sentient' in Peter and Bucephalus can be saved only if the grade of 'being sentient' in Peter is formally distinct from the grade of 'being rational'. Moreover, the rational soul is also the first principle of sensory operations. Therefore, the rational soul must formally be also the sentient soul.[138]

Leaving aside the issue of how Suárez precisely understands this formal distinction in *CDA*, why does he insist on his preference for formal containment over eminent containment? Why does it make a difference to him at all? Seen from the perspective of *TDA* and from the later Thomist critique of Suárez,[139] the reason could be his skeptical approach to the possibility of an overly strong 'elevation' of the less perfect grades *within their own genus* through the more perfect grade's 'eminent presence'. If the theory of the eminent inclusion of the lower souls in the higher one held, the vegetative soul in an olive tree and a horse and the sensitive soul in a horse and Peter would have to be specifically different. If this were the case, in all these instances the lower souls 'intra lineam vegetativi' and 'intra lineam sensitivi' would have to constitute an atomic species. Clearly, this would prevent them from being contracted by a higher difference. If a containing soul had the eminency to perfect a lower soul and its operations in their own grade, the vegetative powers of an olive tree would have to specifically differ from those of a horse and the sensory powers of a horse would have to be specifically different from Peter's sensory powers. The vegetative acts of an olive tree and those of Bucephalus and Peter, and the sensory operations of Bucephalus and Peter would have to be considerably different. However, Suárez stresses, this cannot be the case because vegetative and sensory powers have the same object, namely food, colours, sounds, etc. As I will show in Chap. 5, Suárez capitalizes on this model highlighting the formal containment of the lower grades in the higher grade in the context of his discussion of the interior senses. In this context he takes a critical stance against the Thomist view according to which Peter's cogitative power and memory (reminiscence) are perfected and elevated in their operations due to their participation in the intellectual power and due to the eminent presence of the rational soul in them.

---

independently of reasoning but without the possibility of separating one of its extremes through divine intervention. For Suárez's acceptance of *distinctio formalis* in *CDA*, see also Shields (2014, 201–2).

[138] *Ibid.*, n. 5, 340. Yet in *DM*, as said above, in line with the overall critique of *distinctio formalis* or its reduction to the conceptual version *cum fundamento in re*, Suárez seems to advocate the view that the lower souls are *eminently* contained in the rational soul. Through this containment they seem to be also somehow elevated (see *DM* disp. 15, s. 10, n. 26, 544). In this aspect I observe a doctrinal shift between *CDA* and *DM*.

[139] See *TDA* book 1, ch. 6, n. 7, 503–4. For a critique of Suárez formulated by John Poinsot, which is based on the Thomist's emphasis on eminent containtment, see *Cursus philosophicus Thomisticus*, vol. III, pars 4: De ente mobili animato, q. 1, art. 5, 48–50.

2.5 Indivisibility of the Soul and Its Holenmeric Character

Suárez's Unicity View is supplemented by his theory according to which not only the rational soul but also the soul of perfect animals, such as horses and dogs, is indivisible and exists as a whole in every part of the body it informs.[140] The case of the rational soul does not pose a difficulty. Since it is subsistent, incorporeal and immortal it cannot be extended and divisible.[141] An amputation of a hand is not an amputation of a part of the rational soul. Through an amputation of a hand the soul is not divided or truncated. On the other side of the spectrum, equally unproblematic, are the souls of plants and imperfect animals, such as rainworms and centipedes. As the possibility of grafting shows, a grapevine can be divided into parts which can each continue to live and be nourished. The division of rainworms and centipedes into two parts is at least for a while followed by movement of the two bodily parts. Contrary to the rational soul, their souls are extended and divisible.[142]

A more controversial case is the one of the souls of perfect animals whose bodies are perfectly organized and have the same sensorial functions as we humans have. Regardless of a general comparative significance of these souls for the rational soul, the case of these perfect animals' soul is important since it again points to the systematical importance of the single soul's coordinating function. According to the first view of the character of perfect animals' souls, which is regarded in *DM* as more probable,[143] the souls of these animals (like those of imperfect ones) are extended and divisible, since in their generation and being they depend on matter. They are educed from the potency of a pre-existing matter. They cannot exist without being subjected in matter since they are co-extensive with the body. If they were indivisible they would also have to be immaterial. According to this (anachronically speaking) 'Cartesian' argument, indivisibility entails immateriality as immateriality entails indivisibility. This view can be defended with a set of empirical arguments based on the phenomena of divisibility. Already Aristotle referred to a tortoise that continued to move and live when its heart had been

---

[140]For an analysis of the issue of the (in)divisibility of the souls in Toledo, the Coimbrans and Suárez, see Des Chene (2000, 169–189).

[141]*CDA* disp. 2, q. 7, n. 2, t. 2, 2–4.

[142]*Ibid.*, nn. 3–5, t. 2, 4–8.

[143]In his *editio princeps* of *TDA* Balthasar Alvarez inserts the following paragraph in which he informs the reader about Suárez's change of opinion: 'Sententiam hanc de indivisibilitate animae perfectorum viventium [. . .] mutaret procul dubio ipse auctor [. . .] Quippe iam olim in Metaphysica, hanc retractans quam prius dictarat, oppositam dixerat probabiliorem, disp. 15, sect. 10, n. 32'. More precisely, it is in *DM* disp. 15, s. 10, n. **31**, 545 [bold is mine] where Suárez states: 'Nullam formam materialem esse vere ac proprie indivisibilem'. Des Chene claims that with this later assertion Suárez paves the way 'toward Descartes' view that everything in this world, except the [rational] soul, is extended' (Des Chene 2000, 180). He notes that the idea that something that feels and senses can be divisible (Suárez's view in *CDA* and *TDA*) was utterly inconceivable for Descartes (*ibid.*, 199–200).

removed.[144] Although its main part, the heart, had been removed, the remaining whole continued to live at least for a short time.[145] As suggested above, in *CDA* Suárez embraces the opposite view in which he endorses a parity between the rational soul and the souls of perfect animals: Like the rational soul, the soul of perfect animals is indivisible and unextended. Their souls do not differ in this aspect. Suárez introduces two main arguments for this conclusion, the last of which is symptomatic for his approach.

1. The first reasoning has the following assumption: The existence of a material yet indivisible soul is a possible state of affairs. Suárez denies, anachronically speaking, the 'Cartesian' equivalence between indivisibility and immateriality. Although he agrees that immateriality entails indivisibility, he embraces a view according to which the opposite implication does not hold. Ontological dependence on matter is compatible with soul's indivisibility.[146] In order for a substantial form to be ontologically independent something more is necessary than mere indivisibility. A spiritual accident such as an angelic cognitive act is an indivisible entity and yet it is dependent on the angelic power (intellect) and substance in which it inheres. Accordingly, it is proper to say that indivisibility can be saved without independence of matter.[147] The very argument that perfect animals have indivisible material souls, a possible metaphysical item, derives from Suárez's application of two 'cosmological' (Neoplatonic) principles, which are usually called 'The Principle of *Plenum Formarum*' and 'The Principle of Continuity in the *Scala Naturae*'.[148] According to the first of these principles it holds that given the existence of divine 'fertility' and divine goodness all possible perfections in the universe have to be actualized. Divine goodness would not be sufficiently manifested if one of the possible perfections were not created. The second principle states that the highest part of a lower perfection touches the lowest part of a higher perfection.[149] Through the contact of the highest part of a

[144] Aristotle (2000), *On Youth and Old Age*, ch. 2, 468b14–6, 417–9.

[145] *CDA* disp. 2, q. 7, n. 7, t. 2, 10–2.

[146] From the Cartesian point of view it can also be said that this advocacy of the indivisibility of animal souls deprives Suárez of a potent argument for the immortality of the human soul. For this, see Des Chene (2000, 189).

[147] *CDA* disp. 2, q. 7, n. 10, t. 2, 16–8.

[148] For these rationalist principles, see Lovejoy (1936) and Forigari (1968, 325–6). As regards Suárez's application of these principles in the context of his reasoning for the immortality of the rational soul (in *CDA* disp. 2, q. 3, n. 19, t. 1, 186), see Heider (2015b, 108–10). For this rationalist method in the Renaissance philosophy of mind, see Perler (2019, 28–31). Compared to Aquinas, however, these principles do not find such broad application in Suárez. For this lesser application or rather non-application, see 5.3.1.

[149] Suárez does not employ this principle in other cases, such as the internal senses. For this principle, see Dionysios the Areopagite, *On the Divine Names and the Mystical Theology*, 1920, ch. 7, § 3, 79: 'Nevertheless, as I said, we must draw this knowledge of Wisdom from all things; for wisdom it is (as saith the Scripture) that hath made all things and ever ordereth them all, and is the Cause of the indissoluble harmony and order of all things, perpetually fitting the end of one part

lower level with the lowest part of a higher level two distinct levels are mediated without having to assume 'a missing link'. With respect to the first principle the argument states that indivisible material forms have to be posited since they represent *a* perfection. In fact, this perfection marks the supreme grade in the hierarchy of material forms. The material forms of perfect animals approach at most the genus of immaterial entities. We would have to grant a discontinuity in the *scala naturae* if this kind of soul were not included in the hierarchical order of the forms in the universe. Without these material indivisible forms dependent on the material subject the ontological transition from a lower level (material divisible forms) to a higher level (immaterial indivisible forms) would lack a medium and would become discontinuous.[150]

2. The second argument for the indivisibility of the sentient soul of perfect animals, which Suárez again evaluates as *the best* argument for the existence of the single *indivisible* soul, is based on the abductive inference from observable phenomena.[151] We can observe the same harmony and interconnections between the organs of perfect animals as we find in our bodies.[152] Perfect non-rational animals have the same organs with the same functions as we do. The cognitive virtue resides in the brain, the vital force in the heart and the vegetative *virtus* in the liver. There is an intimate interconnection between them to a degree that if one part is injured, the other is moved by that injury too. At first an injury or its cause in a foot is sensed by the touch, then phantasy located in the brain apprehends it as something harmful and commands an adequate movement to the feet, which is followed by a jerk and an escape. This interconnection would not be possible without the ontological assumption of a single indivisible soul (the identity of the soul must be followed by its indivisibility) which exists in all those animated parts of the body. If the soul were extended and divisible and if one part of the soul were in the foot and another one in the brain, how could the part of the soul in the foot apprehend and 'obey' what is commanded by the part residing in the brain? This 'obedience' can be explained only if one assumes a governing indivisible soul in both parts which is in charge of the mediation and coordination between them. Beside this argument from interconnection, in analogy to the argument for the unicity of the soul, Suárez formulates an argument 'from the phenomenon of interferences'. It is evident that the operations of various organs interfere with each other. A great pain in a tactile organ, say in a hand, can affect or restrain the activity of sight. An intensive attention in the operation of one power leads to a lesser attention or to weariness in the other. These interferences would not be possible if different parts of the soul existed in diverse parts of the body. If this were the case, one part would not be affected through the activity of another. It

---

unto the beginning of the second, and thus producing the one fair agreement and concord of the whole'.

[150] *CDA* disp. 2, q. 7, n. 10, t. 2, 20.

[151] *Ibid.* n. 11, t. 2, 20.

[152] Of course, the same argument can be applied also to the rational soul.

would continue with its activity. Such interconnections and interferences cannot be observed in plants and imperfect animals. A part of a plant can be harmed and even damaged while a distinct part can continue to perform its proper operations.[153]

The singleness and indivisibility of the rational soul and the sentient soul of perfect animals find an important corollary in Suárez's censure of the doctrine of the plurality of substantial *partial* forms. What does this doctrine teach? According to this theory there exist specifically distinct forms of the heterogeneous parts of a living body, whose operations are often mutually contrary, which dispose this or that part of the body for the ensuing reception of a total substantial form. These partial souls must be posited since the total substantial form informs distinct heterogeneous integral parts of an organic body, such as the heart, which is hot, and the brain, which is cold, etc. These distinct parts cannot be conceived without the assumption of partial substantial forms that inform them.[154] Besides this theory of partial substantial forms, Suárez also introduces Plato's theory of the plurality of souls, namely his famous view according to which there are three souls in a human being located in distinct parts of the human body, namely the concupiscible soul in the liver, the irascible soul in the heart and the rational soul in the brain. The last soul occupies the highest bodily position and governs the former two.[155]

In line with his embracement of the Unicity Thesis Suárez declines these views too. In living beings that have an indivisible soul there is only one soul informing all the parts in which vital operations exist.[156] If bodily parts were informed by partial forms, there would be two ways how to make sense of their informing. The parts actualized by these partial forms would be at the same time either informed or not informed by the total substantial soul. Obviously, both options are implausible. The first one would entail the plurality of (at least) two substantial forms informing the same part of the organic body. As we know, this is a view consistently denied by Suárez throughout *CDA* and *DM*. Regarding the second option, there are again two alternative ways to understand it. These heterogeneous parts informed by partial forms and *not* informed by the total soul would be either connected or not. If they were connected, the soul, which would be integrated from them, would constitute a divisible whole. However, this view has already been rejected. If they were disconnected, every part would have to constitute its own living supposit. An animal would be composed of a plurality of disconnected parts. As such it would form an accidental unity. Replying to the second theory, namely to Plato's view, Suárez asserts that if the indivisible form existed only in a single part, say in the brain, the

[153] *CDA* disp. 2, q. 7, n. 11, t. 2, 18–20.

[154] *DM* disp. 15, q. 10, n. 29, t. 1, 544; *CDA* disp. 2, q. 8, n. 5, t. 2, 32–4. Divisible souls, again, are not controversial. They are composed of parts of diverse 'rationes'. The soul's part informing a leaf of a tree cannot constitute its root or trunk. See *CDA* disp. 2, q. 8, n. 2, t. 2, 28–30.

[155] *CDA* disp. 2, q. 8, n. 4, t. 2, 30–2. See Plato (1997), *Rep.* IV 434d–441c, 1066–73. As regards Plato's tripartite soul, see en lieu of other literature Szlezák (1976).

[156] *CDA* disp. 2, q. 8, n. 6, t. 2, 34.

question would be whether this part is truly organic or not. If not, the soul would cease to be the act of an organic body. In fact, it would cease to be a soul at all. However, if it were organic, the soul would have to inform all the subparts of this part of the organic body (the brain). But if it did, why could it not also inform the other parts of the *whole body*? As informing a single part it could not be called the form of the whole (*forma totius*). It would rather be a mover of the body than the *forma informans*.[157] Despite his abovementioned inclination to this theory, Suárez *ex professo* counters it. Last but not least, Plato's view of the plurality of souls can be rejected by employing the abovementioned abductive reasoning again: Consonant operations that are realized in diverse parts of the body can issue only from a single soul. They are made possible only if the single soul as a governing force informs all the parts of an organic body. The existence of various souls in different body parts would make such consonance and harmony impossible.[158]

As a corollary to his theory of the indivisibility of the soul Suárez states that the indivisible soul exists not only whole in the whole (*tota in toto*) but also whole in any part of the body (*tota in qualibet parte*). The soul informs the organic body *holenmerically*. It might be said that it is similar to a hologram that exists as a whole in all of its small parts. However, unlike a hologram, the indivisible soul does not pertain to the category of quantity. It is not indivisible like a geometrical point. Following Aquinas, Suárez says that the indivisible soul transcends the accident of quantity.[159] Unlike the divisible souls in which only a part can be in each part, the indivisible soul does not have a quantitative totality. Its way of being present in the body is not compatible with material and spatial presence, which is the kind of presence we are most familiar with. It is not a whole in all of its parts in the sense of the totality of powers either. It is not this indivisibility because the indivisible soul does not have all of its powers in all of its parts. It has sight only in the eye, hearing only in the ear, smell only in the nose, etc. It is whole in every part of the body, essentially because it does not get divided in them in the manner of a composite that gets divided into matter and form, or of a *definiendum* which is divided into the defining parts. Its totality in all the bodily parts it informs is entitative or essential.[160]

The soul's holenmeric character is not without difficulty, however. How is it possible that the single indivisible soul informs one organic part according to one grade of life and another part according to a distinct grade of life? While some parts of the body are organs of sensitive operations (e.g., eyes and ears), others are organs of vegetative functions (e.g., bones) and some even have no vital functions (e.g., hair and nail). It seems that this manifold informing of the body according to the distinct

[157] *DM* disp. 15, s. 10, n. 36, t. 1, 546–7.

[158] *CDA* disp. 2, q. 8, n. 8, t. 2, 36–8.

[159] Aquinas, *De spiritualibus creaturis*, a. 4 ad 16: '[...] anima non est indivisibilis ut punctum habens situm in continuo, contra cuius rationem esset in loco divisibili esse. Sed anima est indivisibilis per abstractionem a toto genere continui; unde non est contra eius rationem si sit in aliquo divisibili toto'.

[160] *CDA* disp. 2, q. 8, n. 16, t. 2, 48–50.

grades will after all have to result in positing diverse partial substantial forms.[161] Suárez does not think so. He argues that the explanation of the various organic operations in the various organs lies, in part, in the various accidental dispositions.[162] The simple and indivisible soul must inform its body differently since there are distinct dispositions in the foot and in the brain. In line with what has been said in 2.1, Suárez replies that in the order of material causation accidental dispositions, such as the cold in the brain or the hot in the heart, can precede the introduction of the substantial form. In line with his understanding of 'organic body' introduced in 2.1, he adds that these dispositions are only a part of the whole story. The diversity of organic operations can be fully explained only if the soul and its informing is considered too. The soul as a final cause is the reason why there are diverse material dispositions in diverse parts of the body. Although the sentient soul of perfect animals or the rational soul in humans are indivisible in character, their souls function as *equivocal* principles generating different kinds of operations in various parts of the organic body. We may conclude that for Suárez the variety of heterogeneous parts comes from the variety of the *modi informandi* of the soul, which he conceives as an equivocal efficient principle.[163]

2.6 The Powers as Really Distinct Entities

Another topic related to a broad metaphysical One-Many cluster of issues in Suárez's psychology, which is relevant for his theory of perception, is the problem of powers and their distinction from the soul. Contrary to the classical early modern philosophy represented by Descartes, Locke et al., Suárez's *scientia de anima* is essentially a psychology of powers.[164] With the exceptions of *CDA* disp. 1 and 2 devoted to the substance of the soul, all the other disputations explicitly deal with the topic of powers from either a general or a particular point of view. As I have already indicated, the notion of a real distinction between the soul and the powers is directly correlated with Suárez's theory of the single governing and coordinating soul in which the powers are rooted. More precisely, the theory of the single efficient soul is a systematic consequence of Suárez's theory of the really distinct powers. The latter theory seems to have been implicitly assumed by Suárez from the very beginning of *CDA*.

[161] *Ibid.*, n. 1, 28.

[162] John Buridan embraced the view that the distinction of organic operations lies *only* in the material distinction between the various organs. The eye and the foot are informed by the same indivisible soul. The fact that the eye sees and the foot does not see is given only by a suitable complexion of the former and an unfitting complexion of the latter. For this Buridan's view, see Park (1988, 478).

[163] *CDA* disp. 2, q. 8, n. 14, t. 2, 44–6.

[164] For Suárez's theory of powers, see Perler (2013, 2015).

The *status quaestionis* of *CDA* disp. 3, q. 1, where the topic of the nature of the distinction between the soul and the powers ontologically conceived as qualities is analysed, is determined by two opposing doctrinal statements. On the one hand, it is a nominalist stance methodologically inspired by 'Ockham's Razor'. According to this stance, the powers are really identical with the soul, being a simple thing. The various operations of the soul and the composite amount to distinct roles played by the same soul (or souls, if the Plurality View is advocated), or the same substantial composite in the case of material operations such as perception. As William of Auvergne (ca.1180/90–1249) says, one and the same soul, like one and the same person, can take various 'offices' or functions. The same soul takes different roles by displaying different ways of acting without having to assume a plurality of really distinct *homunculi* producing distinct activities.[165] In fact, Aristotle's abovementioned second definition of the soul seems to be far from necessitating the existence of these really distinct powers. If the soul as an equivocal cause can produce faculties of various kinds, why could it not also *directly* produce these processes? This seems to be *a fortiori* the case, if one considers that for Suárez the soul generates proximate causal influence in vital acts.

On the other hand, however, there is a view, advocated by Aquinas and Thomists, according to which this account cannot be right since it is contradictory for created souls and substantial forms to be immediately operative. Considering the imperfection of creatures, they cannot be operative without really distinct instruments. They cannot exercise the manifold operations directly through their substance since their causal agency is fully exhausted by their informing the body. Only in God can we see an identity of operations and substance. Only in God is His operation identical to His substance. In this aspect, God with His attribute of simplicity is fundamentally distinct from the essentially compositional creatures. Without really distinct instruments, the human soul or a body-soul composite cannot realize its activities.[166]

Having presented these two antinomic views and before arriving at his conclusion Suárez introduces several arguments for the theory that postulates a weaker (Scotistic) *ex natura rei* distinction between the soul and the powers, which, as we know, is the distinction he seems to admit in *CDA*. The argument goes like this: The definition of vital powers is distinct from the definition of soul. While soul *qua* soul is essentially defined through its *aptitudo* to the body, powers *qua* powers are characterized through their respect to operations. Assuming that these definitions are *real* definitions, they necessarily lead to accepting the *ex natura rei* distinction in things. If the powers and the soul were not at least formally distinct, one would have to say that the intellect loves and the will understands. If the soul and the powers

[165]For this identity theory in William of Auvergne, see Perler (2015, 100–5). For the identity theory in post-Suarezian philosophy, namely in Pedro Hurtado de Mendoza, and for a comparison of his view with Suárez's theory, see Heider (2019a).

[166]*CDA* disp. 3, s. 1, n. 1, t. 2, 56.

were not *ex natura rei* distinct, by the law of the transitivity of identity the intellect and the will would have to be considered identical.[167]

In the next step, Suárez argues for the *real* distinction, which is the position he embraces. Despite his lengthy reasoning, most of the text bears upon a critique of Aquinas's arguments for the Real Distinction Thesis (henceforth: *RDT*), which, as usual, is the doctrinal background of the Jesuit's exposition of his view.[168] Referring to *STh* 1, q. 54, art. 1 and 1, q. 77, art. 1 Suárez critically evaluates three arguments that he finds in Aquinas. In his first argument Aquinas says:

> [...] since potentiality and actuality divide being and every genus of being, potentiality and actuality have to be referred back to the same genus. And so, if an actuality is not in the genus of substance, then the potentiality corresponding to it cannot be in the genus of substance. Now the soul's operation is not in the genus of substance; rather it is only in the case of God that His operation is His substance. Hence, God's power, which is a principle of operation, is the very essence of God. But as was explained above for the case of angels as well, this cannot be true either of the soul or of any other creature.[169]

Every operation must be preceded by a power that categorially (generically) corresponds to this act. Since operation, whether conceived as act or quality, belongs to the category of accident, therefore the power that performs it must be an accident too. This holds for all creatures including angels. Only in God are we not obliged to posit *RDT*. In all other beings, *RDT* is a necessary consequence of this *a priori* argument.

Suárez does not regard this argument as an instance of demonstrative reasoning. As we can expect from what has been said above, Aquinas's crucial supposition, that is, that no created substance or form can directly exercise efficient causality in its operations, is simply not true. Basically, his reply is based on adopting the opposite claim based on the *proximate* causal efficiency of the soul. This claim is the *cantus firmus* of the entire *CDA*. However, although Suárez agrees that the soul is primarily related to its body, in the same breath he adds that it is secondarily related to its acts. While in the first case Suárez agrees that both 'relata' must be part of the same category (the soul and the body pertain reductively to the category of substance and no 'categorial entanglement' in the form of an inclusion of an accident in the internal make-up of a substance is allowed), categorial uniformity is not required in the latter

---

[167] *Ibid.* n. 6, 62. For Suárez's brusque dismissal of the nominalist view, see also Shields (2013, 366).

[168] Suárez's critique of Aquinas's arguments can be taken as proof of the proverbial saying that when two authors do the same (in this case, embrace the same opinion), namely the real distinction, this does not have to be the same.

[169] *STh.* I, q. 77, art. 1, c., 236: '[...] cum potentia et actus dividant ens et quodlibet genus entis, oportet quod ad idem genus referatur potentia et actus. Et ideo, si actus non est in genere substantiae, potentia quae dicitur ad illum actum, non potest esse in genere substantiae. Operatio autem animae non est in genere substantiae; sed in solo Deo, cuius operatio est eius substantia. Unde Dei potentia, quae est operationis principium, est ipsa Dei essentia. Quod not potest esse verum neque in anima, neque in aliqua creatura; ut supra etiam de angelo dictum est'. Latin quotations from Aquinas's *Treatise on Human Nature* (*STh.* I, qq. 75–102) will be adopted from Alfred Freddoso's translation in Aquinas (2010, here 48).

case, that is, in respect to its operations.[170] Suárez's statement regarding the immediate efficiency of the substantial form is demonstrated by the abovementioned example of the substantial form of water (see 2.2). In line with Aristotle's qualitative and elemental physics, Suárez states that the connatural property of the substantial form of water is coldness. If water is heated by fire and thereby moved outside of its teleological coldness, and if the fire is removed from it, water returns to its connatural quality. The reduction to the original quality is nothing else than the result of the efficiency of the water's substantial form. Although the substantial form of water and the quality of coldness are parts of different categories, the substantial form of water is still related to the quality of coldness as a cause to its effect. But if the soul can produce an act of a different category, the *in re* distinction between the soul and its powers, based on this assumption of categorial conformity, is not necessary. The soul itself can be directly and immediately accountable for producing its operations.[171]

In his second argument Aquinas says:

> Second, this is likewise obviously impossible in the case of the soul. For in its essence the soul is an actuality. Therefore, if the soul's essence were itself an immediate principle of operation, then whatever has a soul would always actually be engaging in [all of the] vital works, in the same way that whatever has a soul is always actually alive. For the soul, just insofar as it is a form, is not an actuality ordered toward any further actuality, but is instead the ultimate terminus of generation. Hence, the fact that the soul is in potentiality with respect to some further actuality is not something that belongs to it because of its essence, just insofar as it is a form; rather, it is something that belongs to it because of its power [. . .] But a thing that has a soul is not always in actuality with respect to [all the] vital works [. . .] For nothing is in potentiality with respect to an actuality insofar as it itself is an actuality.[172]

If the soul were the proximate principle of its operations, it would have to be constantly active in all its vital acts. The soul as the first act of an organic body is continuously in act (it animates all the time a being lives); therefore, if it were directly active in its operations, the same incessant activity would have to occur in operations *ad extra* too. However, this is not true.[173] While we are alive, we do not exercise all our activities incessantly. We do not always think, we do not always see, we do not always desire. Besides, contrary to the powers which can be perfected by

---

[170] *CDA* disp. 3, q. 1, nn. 7–8, t. 2, 64–6.

[171] *Ibid.*, n. 8, t. 2, 66. For this argument, see also *DM* disp. 15, s. 1, n. 8, t. 1, 500.

[172] Thomas de Aquino, *STh.* I, q. 77, art. 1, c., 236–7: 'Nam anima secundum suam essentiam est actus. Si ergo ipsa essentia animae esset immediatum operationis principium, semper habens animam actu haberet opera vitae; sicut semper habens animam actu est vivum. Non enim, inquantum est forma, est actus ordinatus ad ulteriorem actum, sed est ultimus terminus generationis. Unde quod sit in potentia adhuc ad alium, hoc non competit ei secundum suam essentiam, inquantum est forma; sed secundum suam potentiam [. . .] Invenitur autem habens animam non semper esse in actu operum vitae [. . .] Nihil enim est in potentia secundum actum, inquantum est actus'. English translation in Aquinas (2010, 48–9).

[173] Stephan Schmid pointedly calls this view 'The potentialist version of the substantial realist position'. According to this position, the faculties are irreducible primitive entities allowing for unactualized powers. See Schmid (2015b, 154).

operations and habits, the soul cannot be perfected by another act since as the *terminus* of a generative act it *is* already in perfect act. It can be perfected only through additional and really distinct powers that can be then progressively actualized by acquiring various kinds of habits that arise through the production of a number of repeated activities.

Suárez is clear that this argument is not demonstrative either. At most it shows that the formal distinction between the soul and the powers obtains. It states nothing more than that the soul is not a power in the way in which it is the form of the body. It only proves that the soul *qua* formal cause is distinct *formaliter* from the soul *qua* efficient cause. Accordingly, the formal distinction is fairly sufficient to explain the state of affairs that one entity is continuously *in actu* according to one formality, namely the formality of the soul *qua* formal cause, and only at times *in actu* according to another formality, namely the formality of the soul *qua* efficient cause.[174] This can be corroborated by two examples. One and the same accident of heat can formally heat up the subject in which it inheres and only sometimes it can heat up a distinct subject, such as water. One *habitus* can be continuously *in actu* as informing its subject and at the same it can be only at times *in actu* if conceived as a principle of operation.[175]

Finally, Aquinas formulates the third argument as follows:

> Since 'power' or 'potentiality' is predicated with respect to an act, there must be a diversity of potentialities with respect to diverse acts, and because of this it is said that a thing's proper act corresponds to its proper potentiality. Now as is clear from what was said above, every created thing is such that its essence differs from its *esse* and is related to its *esse* as potentiality is related to actuality. But the act to which an operative potentiality is related is an operation. Now the act of understanding in an angel is not the same as his *esse*; nor is any other operation, either in an angel or in any other created thing, the same as the thing's *esse*. Hence, an angel's essence is not his intellective potentiality; nor is the essence of any created thing its operative potentiality.[176]

Powers are distinguished through acts. Diverse acts require distinct powers. However, according to the general premise, being and operating in creatures are distinct acts. Since in all created beings essence and existence are really distinct, operations and being must have distinct principles, which are related to them as their potencies. While the principle of being is the essence or the form that constitutes the essence, the principle of operating must be something different, namely a really distinct power.

---

[174] *CDA* disp. 3, q. 1, n. 10, t. 2, 66–8.

[175] *Ibid.* n. 11, 68.

[176] Thomas de Aquino, *STh.* I, q. 54, art. 3, c., 47: 'Cum enim potentia dicatur ad actum, oportet quod secundum diversitatem actuum sit diversitas potentiarum: propter quod dicitur quod proprius actus respondet propriae potentiae. In omni autem creato essentia differt a suo esse, et comparatur ad ipsum sicut potentia ad actum, ut ex supra dictis patet. Actus autem ad quem comparatur potentia operativa, est operatio. In angelo autem non est idem intelligere et esse: nec aliqua alia operatio aut in ipso aut in quocumque alio creato, est idem quod eius esse. Unde essentia angeli non est eius potentia intellectiva: nec alicuius creati essentia est eius operativa potentia'. English translation in Aquinas (2018).

Following Scotus, Suárez denies this argument by claiming that diverse acts require at most a formal distinction in the powers, since the proposition 'Powers are distinguished through acts' concerns the specific distinction, not the real distinction.[177] Moreover, it is not necessary to consider the distinct principle of operating and being since to cause heat efficiently derives from 'being hot'. However, both originate in the same form. Therefore, the same principle or form can be the principle of 'being hot' and 'causing heat' in a distinct subject.[178]

After rejecting Aquinas's three arguments Suárez formulates a positive argument for his *RDT*. Actually, his positive reasoning boils down to what can be called a 'probabilistic argument', appended by an illustrative example. Although he admits the possibility of the existence of a perfect creature whose powers and essence differ only formally,[179] in a link to the spirit (not *litera*!) of Aquinas's reasoning, in which the imperfection and limitation of the substantial forms of created beings form a fundamental axiom of *RDT*, Suárez argues for a real compositionality of the soul and the powers. This imperfection of finite and created substantial forms and composites can be explained only by positing *RDT*.[180] It cannot be the case that the soul as such is proximately operative as a total cause of its operations. It is proximately operative via the really distinct instruments. Fire can heat up water only by virtue of its necessary accident of heat. A stone can fall only due to its quality of weight. A magnet can attract a piece of iron only by means of its power to attract. Besides the general argument 'from the distinction between the creator and creatures', Suárez refers to the phenomenon of the organic and operational variety of the external senses. Their organic and operational diversity point to real distinctions between the particular sensory modalities. Since the sensorial organs are given to animals for the sake of the powers, real organic diversity entails a real distinction between the powers. But if they are distinct *realiter*, they must be distinct *realiter* also from the soul. If they were not, by the law of transitivity they would have to be identical.[181] Moreover, considering the human sense powers, if these faculties were identical with the rational soul, they would have to be subjected in a spiritual entity. The spiritual soul would have to be their subject and material cause. This, however, is not consistent with their material character due to which they cannot be functional after the soul is separated from the body.[182]

Suárez's critical stance to Aquinas's arguments for *RDT* is appended by a corollary in which his version of *RDT* goes significantly beyond that of Aquinas.

---

[177]For Suárez's theory of the specifications of the powers based on their acts and formal objects, see *CDA* disp. 3, q. 2. For the application of this tenet in the case of the issue of the individuation of the sensory modalities, see 4.1.

[178]*CDA* disp. 3, q. 1, n. 13, t. 2, 70.

[179]Although such a creature would have to be extraordinarily perfect, there is no contradiction in its existence. See *CDA* disp. 3, q. 1, n. 18, t. 2, 76–8.

[180]Cf. also *TDA* book 2, ch. 1, n. 5, 573–4; *DM* disp. 18, s. 3, n. 19, t. 1, 621.

[181]*CDA* disp. 3, q. 1, n. 14, t. 2, 70–2.

[182]*Ibid.*, q. 3, n. 4, t. 2, 118–20.

Unlike Aquinas and Cajetan,[183] in connection with his concept of *distinctio realis* characterized by the mutual separability of extremes, Suárez endorses a view according to which God *can* separate the soul from the powers and the powers from the soul.[184] It is not only that the soul can exist without its powers since God can prevent their emanation from the soul and conserve the 'naked' soul (removing its causal concurrence)[185] but also (no matter how odd it may sound) that a power can exist without its subject. A lesson from the Eucharist teaches us that the property of quantity can be separated from the substance of bread. Given the real distinction and *res*-character of the accident of quantity (and of some qualities too), why could it not be possible for faculties to exist without the soul too?[186] Despite this metaphysical possibility, a separated intellect or another separated vital power could not operate. Unlike a heat miraculously separated from fire, which can still heat air, the intellect cannot exercise its operations without a connection to the soul. As we have seen in 2.2, vital powers and their operations are essentially dependent on the soul's proximate causal influence.

We may conclude that Suárez's theory of the powers, which are *realiter* distinct from the soul, is correlated with his theory of the harmony of the powers rooted in the same soul. Although Suárez nominally adheres to Aquinas's view of the real distinction between the soul and its capacities, he both rejects his arguments and revises his notion of the faculties themselves. Aquinas's arguments are denied by virtue of the Jesuit's rejection of the statement according to which the soul cannot be the proximate principle of vital operations. As we shall see further in Chaps. 3 and 5, it is through attention and coordination of the really distinct powers' acts that the

---

[183]Thomas de Aquino, *STh.* I–IIae, q. 110, art. 4, ad 4, 315: 'Ad quartum dicendum quod, cum potentiae animae sint naturales proprietates speciem consequentes, anima non potest sine his esse'. Cajetan, *In STh.* I, q. 54, art. 3, XVIII, 50: 'Non infert etiam quod Deus possit eas [powers] separare, quia potest supplere vicem causae efficientis. Quoniam hoc intelligitur de efficiente proprie, scilicet per operationem mediam: tale autem non est substantia respectu suae passionis [. . .]'

[184]Dominik Perler and Christopher Shields take Suárez to be saying that faculties can never exist without a substantial form upon divine intervention and that there is only a one-way ontological independency, i.e., the independency of the soul upon the powers. For this reading, cf. Perler (2015, 130); Shields (2014, 207). However, as Perler himself says in note 68, Suárez in his earlier Vivès edition draws a parallel to the Eucharist in which the accidents of quantity and quality can also be supernaturally separated from the substance. It seems to me clear that also in *CDA* Suárez advocates the position according to which the intellect can be separated even though as being separated it cannot be operative since it ceases to be a vitally operating power: 'Sed quaeres rursus, *utrum intellectus separatus possit intelligere*, sicut calor calefacere. Respondetur: Non posse, quia intellectio est actio vitalis, et ideo intrisece pendet a primo principio vitae, sine quo esse non potest'. [my italics; for this see also 2.2]. Apparently, this question *assumes* the very possibility of the existence of a separated intellect (*CDA* disp. 3, q. 1, n. 22, 80–2). Moreover, this seems to be paralel to Suárez's view introduced in 2.2 according to which vital acts *qua* qualities can be created by God in Peter's mind without Peter's vitally operating through them.

[185]For this, see the next section.

[186]*CDA* disp. 3, q. 1, n. 21, t. 2, 78–80.

soul is immediately inherent to these activities.[187] At the same time, unlike Aquinas, Suárez conceives these powers as *res* or encapsulated modules which cannot be causally interactive. His understanding of the powers and the soul as qualities-*res* leads Suárez to the (not entirely intuitive) claim according to which divine absolute power can conserve both *res* independently of each other.

2.7 The Soul's Efficiency and Natural Resulting of the Powers

The powers as really distinct *res* constitute a thick particular[188] or an ontological aggregate (*unum per aggregationem*)[189] with the soul or composite. Without the soul's proximate causal influence in the powers and their operations the soul would be insulated from their acts.[190] I do not entirely share the opinion of some scholars according to which in *CDA* and *TDA* (unlike *DM*) Suárez espouses the view that the soul is *only* the *remote* cause of activities.[191] As it has been said, the soul's efficiency permeates all of the vital acts and guarantees the functional unity in our psychic life.[192] This essential efficiency of the soul, through which it presides over the powers and operations,[193] can be corroborated also from elsewhere. It is above all Suárez's theory of the natural resulting or flowing of powers from the soul

[187]As we shall see below, this view also systematically determines other Suárez's theories concerning the principles of the cognitive act (3.4), perceptual same-order self-awareness (3.7) and perceptual judgment (3.8).

[188]For the term 'thick particular', see Pasnau (2011a, 101–2).

[189]*CDA* disp. 3, q. 3, n. 12, t. 2, 132.

[190]Des Chene (2000, 146) takes this insulation from immediate efficacy in the world to be a disadvantage of the Thomistic model.

[191]For this reading, see Rozemond (2012, 164) and Perler (2013, 17, note 22). In this issue I actually do not see a significant shift in Suárez's views between the two texts. Perler's justification of this interpretation does not seem to be right. In note 22 he quotes the following passage: '[…] potest ergo concedi ad has potentias esse aliquo modo actionem per se […]' (*TDA* book 2, ch. 3, n. 13, 584) and argues that 'This means that they [powers] are the immediate causes of activities' (*ibid.*, 17). However, the context of the quoted passage is different. It does not concern powers as efficient causes or principles but powers conceived as *endpoints* of the soul's efficient causality in a process called natural resulting (*resultantia naturalis*). Moreover, when Perler says that the situation in *DM* is different since there 'he insists that the soul coordinates all the faculties and that it brings about acts of attention, thereby acting as immediate cause' (*ibid.*, 17, note 23), the same (as far as I can see) can be said to hold for both *CDA* and *TDA*.

[192]For the soul as providing unity by being the immediate efficient cause, see also De Boer (2017, 422).

[193]For this unifying and dominating role of the soul over its operations, see also *DM* disp. 15, s. 1, n. 7, t. 1, 499–500.

(*resultantia naturalis*) which (albeit as such not an instance of vital operating[194]) provides additional evidence of the systemic importance of the soul's efficient causality in Suárez's overall project of a *scientia de anima*.[195] In this section, I aim to show that Suárez's view of the natural flowing of powers from the soul ought to be seen as a theoretical consequence of his reified version of *RDT*.[196]

Typically, Suárez starts from a conclusion that he will later reject. This view is here, as elsewhere, represented by the view of Aquinas and the Thomists. It seems that the soul can never be the efficient cause of powers since 'the one who gives the form gives what follows upon the form' (*qui dat formam, dat consequentia ad formam*).[197] He introduces several arguments for this view. First, a generator who educes a substantial form from the potentiality of matter, or God who introduces the rational soul into organized matter, at the same time gives the soul and the essence of the composite its instruments or powers. Without them the soul would be incomplete and incapable of activity. Since nature does not do anything in vain, it must endow the soul with powers too. Second, the soul cannot be the cause of powers since if it were there would be two causal actions, rather than a single action. This is redundant. Third, if the soul were the immediate cause of powers, Aquinas's principle according to which a created substance can never be the immediate principle of accidents would have to be denied.[198] Fourth, if the soul were the efficient cause of its faculties, it would have to be able to restore them after the separation. Once the visual power got damaged or removed, a new power of sight would necessarily spread out from the soul. But such restoration is far from our experience. Fifth, the attribution of causal efficiency to the soul would come to be nothing else than an instance of the agent's (soul's) *self*-causation. However, both Aristotle[199] and

[194]Suárez distinguishes two kinds of internal efficient causality. One is proper, the other one is improper. While proper efficient causality concerns the soul's efficiency toward its vital acts, the improper one refers to natural resulting such as the emanation of powers from the soul. For this classification, see *DM* disp. 34, s. 7, n. 3, t. 2, 412.

[195]As Eleuterio Elorduy (1963) in his introductory study on the topic of natural emanation or natural resulting shows, the notion of natural emanation or resulting has a wide and multiple application in Suárez's philosophy, theology, moral and juridical sciences.

[196]For the claim that Suárez's reification of powers is correlated with his deep transformation of the traditional concept of form, see also Perler (2013, 136–9).

[197]This principle can in fact be found in many places of Aquinas's writings. For instance, in his *In libros Aristotelis De caelo et mundo*, book 3, lect. 7, n. 8, 252 Aquinas says the following: '[...] motor gravium et levium est generans, qui, dum dat formam, ex consequenti dat motum naturalem, sicut et omnia accidentia naturalia quae consequuntur formam: et sic generans causat motum naturalem mediante forma'.

[198]In *STh*. I, q. 77, art. 1, arg. 4, 236 Aquinas formulates an objection to his thesis, viz. that the soul's powers are not the soul's essence. In his argument for the conclusion that the essence of the soul amounts to its powers Aquinas says that it is so since 'Sed *id quo primo sentimus et intelligimus est anima*, secundum philosophum, in II *de Anima*. Ergo anima est suae potentiae'. In his reply Aquinas says 'Et ideo forma substantialis est primum actionis principium, sed *non proximum*. Et secundum hoc Philosophus dicit quod id quo intelligimus et sentimus, est anima' [my italics] (*ibid*., 237).

[199]Aristotle, *Physics* VII.1, 241b24–242a14, 207.

Aquinas claim that nothing can be moved by itself and whatever is moved is moved by another. As can be seen in their proofs for the existence of the unmoved mover, the axiom 'Whatever is moved is moved by another' must be considered as universally valid.

Suárez is well aware that Aquinas and Thomists follow these arguments. Yet, he is also sure that Aquinas after all concedes some efficiency to the soul with respect to its powers. He refers to this passage:

> [. . .] actuality is found in the subject of an accidental form prior to being found in the accidental form, and so the actuality of the accidental form is caused by the actuality of the subject. Hence, insofar as the subject exists in potentiality, it is receptive of the accidental form, whereas insofar as it exists in actuality, it is productive of the accidental form [. . .] Hence, it is clear that all the soul's powers—whether their subject is the soul by itself or the composite—flow from the soul's essence as their principle. For it has already been explained that an accident (a) is caused by its subject insofar as its subject exists in actuality and (b) is received in its subject insofar as its subject is in potentiality.[200]

The soul, insofar as it is the subject in actuality, is the cause of the existence of its powers, while insofar as it is in potentiality it receives them. In this way it seems that it can be indicated as 'self-causative'. However, as Suárez notices, Aquinas is much less clear about the precise character of this efficiency. In *STh*. I, q. 77, art. 6 titled 'Do the powers of the soul flow from its essence?' Aquinas introduces the following objection related to the specification of natural resulting:

> 'Emanation' names a certain movement. But as Physics 7 proves, nothing is moved by itself—except perhaps by reason of a part, as when an animal is said to be moved by itself because one part of it is the mover and another part the thing moved. Again, as De Anima 1 proves, the soul is not moved. Therefore, it is not the case that the soul in itself is a cause of its own powers.[201]

In his reply he says:

> The emanation of the proper accidents from a subject is not through any sort of transmutation but through a sort of natural resultancy, in the manner in which one thing naturally results from another, e.g., color from light.[202]

---

[200]Thomas de Aquino, *STh*. I, q. 77, art. 6, c., 246: '[. . .] actualitas per prius invenitur in subiecto formae accidentalis, quam in forma accidentali: unde actualitas formae accidentalis causatur ab actualitate subiecti. Ita quod subiectum, inquantum est in potentia, est susceptivum formae accidentalis: inquantum autem est in actu, est eius productivum [. . .] Unde manifestum est quod omnes potentiae animae, sive subiectum earum sit anima sola, sive compositum, fluunt ab essentia animae sicut a principio: quia iam dictum est quod accidens causatur a subiecto secundum quod est actu, et recipitur in eo inquantum est in potentia'. For the translation, cf. Aquinas (2010, 59).

[201]*Ibid*., arg. 3, 245: 'Praeterea, emanatio quendam motum nominat. Sed nihil movetur a se ipso, ut probatur in VII libro *Physic*.; nisi forte ratione partis, sicut animal dicitur moveri a seipso, qui una pars eius est movens et alia mota. Neque etiam anima movetur, ut probatur in I *de anima*. Non ergo anima causat in se suas potentias'. For the translation, see Aquinas (2010, 58).

[202]*Ibid*., ad 3: 'Ad tertium dicendum quod emanatio propriorum accidentium a subiecto non est per aliquam transmutationem; sed per aliquam naturalem resultationem, sicut ex uno naturaliter aliud resultat, ut ex luce color'. English translation in Aquinas (2010, 59).

The powers cannot arise from the soul through a transmutation or motion since the soul cannot move itself. There must be an external agent that in respect to a moved thing is already in act and that moves the soul and through the soul consequently brings about the powers. In my opinion, a plausible reading of this *passus* based on the puzzling example of colours resulting from light is proposed by Robert Pasnau. If colours are real qualities that exist even in total darkness (the realist view of colours), light cannot be the cause of their existence. It can cause only an attribute of them, namely their visibility. Equally, the soul cannot be the cause of its powers; it can be only a 'channel' or 'medium' through which an *external* agent (a progenitor) brings the faculties into existence.[203] In his exposition of Aquinas's reply, Cajetan does not abandon the soul's quasi-efficient contribution in the genesis of powers. Nevertheless, he distinguishes between two types of cause. One is a cause that proceeds *through a mediate action*; the other is a cause that runs *without a mediate action* (via *a mere natural resulting*). It is only in this second sense that powers can be said to be caused by the soul. Accordingly, substance can be the cause of its *passio* only 'per modum naturalis sequelae'.[204]

Suárez leaves no doubt that this view is false and that the only way of how to conceive the genesis of powers is to frame it in the manner of the soul's proper causal efficiency. He defines this efficiency as follows: *A* is an efficient cause of *B* iff *A* is an extrinsic principle from which *B* emanates by the mediation of an *action*.[205] It is the mediation through *action* that distinguishes efficient causality not only from intrinsic causality, in which the effect emanates from a cause through intrinsic unification (these causes communicate their own *esse* as parts to a whole, that is, the hylomorphic compound), but also from final causality, in which a cause is causally

[203]Pasnau (2002a, 269–70) says that 'Aquinas takes for granted that colors are qualities belonging to objects. A source of light does not give existence to color, in Aquinas's view, but instead actualizes (illuminates) the medium (the intervening air), enabling color to produce a likeness of itself in the medium, and thereby to enter the eye. Color, then, is not produced by the motion of a light source'. According to this reading for Aquinas the powers are not produced by the soul at all. For a different reading according to which Aquinas admits that the powers are produced by the soul but not by a change, see Ehret (2017, 116–8). According to Ehret when Aquinas speaks about colours here he has in mind the sensible species of the colours that represent and manifest the sensible. Since colour *qua* species produced by light does not determine the receiving subject, the emanated powers do not determine their own subject, namely the soul, since the rational soul is rational even before the rational power's emanation. It is rational in its essence. 'As colors in the air, powers do not represent a distinctive feature, they just make manifest what things are [...] emanation is taken into account as a specific kind of process, namely a non-physical process that does not imply change, and thus avoids any risk of self-causation' (*ibid.*, 118).

[204]Cajetan, *In STh* 1, q. 77, art. 6, VIII, 247.

[205]For a detailed analysis of this definition of efficient cause, see Freddoso in Suárez (2002, xliv–l). As Freddoso notes, for theological reasons to accommodate metaphysical changes such as creation, conservation and concurrence, this definition is a substantially reconstructed version of Aristotle's definition of efficient cause as 'The source of the first beginning of change or rest' from *Metaphysics* V.2, 1013a30–1, 211. For a detailed analysis and application of this definition of efficient cause to immanent actions, see 3.5.

active only metaphorically insofar as it functions as an end.[206] The powers must be caused through the soul's special action since no other way of their *fieri* is intelligible. Without the soul's proper efficient causation such production would come to be nothing more than a mere logical inference (*illatio*).[207]

Given that the soul and the powers are really distinct things, they must be caused by two causal actions since they make two distinct effects. Causal action terminating in the soul[208] must be distinct from action terminating in the powers. An action that is per se one can terminate only in an effect that is per se one. However, as said, the soul with its powers constitutes an ontological aggregate or a thick particular. Such an aggregate cannot be the terminus of a single causal action. There must be two distinct actions that cause two distinct effects. While the soul does not efficiently concur in the production of itself, it causally concurs in the production of powers.[209] A similar and more telling example of this strictly causal exposition of natural resulting is related to the flowing of *passiones* from forms that are temporally separated from the production of substantial forms. One of them is the recurrent example of the substantial form of water and its efficiency during its reduction to its connatural coldness. The emanation of this coldness is temporally separated from the production of the substantial form of water. Water must first be heated and only then can its substantial form start to cause the process of cooling. Suárez is, however, clear that it actually makes no difference whether this natural emanation is temporally separated from the production of a form, or not. In order to show that there are two distinct actions even in the natural simultaneous production of the soul and its powers Suárez introduces this thought experiment: God can create a soul and prevent or block the emanation of powers from it. He can create and conserve the 'naked' soul without its powers. Although this state of affairs is nomologically speaking impossible, as such it involves no logical contradiction. Although according to the

[206]For this see *DM* disp. 17, s. 1, n. 6, t. 1, 582. For *motio metaphorica*, see also *DM* disp. 23, s. 4, n. 9, t. 1, 862. This metaphorical motion does not mean that the agency of a proposed object is not real. It only means that it is not that of efficient causality (*DM* disp. 23, s. 1, n. 14, t. 1, 847).

[207]*DM* disp. 18, s. 3, n. 6, 616–7. The denial of the proper causal role of the soul was a firm agenda in all post-Suarezian Thomists. See Domingo Báñez, *Scholastica Commentaria in Primam Partem Angelici Doctoris S. Thomae. A Qvaestione LXV. vsque ad CXIX, & vltimam*, In *STh.* I, q. 77, art. 6 (Douai: Borremans, 1614), pp. 199–202; Ioannes a S. Thoma O. P., *Cursus philosophicus Thomisticus,* vol. III, *Naturalis philosophia*, p. 4: De ente mobili animato, q. 2, art. 2, 61–74; and his vol. II, *Naturalis philosophia*, p. 3: De ente mobili corruptibili, q. 12, art. 4, Appendix: *De dimanatione*, 267–270; Antonius a Matre Dei, *In octo libros Physicorum Aristotelis, Disputationes*, disp. X, q. 7, 216–28, in *Collegii Complutensis Discalceatorum*. Antonius a Matre Dei (1583–1637) was one of the two Carmelites from the University of Alcalá de Henares (*Complutum*) who wrote the four-volume *Cursus integrum philosophicum*. The other one was Michael a SS. Trinitate (1588–1661). For a comparison of the different analyses of the powers' natural emanation in Suárez and Poinsot, see also Des Chene (1996, 157–61).

[208]*CDA* disp. 3, q. 3, n. 11, t. 2, 130. Des Chene (2000, 150) points out that such a conclusion would be egregious for Descartes for whom not only it is impossible to conceive the soul (*res cogitans*) without the powers but even without the activity of thinking.

[209]*CDA* disp. 3, q. 3, n. 7, t. 2, 122.

natural order the powers, like other necessary accidents (*passiones*), are inseparably connected with the substantial form or composite, they can be separated from their subject through a divine intervention since both are *res*.[210] It is only according to our general mode of expression (*communis modus loquendi*—this expression can be taken as Suárez's verbal 'libation' to Aquinas) that it can be said that a principal agent produces a form together with its powers.[211] Since the goal of the agent is to produce an integral thing with all of its powers, the agent is, absolutely speaking, to be regarded as the efficient cause of the whole aggregate including the powers. Even though the agent produces immediately only the form or the soul, he can be said to radically (*radicaliter*) co-produce all its powers too.[212]

Considering the abovementioned arguments for the opposite view, Suárez replies to the first by referring precisely to this radical co-production due to which it can be said 'the one who gives the form gives what follows upon the form'. And, again, it is only because of this general mode of expression (that does not exactly correspond to the metaphysical facts) according to which we can say that there is only a single action and not two. Suárez also agrees with the fourth argument, sc. that after their separation the intellect and the will would be immediately restored through the soul's efficiency since this separation in no way diminishes the soul's causal power. However, he denies this in the case of the sentient powers which are subjected in the composite. The fact that they cannot be restored is not due to a decrease in the soul's causal power after separation. The reason is that the sentient powers can be damaged or removed only if their organs are damaged and removed. However, the soul cannot restore the organs just as it cannot regain an amputated hand.[213]

As I have already indicated, Suárez's endorsement of the notion of self-causation seems to have been importantly motivated by his libertarian theory of free will, based on the notion of indifference, according to which only the will and no other external force can be conceived as an exclusive principle of contingent self-determination.[214] The libertarian concept of freedom is connected with Suárez's explicit endorsement of Scotus's theory according to which one and the same faculty can at the same time be in virtual or eminent act and in formal potency with respect to the same thing.[215] We have seen that it is not only a vital power that can be both in virtual act and in formal potency at the same time with regard to the same act, but in the process of the

---

[210] *Ibid.*, n. 12, 132; *DM* disp. 18, s. 3, nn. 7–8, t. 1, 617–8. These two modalities are the reason why the propositions 'Powers necessarily flow from the soul' and 'The soul can exist without its powers' are not contradictory. For the distinction between these two modalities, see Shields (2013) and (2014).

[211] *DM* disp. 18, s. 3, n. 8, t. 1, 618.

[212] *CDA* disp. 3, q. 3, n. 13, t. 2, 132–4.

[213] *Ibid.*, n. 14, 134–6.

[214] For his libertarian concept of free will see, e.g., *DM* disp. 19, s. 2, n. 15, t. 1, 697–8. Of the recent publictions, see Anfray (2017) and Pink (2018).

[215] For the importance of this axiom in Suárez's theory of free will in the context of his rejection of the Thomistic theory of *praemotio physica* in the famous debate *De auxiliis*, see Piñon (1975, 343–4).

natural emanation of powers the substantial form or soul also complies with this principle.[216]

The exclusive role of the soul as the efficient cause of powers is corroborated by Suárez's denial of Aquinas's view according to which the less perfect powers emanate from the perfect powers.[217] The denial of this metaphysical Neoplatonic descendance from the most perfect to the least perfect (from ἕν to ὕλη) is symptomatic for his theory of a-causal mediation among the operations of various powers. In connection with this inter-power emanation, Aquinas claimed the following:

> [. . .] what is closer to the first thing is in some sense a cause of what is more remote from the first thing. But it was shown above that there is a multiple ordering among the powers of the soul. And so one power of the soul proceeds from the soul's essence by the mediation of another power. But since (a) the soul's essence is related to the powers both as an active and final principle and also as a receptive principle [. . .] since (b) that which is an agent and an end is more perfect, whereas that which is a receptive principle is as such less perfect, it follows that those powers of the soul which are prior in the order of perfection and nature are principles of the other powers in the manner of an end and *an active principle* [. . .] *the sensory power is a certain deficient participation in the intellect*, and so according to its natural origin it in some sense arises from the intellect in the way that what is imperfect arises from what is perfect.[218]

Suárez entirely rejects this view. He labels this view, as he does in many other cases, with his euphemistic evaluation 'difficilis'. How could vegetative and sentient powers emanate from the soul through the intellect? The process should rather go in the opposite (evolutionary) direction, namely from the less perfect to the more perfect. Moreover, the flowing of one power from another is not necessary for

---

216On this voluntaristic basis Suárez also rejects the universal validity of the abovementioned Aristotelian-Thomistic axiom 'Whatever is moved is moved by another'. For Suárez's explicit critique of this view, see *DM* disp. 29, s. 1, n. 7, t. 2, 23: '*Omne quod movetur ab alio movetur* adhuc non esse satis demonstratum in omni genere motus vel actionis; nam multa sunt quae per actum virtualem videntur sese movere et reducere ad actum formalem, ut in appetitu seu voluntate videre licet, et in aqua reducente se ad pristinam frigiditatem [. . .]' For the contrary axiom that Suárez finds in Scotus, see *DM* disp. 18, s. 7, n. 4, t. 2, 632.

217Surprisingly, this important part of *CDA* (disp. 3, q. 3, nn. 16–22, t. 2, 140–50) has been entirely left aside by scholars. In this part Suárez lays out the various kinds of the ordering of the emanated powers. He mentions, among others, the order of duration (one power exists before another); the order of dignity (one power is more perfect than another), the order of generation (one power is generated before another), the order of effective emanation (one power emanates from another). And it is this last ordering that is especially relevant for us here.

218My italics. *STh*. I, q. 77, art. 7, c., 247–8: '[. . .] ita quod est primo propinquius, est quodammodo causa eorum quae sunt magis remota. Ostensum est autem supra quod inter potentias animae est multiplex ordo. Et ideo una potentia animae ab essentia animae procedit mediante alia. Sed quia essentia animae comparatur ad potentias et sicut principium activum et finale, et sicut principium susceptivum [. . .] agens autem et finis est perfectius, susceptivum autem principium, inquantum huiusmodi, est minus perfectum: consequens est quod potentiae animae quae sunt priores secundum ordinem perfectionis et naturae, sint principia aliarum per modum finis et activi principii [. . .] Sensus etiam est quaedam deficiens participatio intellectus: unde secundum naturalem originem quodammodo est ab intellectu, sicut imperfectum a perfecto'. English translation in Aquinas (2010, 61).

securing the necessary subordination of powers, such as that of the sensory appetite to the rational appetite. This can well be secured by the powers' rootedness in the same soul.[219] In sum, while the qualitative difference in the perfection of the human and the nonhuman interior senses in Aquinas is based on the assumption that the former emanate from the intellect and the latter do not, Suárez is clear that the former are such just because they are rooted in a more perfect soul and not due to this emanation.[220]

Suárez's denial of metaphysical inter-power descendance is peculiar to his approach.[221] His theory of the emanation of all the powers directly from the soul is an important metaphysical underpinning of his view of the soul's mediation between the acts of the various powers, which also grounds his theory of the production of the imaginative species in the internal sense.[222] There is no direct causal mediation between the acts of the really distinct modulary powers, such as between the external and the internal senses, between the acts of the internal sense and the intellect, and between the acts of the cognitive powers and the appetitive powers in general. The only mediation can go a-causally and indirectly through the soul's coordination.

[219] *CDA* disp. 3, q. 3, n. 18, t. 2, 144–6.

[220] *Ibid.*, n. 20, 146–8.

[221] For the systematic impact of this theory on Suárez's denial of the 'intellectualized' internal senses, see 5.3.1 and 5.3.2.

[222] For this, see 5.1.

Chapter 3
External Senses I

In this chapter, I present Suárez's doctrine of perception by combining his general exposition with his 'first-order *in particulari*' analysis of external sense perception presented in *CDA* disp. 6, in which he focuses on the external senses without considering their specific modalities. The 'second-order *in particulari*' treatise of their individual modalities will be discussed in Chap. 4. This part has nine sections. First, I present Suárez's arguments for the existence of the sensible species both *in medio* and in the powers. Second, I bring in Suárez's views related to the issues of the nature and the origin of these sensible species. Third, I outline Suárez's ontology of the perceptual act. Fourth, I lay out his theory of the causes of perceptual acts. Fifth, I explicate Suárez's view of the endpoint of perceptual act. Sixth, I present Suárez's theory of selective attention and its division that is closely connected with his theory of the causes of perceptual operations. Seventh, I introduce his doctrine of perceptual self-awareness. In a link to the Aristotelian-scholastic theory of three kinds of sensible objects and Suárez's theory of perceptual judgment, I finally explicate his views related to illusory perception at the natural level (Sect. 3.8) and at the supernatural level of a hypothetical divine intervention (Sect. 3.9).

3.1 Necessity of the Sensible Species

In *CDA* disp. 5, q. 1 Suárez starts by asking whether a perceptual power must be connected with a sensible object to produce a perceptual act. As he notes, refuting this connection is endorsed by those who deny the existence of the intentional (sensible) species. Therefore, the problem really is whether such intentional

D. Heider, *Aristotelian Subjectivism: Francisco Suárez's Philosophy of Perception*, Studies in the History of Philosophy of Mind 28,
https://doi.org/10.1007/978-3-030-67341-3_3

species exist at all.[1] The negative position can take different versions according to the number of items denied. These can be the species that are the principles of the external senses (of all, or only of some), of the internal senses, of the intellect, or even of all of the above. Such an eliminativist position in its various versions was advocated by some philosophers in favour of the extramission theory (see Sect. 4. 2.4) and by some late medieval philosophers who regarded the existence of the sensible likenesses as problematic and entailing an implausible representationalism. Although Suárez does not mention Peter Olivi, two brief arguments for the eliminativist view point to the Franciscan's implicit doctrinal presence.[2] According to the first argument, the existence of the causal principle prior to the perceptual act, sc. the species, is not necessary because the object's epistemically relevant function is only *terminative*. The object can be nothing else than a terminative cause of perception to which the active perceptual power is intentionally stretched. These objects cannot function as motive causes since they are not capable of acting in an epistemically relevant way on the soul. Being on a lower ontological level, the corporeal affections received from the external objects cannot actualize the soul's powers.[3] Second, intentional unification through the sensible species is impossible, since postulating this principle does not make sense. A visible object can be joined to the visual power owing either to real being, or to intentional being. But neither of these is possible. Starting from the second alternative, it makes no sense to postulate likenesses of objects having intentional being. As such they can be only 'ficta'. The first option also results in an implausible consequence, since (as stated by Aristotle) if an object in its real being is placed on the eye, it necessarily impedes vision.[4] In other terms, such an entity necessarily inserts a 'veil' between the object and the perceiver. However, if this were the case, the intuitive cognition of external objects existent here and now would be compromised. Accordingly, in order to have an intuitive cognition, a

[1]In his *De Deo uno et trino* Suárez names six functions (*munera*) of the intelligible species that can be *mutatis mutandis* extended to the sensible species as well: (1) to inhere in the intellect; (2) to constitute the intellect in the first act or to inform it; (3) to unite the object with the power; (4) to produce an act of understanding; (5) to represent an object; and (6) to determine the power to this or that kind of act. For this account, see Suárez (1856a), *Opera omnia*, vol. 1, *De deo uno et trino*, book 2: *De attributis Dei negativis*, ch. 12, nn. 5–13, 87–9. All these functions will be discussed *passim* below.

[2]*CDA* disp. 5, q. 1, n. 1, t. 2, 284.

[3]For recent interpretations of Olivi's criticism of the existence of the sensible species and an overview of his active theory of perception based on the Augustinian axiom that the ontologically higher (spiritual) cannot be produced from the ontologically lower (corporeal), see Perler (2004, 109–27), Pasnau (1997, 168–81), Tachau (1988, 39–54); Toivanen (2013b, 141–61); Lička (2017, 103–10). For this causal axiom based on the impossibility of ascendant causality and its application to sensory perception, and for the limitation of the causal role of an object in the production of perception in Augustine and medieval Augustinianism, which, as we shall see below, is operative also in Suárez, see Silva (2014a, 2014b). For Olivi himself, see Petrus Iohannis Olivi, *Quaestiones in secundum librum Sententiarum*, vol. II, qq. 49–71, q. 58, 414–5; *Quaestiones in secundum librum Sententiarum*, vol. III, qq. 72–118, q. 73, 87.

[4]Cf. Aristotle, *DA* II.7, 419a12–4, 107.

distant object, as Ockham argues, must act on the external senses immediately, that is, without the mediation of a species. The object must be able to act at a distance. As is well-known, for Ockham such acting was not limited to perception: a magnet acts on a distal metal; the Sun illuminates and heats the air near the Earth's surface without illuminating and heating the planets in between etc.[5]

In line with his metaphysics of efficient causality, Suárez rejects this Ockhamist possibility of acting at a distance.[6] Not only that the visible objects cannot act on the power at a distance, but no transitive cause as such can produce an effect without passing through a medium up to its *passum*. Material causes and effects must meet the necessary condition of mutual contact.[7] Consequently, Suárez concludes, the conjunction of the cognoscible object and the cognitive power is necessary for *all* kinds of cognition.[8] In other terms, the intentional species must be posited in all kinds of cognition as an indispensable (co-)principle in the process of the cognitive acts' production. Without the power's reception of this principle the production of a perceptual operation cannot be explained. In order to refute the abovementioned objection from the fictive character of the species, Suárez states that an intentional species refers to a *representative* form (as such this form is in its essence 'carried away' to something else) that does *not* rule out its *real* being (conceived as imperceptible, though). It is called 'intentional' since it is the principle of a cognitive act called intention or second act. He is sure that an object cannot operate only as an endpoint of cognition but constitutes a necessary motive principle in the first and receptive phase in the overall genesis of the production of perception.[9] Since

[5]Ockham, *Quaestiones in librum tertium Sententiarum (Reportatio)*, q. 2, 56–7. For Ockham's critique of the species theory and his theory of action at a distance, see Maier (1967, 433–44); Pasnau (1997, 162–7) and Lindberg (1976, 139–42).

[6]For the impossibility of *actio in distans*, see *DM* disp. 18, s. 8, t. 1, 650–68. For an analysis of Suárez's view according to which the *propinquity* of cause and effect is one of the necessary conditions of efficient causation, see Des Chene (2012).

[7]Suárez explains that 'an efficient cause's natural mode of acting is, as it were, through a continuous line of action from what is nearby to what is distant' (*DM* disp. 18, s. 8, n. 20, 657–8; for this translation, see Suárez, 1994, 195–6). But he also says that no conclusive argument based on the nature of the efficient cause can be given. This is why his arguments for this upshot are based largely on experience and thought experiments. One of these is the thought experiment with the void which is postulated as hypothetically existing. If *per impossibile* a void were inserted between the efficient cause and the patient, the agent could not affect the distant patient. No action across an empty space, no emission of sensible species from a visible object would be possible (*ibid.*, n. 14, 655). The same happens, as we know from experience, if space is filled with an opaque body (*ibid.*, 15). However, when pressed to formulate an a priori reason for this impossibility, Suárez comes with this reasoning: In order to exercise its causality, every act involves a connection with the power because the agent is the *act* of the patient and the patient is in potency to its action. As he says, this is not a consequence of the agent's imperfection, but rather it is part of the act's perfection (*ibid.*, n. 20, 658).

[8]*CDA* disp. 5, q. 1, n. 3, t. 2, 286: 'Sit tamen prima conclusio: unio obiecti cognoscibilis cum potentia est necessaria in omni cognitione'.

[9]*Ibid.* n. 3, 286–8. In *DM* disp. 35, s. 4, n. 18, t. 2, 464 when talking about the intelligible species Suárez presents the following general criterion for introducing the intentional species: 'Species intelligibilis poni solet, vel ut objectum intelligibile in potentia fiat intelligibile in actu, vel ut

cognition amounts to a process of *assimilating* the perceiver to the perceived object, the sensible species must be posited. And while it can be argued that such assimilation can be accomplished through a mere natural causality, Suárez is adamant that this cognitive assimilating must be mediated through the intentional species, which is the only vehicle transmitting information from the object to the perceiver.[10]

Beside this basic a priori argument from the assimilative character of cognition, Suárez formulates several a posteriori arguments which are largely taken from premodern catoptrics (the optical study of mirrors and phenomenon of reflection).[11] Most of the space is devoted to the 'mirror experience'. Normally, we cannot see an object standing behind us (*a tergo*). We can see it only if it is situated in our visual field. However, albeit an object standing behind us cannot be perceived directly, its presence can be mediated by a mirror situated in front of us. The mirror can mediate a new presence of the object, namely behind the mirror, to us. Its presence can be literally *re*-presented. But how can we explain this *re*-presentation? Following the tradition of the medieval *perspectivi*, Suárez claims that the only way how to explain this is to assume the existence of visual likenesses that are rectilinearly multiplied from the object toward the mirror from which they are then reflected back to the observer's eyes where, if all physiological and psychological conditions on the side of the percipient are met, it (co-)causes a perceptual experience.[12] In this context

---

objectum quod erat separatum aut distans et improportionatum, conjungatur vel proportionatur potentiae'. The statement about the distance and disproportionality between the object and the power can be applied also to the external senses. Ironically, after Suárez this criterion was taken up by some Jesuits, such as Pedro Hurtado de Mendoza, who subjected Suárez's species theory to criticism and denied the existence of the sensible species for the lower external senses (smell, taste and touch). For this, see Heider (2019a). In any case, for Suárez there seems to be only a single case in which such an intimate union of the object with the power exists that it makes the existence of the intentional species dispensable, viz. the *visio beatifica*, even though even in this beatified vision these species are possible. For this, see Francisco Suárez (1856a), *Opera omnia*, vol. 1, *De deo uno et trino*, book 2: *De attributis Dei negativis*, ch. 12, nn. 14–21, 89–91.

[10] *CDA* disp. 5, s. 1, nn. 4–5, t. 2, 288–90. For a characterization of assimilation in the Aristotelian cognitive model as entailing the objects' causality upon our senses through the emitted species, see Silva and Thörnqvist (2019).

[11] Beside this 'mirror experience' Suárez adduces also examples connected with perceptual errors: A stick immersed in water appears to be broken; a coin immersed in water seems to be bigger than it really is. In Suárez's view these experiences are possible only due to the emission and refraction of the sensible species in the medium (*CDA* disp. 5, s. 1, nn. 4, t. 2, 290). These examples were used also by immediate critics of Ockham's species-less theory, such as John of Reading († 1346); cf. Tachau (1988, 174–9).

[12] According to Durandus of Saint-Pourçain we see the sensible species of the object on the mirror's surface because the sensible species are of the same sort as the sensible objects represented by them. The sensible species of red is red and the species of light is light. In the next section we will see that Suárez rejects this view. For him the intentional species is not the object known but only a medium through which we apprehend the object. If the species were seen in the mirror, it would have to be visible in the medium (e.g. the air) too. This is not the case. He shows that in the mirror we see the object itself in its depth and we see it at a distance from the mirror in which it exists. An object at a great distance from the mirror situated close to the perceiver is seen at the distance in which it exists, and not closely in the mirror. If the species were not a sortally distinct item from the sensible, it

Suárez denies an alternative explanation based on the theory of extramission according to which the eye sends forth the visual rays reflected from the mirror and attaining the object seen.[13]

In the second part of *CDA* disp. 5, q. 1, Suárez raises a second question important for his other doctrinal decision-makings. Are the sensible species, multiplied *in medio* up to the perceiver, received in the power, or in the organ? If the latter were the case, the species could not be an intentional *form* of the sensibles and the epistemic role of the sensibles would seem to be rather restricted to a mere 'excitation' of the soul. In his doctrinal survey, Suárez mentions two Greek commentators on Aristotle, namely Simplicius and Philoponus, who advocated this view. In his *Commentary on Aristotle's On the Soul* II.5, Philoponus says:

> [. . .] the apprehension of sense-objects is nothing other than discernment of the affection produced in the *sense-organ* by the sense-objects. For when the sense-organ is put into this or that state by the sense-object the power in it apprehends the affection which occurs in its own subject.[14]

This view seems to be similar to Augustine's tenet from *De quantitate animae*, in which he defines sense perception as 'an effect on the body of which the soul is not unaware'.[15] Setting aside the issue of the complicated harmonization of his accounts of extramission and intromission, Augustine regards the soul's vision as primarily focused on the affections of the *body* (*passio corporis*) caused by the corporeal affections. This view is (at least verbally) rejected by Suárez. In his second conclusion of *CDA* disp. 5, q. 1 he embraces the view of Aristotle[16] and Aquinas,[17] namely that the object is united with the power through the species that inheres in the *power*. Two reasons for this conclusion are mentioned. First, there is no logical contradiction in the species's inherence in the power. The intelligible species inheres in the angelic intellect and in the separated human intellect where there is no bodily organ. Second, the intentional species is posited for the sake of the power's actualization,

---

would have to be seen on the mirror's surface. But this does not comply with our experience (for this, see *CDA* disp. 5, q. 2, nn. 14–6, 314–6). This *a posteriori* argument for the existence of the sensible species does not seem to be entirely satisfactory, though. It does not demonstrate why a more parsimonius exposition based on the species-less causal action of an object would not suffice to explain the observation. At the same time it leaves unexplained why the perceiver does not see the object at the very place where it exists, namely behind the perceiver, and why it is seen behind the mirror. For the issue of what is actually seen in a mirror with a presentation of two distinct replies (models)–the multiplication account, according to which we see an image constituted by the mirror (the theory endorsed by Albert the Great), and the unification account, endorsed by Suárez as well, according to which we see an appearing real object (Peter Auriol)–, see Lička (2019).

[13]For his rejection of the theory of extramission, see Sect. 4.2.4.

[14]My italics. John Philoponus, *On Aristotle's On the Soul 2.1–6*, 100.

[15]Augustine, *The Greatness of the Soul*, ch. 23, 41.

[16]Aristotle, *DA* II.12, 424a17–9, 137: '[. . .] sense is that which is receptive of the form of sensible objects without the matter [. . .]'

[17]Thomas de Aquino, *STh.* I, q. 14, art. 1, c., 166: 'Sensus autem cognoscitivus est, quia receptivus est specierum sine materia [. . .]'

not the organ's. However, this does not mean that the organ cannot serve as a *medium* of this reception. The organ is essential for the reception of the species in the power since it mediates this reception in the power.[18] The sensible species cannot be 'cognitively' received in the power if a *temperamentum* of the organ is not proportional to the received species. This is also why Suárez devotes so much (in fact, much from the disp. 7) to the problem of the organs of the external senses and their correlation with their proper objects.

3.2 Nature and Origin of the Sensible Species and the Agent Sense

After answering the question 'An sit species sensibilis?' Suárez approaches the query 'Quid sit species sensibilis?'. Given the general problems connected with the intentional species brought up by critics of the species theory, an inquiry into the nature of the species seems to be crucial. In *CDA* disp. 5, q. 2 Suárez raises four questions: (1) Is the sensible species substance, or accident?, (2) Is the sensible species specifically (in kind) the same as the represented object?, (3) Is it spiritual, or material?, (4) What is the relationship between the entitative and the representational aspect of the intentional (sensible) species?[19]

Regarding the first question, Suárez affirms that the sensible species is an accidental quality pertaining to the kind of disposition which inheres in the power through the medium of the organ.[20] Contrary to the species of the internal senses, the sensible species of the external senses depends (in the natural order) in its being and becoming (*in fieri et conservari*) on the sensible quality represented by it.[21] In order to actualize the sensory power, the species does not need to be a substance, such as the Epicurean effluences of 'eidola' or the Democritean configurations of atoms.[22] Since species are accidents, their union with the power can be only that of an accidental unity composed of subject and accident.

True to his contrasting thinking, Suárez presents a rival position according to which the unification of the object and the power through the sensible species does not result in an accidental unity. On this view, endorsed by Thomists headed by Cajetan, the connection of the sensible species and the power is supposed to bring about a unity considerably 'tighter' than the accidental unity of a subject and an accident. This unity is even more intimate than the unity of a matter-form composite.

---

[18] *CDA* disp. 5, q. 1, n. 6, t. 2, 292–4.

[19] *Ibid.*, q. 2, n. 1, t. 2, 296.

[20] *Ibid.*, n. 23, t. 2, 324.

[21] As I will show in Sect. 3.9, this dependence can be 'affected' by a (hypothetical) divine intervention: God can cause the external senses' (abstractive) cognition of an absent or nonexistent object by implementing the species in the observer's eyes.

[22] *CDA* disp. 5, q. 2, n. 2, t. 2, 296–8.

While for Suárez this second composition results in the production of a new thing (*a tertium quid*), for Cajetan no such *tertium quid* is established. Following the assimilativist view of perceptual act he claims that the perceptual power identifies *intentionally* with the cognized thing.[23] Suárez is quick to say that the mode of union of the species and the perceptual power cannot differ from the aggregative union of accident and substance. The power and the species can constitute only an accidental unity.[24] Following Aquinas,[25] Cajetan spells out his claim by employing this distinction: the intentional species (in fact, he speaks about the intelligible species but the reasoning applies to the sensible species too) comprises two distinct aspects: (a) the species conceived as an inherent form, and (b) the species conceived as an image of the object.[26] While the species in the meaning of (a) inheres in the power and constitutes an *unum* per accidens with the subject, the species in the meaning of (b) constitutes a unity that is more 'intimate' than a substantial unity.

Suárez rejects this distinction as fictitious. In answering the abovementioned fourth question that is related to the first question he claims that the species *qua* (a) cannot be considered without the aspect (b). These two aspects cannot be separated.[27] But why does Suárez resist this distinction? It seems to me that he does so in order to highlight the real distinction between the first-act reception of the intentional (sensible) species and the second-act perceptual operation conceived as a distinct unit, sc. quality, which, as we shall see in Sects. 3.3 and 3.5, is a *res*-accident. If he admitted Cajetan's claim that the species is received in the power only according to its representational being or the pure perceptual content transmitted by the species, this reception would come to be identical with the perceptual act. An act of perception would become the *formal* effect of the sensible species. However, as we will see below, cognitive operation cannot be the formal effect of the received species since the reception of the species can and often does occur without the production of a perceptual act. If a perceptual state were the formal effect of the reception of the species, it would have to be inseparable from this reception. The reception of the species would necessarily produce a perceptual experience. Even more importantly, if perception were produced in this way, it would not be an immanent and vital operation; rather, it would become an effect of transitive operation.[28]

Albeit the species pertains to the accident of quality, in the reply to the second question Suárez states that it is *not* of the same order and of the same kind as the sensible object of which the species is a representation. Following Aquinas[29] Suárez

[23]Cajetan, In *STh.* I, q. 12, art. 2, XV, 118.

[24]*CDA* disp. 5, q. 2, nn. 4–5, t. 2, 300–2.

[25]See especially his *Summa contra gentiles*, book 3, ch. 51, 140.

[26]Cajetan, *In STh.* I, q. 12, art. 2, XIV, 118.

[27]*CDA* disp. 5, q. 2, n. 23–4, t. 2, 322–4.

[28]*Ibid.*, n. 6, 304–6.

[29]Thomas de Aquino, *Quaestiones disputatae de veritate*, q. 8, art. 11, ad 3, 256: '[...] inter cognoscens et cognitum non exigitur similitudo quae est secundum convenientiam in natura, sed secundum repraesentationem tantum. Constat enim quod forma lapidis in anima est longe alterius naturae quam forma lapidis in materia; sed inquantum repraesentat eam, sic est principium ducens in cognitionem eius'.

asserts that the intentional species can be the *intentional* vehicle of perception only if it is *not* of the same nature as the represented item. This difference in kind is expressed by the term '*spiritual* likeness'. This is also what Aristotle means when he says that the sense receives the forms *without* the matter.[30] The visual species of a white colour is not and cannot be white. Clearly, when emitted from the object the species of white colour does not affect the medium such as the air *realiter*.[31] The air does not turn white through the presence of the species of white. When we see the white colour of snow, we do not see the coloured (white) air. In fact, if the air turned white, no other species, such as that of black, could occur in the same part of this medium. Also, if on the reception of the species of white the eye turned white, it could not be receptive of the species of black colour. The former would preclude the latter from being in the same place. As Suárez says, this very possibility of the coexistence of white and black colour in the eye is based on the fact that sensible species do not have their contrary opposites and thus they do not expel each other from a place. They can represent two distinct things in the eye at a time. Moreover, if the species were the same in kind as a colour, no sensation could be produced because, as it is said, 'what is placed on the sense organ should be imperceptible'.[32]

Suárez finds doctrinal opponency to this conclusion in the theory of Durandus of Saint-Pourçain.[33] This *enfant terrible* of the Dominican Order was well aware of the problems befalling the sensible species.[34] In order to make things less complicated, Durandus says that the sensible species *is* specifically the same as the sensible object represented by it. It differs from it only in the mode of being. At one time it is an entity existing in a sensible object, at another time it exists in the medium and the organ. Durandus embraced this view because he was not clear as to what a species's essential distinction from a sensible object might be like. We could say today that if it is a real entity, it should be observable at least by a microscope. Suárez presents

---

[30] *CDA* disp. 5, q. 2, n. 8, t. 2, 306–8. Aristotle, *On the Soul* II.12, 424a18–22, 137: '[. . .] sense is that which is receptive of the form of sensible objects without the matter just as the wax receives the impression of the signet ring without the iron or the gold, and receives the impression of the gold and bronze, but not as gold or bronze [. . .]' Symptomatically, when referring to this passage Suárez never mentions the second part of this passage about the reception of the signet ring in the (passive) wax.

[31] Of course, in this colours differ from other sensibles, such as sounds or odours, that can affect the medium *realiter* too. For this, see Chap. 4.

[32] Aristotle, *DA* II.9, 421b17–8, 123: 'That what is placed on the sense organ should be imperceptible is common to all senses'. For Suárez's substantial relativization of this statement, see Sect. 4.5.4.

[33] I cannot share Jörg Telkamp's interpretation according to which '[. . .] Durandus is Suárez's main source, with whom he disagrees only occasionally'. From my reading Suárez often takes Durandus's views as opponent's opinions. See Tellkamp (2012, 7). For Durandus's view of the sensible species, see Durandi de Sancto Porciano, *Scriptum super IV libros Sententiarum*, in II *Sent*., d. 3, q. 5, 166–7.

[34] Anneliese Maier (1967, 429) pointedly says that 'Die Theorie des species sensibiles ist ein Lehrstück, das wie kaum ein anderes der scholastichen Philosophie mit Schwierigkeiten und Unklarheiten belastet ist'. For a lucid presentation of Durandus's theory, see *ibid*., 429–31.

Durandus's four arguments. First, the sensible species of white colour must be white since the species is related to white colour in the same way as a luminous object (*lux*) is related to light (*lumen*). Both light and luminous objects, such as the Sun or fire, are corporeal entities. Therefore, the sensible species of colour must be corporeal like the object it represents. Second, *X* can be the likeness of *Y* only if they are specifically the same. Third, the action of the sensible object is *univocal*. The sensible generates a univocal causality. Colour as a real quality cannot emit something that would be of an entirely different order than colour. Fourth, it is not true that the species is not visible. In fact, the species can be seen in a mirror. Peter sees his face in the mirror from the reflection of the species. Therefore, he sees the sensible species.[35]

Suárez's replies to these arguments show how differently from Durandus he conceives the nature of the sensible species. Concerning the first argument he argues by pointing to the dissimilarity between the luminous source and light on the one side and the real quality of colour and its species on the other. While light is a real entity that is specifically the same as the luminous source (light can be seen only through its own sensible species[36]) the sensible species, as said above, is not cognized (it is 'ignota'). It is only that through which (*quo*) a colour is sensed. The species only 'carries' us to the cognition of something else. It refers us to something else. It is not a 'quod' or object in our perception. Second, Suárez finds an essential difference between a real similarity based on the assumption of specific unity (two horses are similar because *in re* they share the same specific or common nature) and a representative likeness that does *not* require this kind of formal unity and which is in a way *primitive*. The dissimilarity between these two can be seen on the example of the representation of a colour through the intelligible species in which no natural likeness can be the case. Third, as he shows in *CDA* disp. 7, although an intentional action often goes hand in hand with a real action, as in the case of heating, they are *not* the same. As we will see further in Sects. 4.2.3, 4.3.2, and 4.4.2, the sensible species can attain more distant places than those reached through natural actions. Fourth, as we have seen in the previous section, it is not true that in the mirror we see the images of sensible species. What we see there are the material objects in their three-dimensional depth. We do not see the objects on the surface of the mirror. Objects in the (flat) mirror are seen at the same distance at which they are located from the mirror. Moreover, it cannot be said that they are seen only according to representational being and not according to their entitative being since these two aspects cannot be separated. The only function of the mirror in this case is to rebound the species back to the perceiver and mediate the cognition of objects.[37]

The reception of a species from a sensible object that differs from it in kind does not mean that the power or the organ need not be deprived of the real nature of its

[35] *CDA* disp. 5, q. 2, n. 13, t. 2, 312–4.

[36] See *CDA* disp. 7, q. 1, n. 1, t. 2, 552–4. See also Sect. 4.2.1.

[37] *CDA* disp. 5, q. 2, n. 13, t. 2, 312–4. For his 'realist reading' of the mirror experience, see *ibid.*, nn. 14–6, 314–6.

object.[38] It may be argued that if the power receives a species that is essentially different from the represented quality, then the power disposed for this reception must be only deprived of species representing the given quality. This is not the case. The overall thrust of the *quaestiones* on the issue of the organ of the external senses in disp. 7 shows that the organs must be corporeally disposed to be able to receive species. The crystalline humour of the pupil must be colourless, transparent and soft, the air in the inner ear must be motionless, the organs of the tactile power must be set in a proper medial position etc. Suárez shares with Aquinas the view that a cognitive power must abstract from matter.[39] The degree of materiality from which the external senses abstract has to do with the absence of the real nature of sensible qualities in the organs. Equally so, Suárez's conclusion about the essential difference between the species and the objects does not foreclose the possibility of a damage to the powers through activity of the objects. A strong noise through a strong movement of air that accompanies the former damages the inner ear, a very bright colour damages the eye by reverberating an intensive light, and an odour through a concomitant hot evaporation causes a pain the sense of smell. Properly speaking, these injuries are not caused by the species. The species always perfects the power in which it is received because it modifies it intentionally and disposes it for the elicitation of a perceptual act. But since intentional action often goes hand in hand with material and natural (tactile) action, a sense can get impaired through its inadequate and violent affection of the latter.[40]

The sensible species' essential difference from the sensibles cannot be regarded as a reason for their *literal* immateriality and spirituality. Suárez's employment of the term 'spiritual' in connection with the sensible species does not mean (much more clearly so than in Aquinas[41]) that they are immaterial entities in a literal sense. In the famous controversy over the ontological status of the sensible species Suárez clearly sides with the materialist understanding of these species. In his reply to the third question he states that contrary to the intelligible species, which is spiritual and indivisible since it is received in the intellect not having a bodily organ, the sensible

[38] *Ibid.*, nn. 27–30, 330–6.

[39] *Ibid.*, n. 29, 332.

[40] *Ibid.*, n. 31, 336–8. For a more detailed description of the relation between natural and intentional actions in the particular external senses, see Chap. 4.

[41] As is well known, the question of whether sensible species were literally material or immaterial entities for Aquinas has been widely discussed by many scholars. For a survey of this debate and the division into two camps regarding the exposition of Aquinas's view of the sensible species, see Pasnau (1997, 42–7, esp. 42, note 20). According to Pasnau (2004, 42–60), Aquinas defended 'a materialist theory of sensation' which he, however, regards as a controversial reading. He notes that for Aquinas sensible species are instantiated within a physical substance and that sensation is an act of a corporeal organ. The reception of sensible species in Aquinas is 'a wholly physical process' (*ibid.*, 46). Pasnau does not see an incompatibility between intentionality and physicality. Aquinas's talk about the immateriality of the sensible species means, like for Suárez, only that the sensible form does not alter the organ according to its natural being. The same reading of Aquinas (and Scotus) can be found in Cross (2014, 32–9). For a different (immaterialist) reading of Aquinas, see, for example, Burnyeat (2001).

species is material and divisible. When it is received in a material and extended power, it must be material and extended.[42] It is received *in medio*, which is also extended and material. As said above, its 'spirituality' points above all to a difference from the physicality or the corporeal character of objects. Intentional alteration of the forms does not comply with natural alteration, but it is compatible with its materiality. Intentional reception cannot be physical or natural since a natural change involves that the recipient of the form becomes an instance of this form. However, this does not rule out its materiality.

Suárez confirms the second attribute of the sensible species, namely its divisibility, through a reference to the mirror experience. In the mirror the visible species are reflected as extended entities. Peter's face is represented by the whole mirror image of his face. If the mirror gets broken into two parts, one part will reflect only a part of Peter's face and the other part of the mirror will reflect only the other part of Peter's face. Neither of the fragments will reflect his whole face. The mirror's fragmentation will be correlated to the fragmentation of the reflected image. For Suárez this indicates that the sensible species representing Peter's face is composed of a multiplicity of divisible parts that rectilinearly represent the various parts of his face. Clearly, if the represented images get fragmented, the sensible species representing the objects must be divisible as well.[43]

But are these species the *formal* likenesses of objects, or do they represent them only *efficiently* (*effective*)? The spiritualist reading of the sensible species seems to lead to a fully-fledged formal representation. On this interpretation, the species is a formal cause whose mere reception *ipso facto* leads to a perceptual act. Suárez's reply to this (fourth) question is not easy to determine. The difficulty can be seen in Suárez's oscillating formulations between two distinct views. While one dominates in *CDA*, the other one is presented in *TDA*. In *CDA* disp. 5, q. 2, Suárez concludes that intentional species are *formal* likenesses of the sensibles.[44] The species, including the sensible species, is posited in the power to substitute for the object (*supplere vicem obiecti*). However, it can substitute for it only if it represents it. And it can represent it only if it is a true, sc. *formal*, similitude of an object. In *TDA* Suárez diverges from this formulation. There he states that the intentional (sensible) species does not represent the object formally but only *efficiently*.[45] Although in *TDA* he notes that the opposite view is also probable, he makes clear that the claim about efficient representation is more probable. Interestingly, his arguments are based on the same premises that are mentioned in *CDA*. If the intentional species were the formal likeness of the object, it would necessarily communicate its effect to the power in which it inheres. But this is at odds with the clear separability of the

[42] *CDA* disp. 5, q. 2, n. 17, t. 2, 316.

[43] *Ibid.*, nn. 17–19, t. 2, 316–320.

[44] *Ibid.*, n. 21, 322: 'Istae species intentionales sunt similitudines formales obiectorum'.

[45] *TDA* book 3, ch. 2, n. 20, 620: 'Species intentionales non repraesentant formaliter obiecta, sed effective tantum'. This difference is observed also by Spruit and Tellkamp. See Spruit (1995, 298, note 120) and Tellkamp (2012, 11, note 33).

reception and production of the perceptual act. Contrary to *CDA*, in *TDA* Suárez seems after all to propose a different reading of 'the semen metaphor'. Semen, to which the intentional species is often compared, does not formally contain the parts or organs of the foetus. It is not possible to distinguish the parts of a foetus or a plant, such as a tree, in the semen. The semen has only a *virtual* power to produce the body with the organs. It is not a formal representation in the proper sense. Analogically, the *species impressa sensibilis* is an inarticulate and obscure representation.[46] Against Durandus, Suárez highlights that the identity between an intentional likeness and an object does not need to be univocal. Unlike the unity in a similarity of extramental objects belonging to the same kind, the representational unity between the species and the objects is more deficient. He compares it to the similarity between a shadow and the object casting this shadow, or between a picture and the object depicted by it. Neither needs to be entirely isomorphic. Unity or similarity between the species and the represented object can be predicated of them only by means of the analogy of attribution.[47] It seems that this conclusion, which is explicit in *TDA*, corresponds better also to the overall thrust of the text and spirit of *CDA*. Also in *CDA* Suárez speaks about the efficient causality which the species generates in relation to the powers. In *CDA* he also frequently employs the biological metaphor comparing the species to the semen received in 'the womb' of a cognitive power. In *CDA* he is equally explicit about the species's virtual representation.[48] In *CDA* he also likens the production of a cognitive act through the subordinated causes to the generation of an offspring in the procreative act of father and mother. Importantly, this reading can also be corroborated by Suárez's theory of the origin of the sensible species.

The topic of the origin of the sensible species is associated with the famous medieval controversy over the agent sense.[49] Although in the theory of the Latin Averroist John of Jandun, one of the most ardent advocates of the agent sense, this sense was posited for the sake of the formation of an act of perception,[50] in *CDA* disp. 6, q. 2 Suárez conceives the agent sense as a hypothetical force that *elevates* the material and imperfect sensible qualities to the level of the more noble (spiritual) sensible species. The fact that the extramental qualities need to be perfected is based on the abovementioned causal axiom, according to which the (lower) material and imperfect sensible objects cannot act on the (higher) vital sense powers. For a sensible species to be able to epistemically act on the power it must be more perfect

[46]*TDA* book 3, ch. 2, n. 22, 621.

[47]*Ibid.*, n. 25, 622.

[48]*CDA* disp. 5, q. 2, n. 6, t. 2, 306: '[. . .] species concurrit effective ad actum; ergo non est causa formalis illius'; for the comparison of species to semen, see, for example, *CDA* disp. 5, q. 2, n. 21, t. 2, 322.

[49]For the following historical background and Suárez's theory, see Heider (2019b).

[50]This is also why Suárez treats this doctrine of Jandun's in the context of the question about the principles of the cognitive act in *CDA* disp. 5, q. 4 (for this topic and Jandun's theory, see Sect. 3.4). For Jandun's theory of the agent sense that – in analogy to the agent and the potential intellect – is a power really distinct from the passive sense, see Pattin (1988).

than the represented qualities. It was Averroes who introduced the *sensus agens* into medieval debates.[51] In his 'long' commentary on *De anima* II.5 he says:

> And one can say that sensibles do not move the senses in the way they exist outside the soul, for they move the senses insofar as they are intentions, since in matter they are not intentions in act but in potency. And one cannot say that this difference occurs by virtue of the difference of subject such that intentions come to be on account of spiritual matter which is the sense, not on account of an external mover. It is better to think that the cause of the diversity of matter is the diversity of forms and not that the diversity of matter is the cause of the diversity of forms. Since it is so, it is necessary to posit an external motor in senses different from sensibles, as was necessary in the case of the intellect. It has been seen, therefore, that if we concede that the diversity of forms is the cause of the diversity of matter, then it will be necessary that the mover be external. But Aristotle was silent about this because it is hidden in the case of sensation and apparent in the case of the intellect.[52]

Although Aristotle did not speak explicitly about this external mover in the senses, this does not mean that he did not take it into account. Although this agent is not that apparent when compared to the agent intellect, that does not mean that it does not exist. Since sensibles cannot immediately affect the sensory powers, it is necessary to elevate them to the level of sensible intentions. We cannot say that the sensibles become intentions by virtue of the subject that receives them, since form in general does not exist for the sake of matter but rather vice versa (in the same way an organ exists for the sake of the power). The emission of stimuli from sensibles must be assisted by a mover, which is not inherent to the perceiver's soul. Only such an *external* mover can secure the necessary reduction of potential sensibles to actual sensibles, as the agent intellect undergirds the reduction of potential intelligibles to actual intelligibles. Only the assistance of an external agent can explain how an ontologically lower entity, that is, a sensible thing, can alter an ontologically higher one, that is, a cognitive power.

Suárez quotes Agostino Nifo who showed that Averroes actually embraced the theory of an external mover of sensible forms that reduces them to sensibles in acts.[53] Nifo aimed to spell out a theory that Averroes had sketched more clearly. What is this mover and how does it operate? For Nifo, a perfect understanding of Averroes's view must be based on three main assumptions related to three factors inherent to each natural change on the part of an agent. The first assumption concerns the form of the agent; the second is related to the matter, in which the natural form exists, since if the physical change is an individual transformation, the existence of matter is necessary; the third element is the universal agent in virtue of which all particular agents are made operative. All these factors are also indispensable in perception. The universal agent is the prime mover, or God, who is ultimately

---

[51] For Suárez's reference to Averroes, see *CDA* disp. 6, q. 2, n. 3, t. 2, 471.

[52] Averroes, *In Aristotelis De anima librum secundum*, 60, 221.

[53] *CDA* disp. 6, q. 2, n. 3, t. 2, 471.

responsible for all particular movements. While the sensible object is the proximate and instrumental cause, the prime mover is the remote and principal cause. Since the universal agent (God) cannot be the cause of particular effects, such as the apprehension of this or that shade of blue, Nifo asserts that this particular causation, which is connected with the particular formal content in the given sensation, is provided by the concrete sensible form. The prime mover is the principal cause of the spirituality of the sensible species. Without being elevated by the prime mover the sensible forms could not intentionally affect the cognitive powers.[54] It is no wonder that according to this theory the sensible species brings a formal or spiritual alteration to the power and not an efficient one, and it is no wonder that the reception of this elevated species is in itself quite sufficient for an actual perception.[55]

However, it was not only Nifo who referred to an external mover in this connection. In at least one of his texts also Aquinas makes use of this 'lifter'. In *De potentia* , q. 5 art. 8 c. he says:

> But we should note that a material substance has two kinds of action. One is by a property of the substance, so that it, being moved, causes movement and acts. The other has action insofar as it attains the order of separate substances and shares part of their way. For example, lower natures were accustomed to share in the property of a higher nature, as is evident in certain animals that share in a likeness of practical reason, which is proper to human beings. But this is the action of a material substance that is not related to the transformation of matter. Rather, it is related to communicating a likeness of its form to a medium by analogy to the immaterial representation received in the senses and the intellect about a thing. And the sun in this way illuminates air, and color multiplies its form in a medium.[56]

This passage makes a number of important points. However, I will restrict the discussion to the issue of the production of the sensible species. In this respect, historically speaking, the passage became defining for Cajetan.[57] For him a thing has

---

[54] Agostino Nifo, *De sensu agente*, in *Librum Destructio Destructionum Averrois commentationes*, 128a–29a.

[55] Nifo (*De sensu agente*, 128b) explicitly says that '[. . .] species et sensatio, ut dicendum, sunt una res simpliciter [. . .]'. Of course, this theory would have to face the objection why the medium in which a spiritualized species inheres is not cognitive, and the sense power is. If this kind of strong intentionality is inherent to an extramental species, a cognitive power will turn out to be, as South says with reference to Aquinas's view, 'a black box' (South, 2001a, 224).

[56] Thomas de Aquino, *Quaestiones disputatae de potentia. A quaestione V ad quaestionem VI*, q. 5, art. 8, c: 'Sed sciendum quod corpus habet duplicem actionem: unam quidem secundum proprietatem corporis, ut scilicet agat per motum (hoc enim proprium est corporis, ut motum moveat et agat); aliam autem actionem habet, secundum quod attingit ad ordinem substantiarum separatarum, et participat aliquid de modo ipsarum; sicut naturae inferiores consueverunt aliquid participare de proprietate naturae superioris, ut apparet in quibusdam animalibus, quae participant aliquam similitudinem prudentiae, quae propria est hominum. Haec autem est actio corporis, quae non est ad transmutationem materiae, sed ad quamdam diffusionem similitudinis formae in medio secundum similitudinem spiritualis intentionis quae recipitur de re in sensu vel intellectu, et hoc modo sol illuminat aerem, et color speciem suam multiplicat in medio'. For the translation, see Thomas Aquinas (2012, 155).

[57] See Cajetan (1939), *Scripta philosophica, Commentaria in De anima Aristotelis*, 253–4.

two actions. The first is the action that runs according to its own nature, the second is the action that participates in the action of a superior agent, which is a separate form. This is why a hot body has twofold (equivocal) action: one that makes a heat in its natural being in the air; the other that creates the intention (sensible species) of heat in the air.[58]

How does Suárez evaluate all these theories that share this ontological elevation of sensible qualities? In its core his evaluation is negative. He denies all theories according to which the origin of the sensible species is to be referred to non-naturalist sources. He does not see what the Cajetanic participation could be. He tentatively argues that it must be a participation in some perfection. This perfection must be higher than the perfection of a produced species. It is not sufficient for the production that a higher perfection occurs only in a participated nature: Suárez says that it must somehow be in the participating thing, namely in a sensible. However, if it is in a participating thing, is it intrinsic to it, or is it only added to it? If internal, this position implies that the overall perfection of the sensibles would be sufficient for the production of the species. If it is somehow added to the sensibles, the theory would make no sense at all since it would hardly be possible to conceive what this added entity actually could be.[59]

Dismissing all these 'Deus ex machina-theories' Suárez concludes that the sensible species is caused only by the sensible object. No agent sense and no external motor are needed for its production. According to this 'naturalizing' statement the sensible species is *not* more perfect than the represented objects since just the opposite holds, namely that a white colour is more perfect than its species. Both are material qualities, but while a white colour is perfect in its kind, its species is only a trace (*vestigium*) of it. The subtlety and imperceptibility[60] to which the species owes its cognitive transmittability through a medium is not a consequence of its perfection, that is, of its *literal* immateriality, since (as said) its 'immateriality' is nothing else than its material subtlety. A major subtlety is only a mode of their entity with which the vestigial character and its imperfection are compatible. For Suárez the sensible species is an unsensed *virtual* representation that stands in need of a further cognitive elaboration and processing before acquiring the expressed form of a full-blown perceptual act, which only can be seen as the formal similitude of an object. Despite its material character, however, Suárez keeps the action of this intentional emission apart as significantly distinct from the causal action whose effect is a corporeal alteration in a *passum*, such as the heating of an object by fire. However, a following objection can be mentioned. With this distinction between two kinds of action in the mind, someone may argue that sensible qualities cannot produce sensible species because such intentional action is a contradictory item. In general, it holds that an agent is active (a) either for the sake of the conservation of its kind, or (b) for the sake of its own perfection. However, the intentional action of

---

[58]For Suárez's presentation of this view, see *CDA* disp. 6, q. 2, n. 4, t. 2, 472.

[59]*CDA* disp. 6, q. 2, n. 5, t. 2, 474.

[60]For its imperceptibility, see *CDA* disp. 5, q. 2, n. 6, t. 2, 306.

sensibles, such as the emission of the species of heat by fire, results neither in the fire's perfection, nor in its conservation. It is only the causal operation of heating that can result in the spreading of its perfection with respect to extension and intension, and thus also to the fire's conservation. In his reply to this argument, which in line with what has been mentioned above eliminates the view of the sensible species that aims to identify intentional action as an action *sui generis*, Suárez says that intentional action generated by sensible qualities themselves is *not* a contradictory item. He denies the universal validity of the abovementioned disjunction raised by an opponent as being a comprehensive description of all the possible goals an action can have. In fact, there are many agents, such as separate substances that only communicate themselves to other objects. They themselves are neither perfected nor conserved by their actions. The Sun illuminates other objects without perfecting itself and without conserving itself. In the same way, beside its efficiency of natural effects, that is, heating external objects, a fire's accident of heat produces its own sensible species by means of which it communicates itself and perfects an (attentive) tactile power. This also belongs to the scope of the quality's *radius operandi*. It is possible to conclude that sensible objects permanently diffuse their sensible species even in the absence of a perceiver. They are of a vestigial character and thus less perfect than the represented sensible qualities. They are caused by the sensibles themselves without the assistance of higher elevating agents. Accordingly, Suárez suggests that the sensible objects's generation of this dual (equivocal) activity should be taken as a primitive state of affairs.[61]

3.3 The Perceptual Act as a Distinct Unit

Having treated the issues of the necessity, nature and origin of the impressed sensible species, in *CDA* disp. 5, q. 3 Suárez approaches a topic related to the entity of a cognitive act which he takes for a distinct unit. The main doctrinal opposition to this view represents an account that denies its distinct ontological character. According to this negative view, a perceptual act is nothing above the reception of the sensible species. Clearly, those who stress the *formal* causality of the species and the *formal* identity of an object in extramental and intramental being are inclined to attribute the total causal activity to the species and tend to identify an act of cognition with the reception of a species. This conclusion, which is ontologically more parsimonious than that endorsing a perceptual act as a unit *sui generis*, is involved in many of Aristotle's and Aquinas's deflationary formulations. The perceptual process is characterized in the formulations as a passive procedure, identical with the activity of the sensible objects: 'For perception is a form of being acted upon';[62] 'The

[61]*CDA* disp. 6, q. 2, n. 6, t. 2, 474–6. In Chap. 4 we will also see that Suárez has several arguments for postulating a distinct intentional action.

[62]Aristotle, *DA* II.11, 424a1, 135.

activity of the sensible object and of the sensation is one and the same [. . .]';[63] '[. . .] the senses are such that their being affected is their very sensing'.[64] A question that advocates of this passivist view of perceptual acts are supposed to answer is why the reception of sensible species entails perception in the sense powers, but not in the air or the mirror, if the species *in medio* are fully intentional items. What makes the sense powers so special that only in them the reception of the species results in perception, while the species' inherence in the air does not lead to it? A common reply has it that this happens due to the immateriality and a certain illimitability of the sensory powers. It is the way of reception and the 'quality' of a *patiens* that makes the sense powers perceptual and distinguishes them from noncognitive subjects, which can receive only corporeal affections coming from objects.[65] In line with this position, Suárez distinguishes between two kinds of reception. While the first is called 'dead', the second kind is 'vital' (*immutatio vitalis*). The first kind is the reception of the species of a sensible *X* in the air, in a mirror, or in a percipient *P* who is not attentive, is asleep, in ecstasy, shock, or after apoplexy. If *P* is inattentive due to some reason, or if *P*'s attention is distracted and captured by something else, then although *P*'s sense organs receive myriads of sensible species representing *X* and other sensibles in the environment, *P* remains noncognitive. However, as Suárez notes, actually all (including representatives of the passivist account) agree that a cognitive act cannot be a 'dead' reception. It must be a vital reception. This vital reception occurs only if the soul is attentive or if it attends to *X*.[66]

Even assuming such attending on the side of the powers, Suárez views this theory as false. If a perceptual act were equivalent to the power's reception of a species, it would be reducible either (a) to the species,[67] or (b) to the power. However, neither of these options is tenable. The first alternative is implausible since it holds that the power is related to an act in a different way than it is related to the impressed species. As we shall see in Sect. 3.5, Suárez endorses a three-phase view of perception or, as Peter King says, a theory of a composite mental representation that cannot be reduced to the simple version of the conformal account of mental representation.[68] The first stage involving the reception of the species is passive, the second one yielding the production of a cognitive act is active, while the third stage characterized by the reception of the act in the very same power by which it has been formed is again passive. As we know, the reception of species cannot produce a cognitive act

---

[63] *Ibid.*, III.2, 425b27–8, 147.

[64] Thomas de Aquino, *STh* I, q. 17, art. 2, ad 1, 220: '[. . .] sensum affici, est ipsum eius sentire'.

[65] For this exposition in the theory of Aquinas, see Tellkamp (2006, 275–89).

[66] *CDA* disp. 5, q. 3, 1, t. 2, 340–2.

[67] This view differs from that which states that the sense power produces a perceptual act conceived as a distinct ontological unit whose total cause is the sensible species (for this view see the next section). Here we have an account that reduces a cognitive act to the more fundamental units. Cajetan is mentioned as a representative of this view : '[. . .] nihil aliud est videre quam esse species visibilis in visu [. . .]' (Cajetan, *In II De anima*, ch. 5, n. 173, 156).

[68] For these two models of intentionality, see King (2007).

since a vital and immanent operation can never originate through a transitive causality of the external objects.[69]

On a different reductionist approach, an act of perception must be reduced to the power. Without naming an adherent of this approach, Suárez seems to be referring to those who endorse a radically activist view of perception freely inspired by Olivi's theory. This reduction is implemented by those who identify cognition with vital attention. This attention or *aspectus* is a virtual stretching of the power to a sensible object which is the terminus of this intentional (attentional) directedness.[70] Employing the ontology of modes, in his *reductio* Suárez argues that this attention will have to be finally construed in this theory as a separable mode of the soul, rather than of a power. All in all, on this view we would have only the soul modified by the various modes of attentions directed to various objects. There would be no theoretical room left for the existence of a perceptual act conceived as a distinct accident of quality. In the reply to this approach, Suárez expands on this *reductio* by presenting three arguments supporting the claim that he has in mind a theory similar to Olivi's view. (1) In line with the abovementioned emphasis on the soul's efficient causality, only the soul is responsible for paying attention. Blind people experience intentional stretching out to visible objects even though their attentional intention can never be, phenomenologically speaking, 'fulfilled'.[71] Attention can exist in a subject whose visual power does not exist at all or is irrecoverably damaged. A blind man could not experience attentional intentionality if it were a mode of the visual power, since this power is not available to him. Suárez pushes this argument further by saying that if this attention were a mode of the soul, even the powers would be come to be redundant. In other terms, if this theory were valid, it would imply the nominalist and the reductionist view according to which the soul or a material composite is identical with the powers. No real distinction between the soul and the powers would be possible.[72] Accordingly, the operations conceived as modes would have to be identical with the soul. They would be nothing above the soul's modifications. This is at odds with the essentially compositional character of creatures. (2) Moreover, if a perceptual act were identical with a mode of the soul, it would have to be an act that is proper to the soul. Accordingly, the subject of seeing would not be a soul-body composite but the immaterial soul. Sense perception would be subjected in the immaterial rational soul and it would become an entirely spiritual process that

[69] *CDA* disp. 5, q. 3, n. 3, t. 2, 344–6.

[70] For this theory of *aspectus* or intentional directedness in Olivi, see Toivanen (2013a, 2013b, 151–61).

[71] In *The Trinity* 9.1.2 (transl. by Hill), 305, Augustine indicates something similar: '[. . .] and though its intention, now that the eyes have been lost, has no sense of the body which it can join in the act of seeing to a body outside, and keep its gaze fixed on it once seen, nonetheless it shows by its very exertions that it neither perishes nor even diminishes with the loss of the bodily sense'. Suárez without doubt subscribes to this Augustinian view.

[72] Interestingly, just this is Olivi's view. For Olivi '[. . .] the soul is nothing but an aggregate of its powers', (Toivanen, 2013b, 163).

could also exist in separated souls.[73] (3) If a perceptual act were reducible to the soul's mode of attention, the sensible species would turn out to be redundant. In a very Olivian manner Suárez says that if a sensible object were present, nothing more would be required for the production of a perceptual act than this attention, since this attention is not and cannot be causally dependent on the previous reception of stimuli. As we read in Olivi, such virtual outreach to sensibles depends on them only terminatively. When this attention or 'aspectus' is 'caught' in the object, the sight causes the following act. Accordingly, the object will not be the motive but only the terminative cause of the perceptual act's content. Useless to say how sharply this dispensability of sensible species conceived as motive causes contrasts with Suárez's abovementioned endorsement of the sensible species.[74] All in all, the only possible conclusion that can be embraced is the following: an act of perception is nothing else than a special *quality* pertaining to a kind of disposition which, as the last perfection of the perceiver, is really distinct from the power insofar as it is informed by the sensible species.[75]

3.4 Moderately Active Perception: The Causes of a Perceptual Act

The abovementioned characteristics of the *species sensibilis* and the theory of perceptual acts conceived as distinct ontological units comes a long way towards Suárez's view of its principles. Not surprisingly, in *CDA* disp. 5, q. 4 Suárez rejects two main versions of cognitive passivism by appealing to the material and 'incomplete' nature of the sensible species and by emphasizing the vital and immanent character of cognitive operations. The first one, which can be called 'stronger', is identical with the abovementioned claim according to which a cognitive power does not produce its act but receives it from the species.[76] The second ('weaker') form recognizes a perceptual act as a unit *sui generis* but claims that the efficient principle is the perceptual power informed by the sensible species, where the *ratio agendi* is the species. In analogy to hot water heating up the immediate surrounding air, where the immediate agent of the heating is not the water but the accidental form of heat, it can be said that the perceptual act is not caused by the power but by the received

[73]For Suárez's critique of the view according to which the senses are immediately subjected in the soul, see *CDA* disp. 3, q. 3, n. 4, t. 2, 118–20.

[74]*CDA* disp. 5, q. 3, n. 4, t. 2, 346–8.

[75]*Ibid.*, n. 2, 342. In this Suárez accepts Scotus's opinion according to which cognitive acts are formally neither actions, relations nor any other category but qualities that pertain to the first kind of the accident of quality, sc. disposition. For this Scotus's view and arguments for it, among other places, see *Quodl.* q. 13, nn. 3–5 (Vivès 25: 508–9) and *Reportata Parisiensia* I, d. 3, q. 6, nn. 2–7 (Vivès 22: 112–5).

[76]*CDA* disp. 5, q. 4, n. 3, t. 2, 352.

accidental species.[77] The basic principle underlying Suárez's reasoning against these views, which he associates with the names of Nifo (the first version) and Aquinas (the second version), is the claim that the species cannot efficiently and immediately concur as a total efficient cause in the production of cognitive vital operations. If it did, perceptual acts would cease to be vital operations and would become transitive acts originating in an external agent. Assuming that the sensible species is only a material trace of the sensible object, as a less perfect entity it cannot produce a more perfect (animate) quality, namely a perceptual state. If the species were the total *ratio agendi*, its reception would necessarily entail the production of a cognitive act. However, this is at odds with our experience. Despite being affected by numerous species, if we are not attentive, we do not perceive. Moreover, diverse people whose organs are equally organically or physically disposed to the reception of the same species can have different perceptions of the same object due to their different attending. As everyone know from introspection, one and the same perceiver often cognizes differently due to varying degrees of attention in the morning and in the evening. With respect to the variable attention, the perfection of our perceptual acts can vary significantly. Accordingly, perceptual acts cannot be caused only by the species but must originate also in the powers.[78]

Suárez does not reject only the passivist account of perception in its various versions. He also rejects Jandun's abovementioned theory of the agent sense, which in the context of the issue of the principles of cognition creates a negative background for his own exposition. Therefore, it is in place to briefly introduce Jandun's theory now.[79] As I have said in Sect. 3.2, the Latin Averroist does not introduce a really distinct power called *sensus agens* to 'elevate' the sensible species from the plane of natural being to the level of intentional (spiritual) being. He does not deem it necessary to posit the *agens* in this field of the production of the sensible species. He postulates it in the process of the act of perception's formation. The necessity to consider this sensory power is rooted in four basic principles.[80] First, everything that is received in a subject depends on an effective principle. A passive potency always presupposes an active potency that reduces the passive faculty to act. Second, the nobility of beings is directly proportional to the nobility of their proper operations. If it holds that an operation is the goal of that operation's subject, one can recognize the perfection of the operation's subject through knowledge of the perfection of its operation. Third, producing an effect is *simpliciter* more perfect than receiving an effect. Activity is nobler than passivity, and a more perfect being is more active than a less perfect being. While separated substances occupy a higher position on the ontological ladder because they are active in moving heavenly bodies, prime matter is the least noble being because it is pure potency. Fourth, as regards sensory

[77] *Ibid.*, n. 4, 352–4.

[78] *Ibid.*, n. 5, 354–6.

[79] For this presentation, see also Heider (2019b, 169–71).

[80] Jandun, *Sophisma de sensu agente*, in Pattin (1988, 128–31).

cognition, the sensible species is less perfect than the perceptual power. The sensible species of a red colour is not more perfect than the visual power. One of Jandun's key reasons for this ordering of powers and the sensible species consists in that the cognitive act is animated (vital), whereas the sensible species is only of a corporeal nature. Leaving aside arguments for the doctrine of the agent sense that are based on the activity of the internal senses, namely that of the cogitative power, the common sense, and the memory, which are all essentially active powers that judge, distinguish and proceed discursively in the field of particular intentions,[81] one of Jandun's main arguments for the theory of the agent sense can be labelled as 'reasoning from the elimination of alternatives'.[82] Obviously, an act of sensation must have its immediate efficient principle. If not the agent sense, it must be one of the following: (1) the soul, (2) the extramental sensible, (3) the sensible species, or (4) the passive sense. Concerning the first option, in line with Aquinas and Thomists Jandun claims that the soul cannot be the immediate principle of the production of sensory operations, since the soul becomes active *only* through the sensory powers. As for the second option, he shows that the sensible, being an entity less perfect than the cognitive power, can in no way be the immediate cause of the perceptive act. Third, if sensible species were the efficient cause of sensation, they would have to be more perfect than the cognitive power since activity is nobler than passivity. This is not the case. Finally, as regards the passive sense, passive potency cannot be the immediate principle of a cognitive act since one and the same power cannot be simultaneously passive and active in the same respect. Beside the first premise it is also Jandun's last premise which Suárez will deny sharply.

The only remaining option for Jandun is to conclude that sensation is caused by the agent sense. When describing the mechanism of sensation Jandun asserts that the first actualization consists in the passive sense's reception of a determination from the sensible species that is emitted by the sensible without an 'ontological lifter'. This reception is a mere disposing reception, which is not yet sensation proper. According to Jandun, sensation consists in the second act, which is then received in the passive sense. While discussing visual perception, he stresses that the eye is a corporeal potency, which puts it on the same ontological level as an inanimate medium such as illuminated air. When disposed by the sensible species, the agent sense, as the perfecting agent, produces an act of sensation.[83] In the third stage this act of sensation is received in the passive sense. Accordingly, the cognitive faculty proper is not the agent sense but the passive sense.[84]

---

[81] *Ibid.*, 226.

[82] *Ibid.*, 227.

[83] *Ibid.*, 159–60.

[84] When explaining the Stagirite's receptivism, Jandun underlines the specific historical context of *De anima* which in Aristotle's days was determined by the dominant theory of extramission, according to which visual perception (at least in its first stage) is based on the primordial emanation of intraocular fire or visual rays toward the sensibles. According to the Latin Averroist's reading, when Aristotle stresses the reception of forms, he does so only to counterbalance this dominant activist approach. Consequently, Jandun writes, what is only briefly indicated in Aristotle, i.e., his

Suárez's criticism of Jandun's theory is primarily related to his theory of the real distinction between the agent and the passive sense, which has to do with his premise that the same power cannot be simultaneously passive and active in the same respect. In Sect. 2.7 and elsewhere we have seen that Suárez cannot be regarded as advocating this view. He adduces two main arguments against Jandun's doctrine. First, by analogy to his rejection of the passivist conceptions, Suárez states that if the Latin Averroist's theory were true, we would have to admit two really distinct powers, one of which, sc. the agent sense, would be productive of the act of sensation, while the other, sc. the passive sense, would be properly cognitive. However, this would violate the immanent and vital character of the cognitive operation (presented in Chap 2) too. Its immanent character requires that the act remain in the *same* power through which it was produced. Hence, such an act cannot be received in a potency that is really distinct from the potency producing the act, even though it resides in the same supposit. Second, Jandun's doctrine after all leaves open which of these two senses is cognitive. Will it be the power that produces the cognitive act, or the power that receives it? Neither of these alternatives is plausible. The passive sense cannot be the cognitive power since cognition is a vital operation. The cognitive act as a vital operation must intrinsically include an active component. The *sensus agens* cannot be the cognitive power either. In consonance with his Aristotelianism, Suárez declares the necessity of the power's determination from the sensible through the species. If cognition involves both assimilation to a cognized thing and its cognitive expression or articulation, the perceptive power can be determined only by receiving the sensible species. Suárez's evaluation of Jandun's theory is clear: *Agere* and *recipere* cannot be separated in such a way that they would be attributed to two different powers. Cognition is formed by one and the same power, which is both passive and active.[85]

A significant doctrinal approximation to a theory that Suárez holds to be true comes with Scotus's conclusion.[86] An act of apprehension is caused by the power and the species, the two being partial causes integrated in the total cause. Vital attention of the soul through its powers is necessary. However, since cognition is an assimilative process the species must be part of the total cause too. Cognition depends not only on attention conceived as a general mode of the soul, but also on the received species. The power needs to be determined in the first act, viz. by the species, in order to produce the second act. In his exposition in *Ordinatio* I, d. 3, p. 3, q. 2 Scotus distinguishes between two kinds of causal concurrence. One is the production 'ex aequo', that is, concurrence without an essential ordering. Two horses drawing a carriage concur equally in the production of the effect which is the motion of the carriage. The other one is concurrence 'non ex aequo', that is,

---

cognitive activism, must be considered no less seriously than the numerous passages manifesting his passivist approach. Therefore, Aristotle's silence about the *sensus agens* is far from implying a denial of this power. For this see, Pattin (1988, 163).

[85] *CDA* disp. 5, q. 4, n. 2, t. 2, 350–2.

[86] *Ibid.*, n. 7, t. 2, 356–8.

causal concurrence with an essential ordering of the causes as superior and inferior. Such concurrence can be twofold. The lower cause concurs only if the higher one moves it. A stick (the inferior cause) is moved by a hand (the superior cause). But if the superior cause does not move the inferior one and the inferior one *by itself* contributes to the effect, we get the second type of essentially ordered causes. As an example, Scotus mentions father and mother who generate an offspring. It is this type which Scotus applies to his theory of the principles of a cognitive act. The power and the species are related as higher and lower causes which both (in a different degree of perfection) concur in the production of cognition. Scotus is clear that they cannot concur 'ex aequo', since if the capacity were powerful enough, it could substitute for the efficiency of the second and causally less powerful cause, which is not possible in the production of cognition. He is also sure that the causal contribution of the species cannot arise from a 'motion' coming from the power. The species has its own irreplaceable causal influence in the production. All in all, the production of cognition corresponds to the second type of essentially ordered causes.[87]

In a link to Scotus's doctrine, Suárez formulates his position in three conclusions. First, with respect to its act a cognitive power has a proper and immediate activity. Since the soul is the principle of vital operations and operations come through its faculties, the powers must be causally concurrent in the production of perceptual acts. Even though the quality of the intentional species and other conditions on the side of the sensible object has an impact on the quality of the perceptual act, a causal contribution of the power cannot be eliminated.[88] Second, it is not only the power but also the species that possesses some activity in regard to an act of cognition. Cognition proceeds through assimilation of the knower to a known thing and assimilation requires a principle. If we want to know something perfectly, we strive as much as possible to form a likeness of the object. This involves the principle of the assimilation, namely the species.[89] Lastly, the integral principle of the formation of a cognitive act comprises both the activity of the cognitive power and the sensible species. Suárez's conciliatory conclusion is formulated as follows: 'The integral principle of the cognitive act is the power informed by the intentional species'.[90]

However, is this formulation not at odds with the abovementioned assumption that a sensible species, being less perfect, cannot immediately concur in the production of a cognitive act? Shouldn't Suárez rather conclude that the sensible species does not efficiently concur in the formation of a cognitive act at all and,

[87]For this, see his *Ord.* I, d. 3, p. 3, q. 2, nn. 495–98 (Vat. 3: 293–5). Although Scotus speaks about intellectual knowledge, the conclusion can be extended also to the principles of perception. For the similarity between Scotus's and Suárez's theories of the principles of a cognitive act, see also González-Ayesta (2012).

[88]*CDA* disp. 5, q. 4, n. 10, t. 2, 360.

[89]*Ibid.*, n. 14, 364.

[90]*Ibid.*, n. 15, 364: 'Principium integrum productivum cognitionis est potentia informata specie'.

consequently, that the only efficiency pertains to the cognitive power? Shouldn't Suárez embrace precisely an extremely activist conception, which Leen Spruit describes as mirroring the *Zeitgeist* of Renaissance philosophy?[91] The brilliance of Suárez's solution is that he is not forced to accept this somewhat extreme opinion. In his reply to the argument that the sensible species cannot concur in the production of a cognitive act (an argument formulated already by Henry of Ghent, ca. 1217–1293, who espoused a theory of active cognition presented by Suárez as a position of radical activism[92]), Suárez starts his complex reply from the basic assumption according to which a cognitive act is a more perfect entity than the intentional (sensible) species. He begins with the following three negative statements, in which he agrees with Henry. He claims that a less perfect thing, the sensible species, cannot concur in the production of a more perfect thing, the cognitive act, whether (a) as an immediate total cause (strong receptivism), or (b) as a principal cause complemented by the cognitive power (weaker receptivism) or (c) as the total instrumental cause (*sic*!). On my reading, this last denial shows that Suárez differs not only from Aquinas and Thomists, but also from Scotus whose theory states that an act of cognition occurs on the basis of the coordination of two partial causes, namely the cognitive power and the species, where the cognitive power is the principal cause and the sensible species is the instrumental cause. The originality of Suárez's thesis lies in his claim that the sensible species does not constitute the total instrumental cause: it only instrumentally co-produces the total instrumental cause in concert with the second partial instrumental cause, sc. the sensory power. Only in conjunction do these two partial instrumental causes constitute the total instrumental cause. Importantly, this total instrumental cause is operated by the principal cause, which is the attentive *soul*. Accordingly, Suárez solves the abovementioned problem of ascendant causality by means of this causal distribution, in which one can discern traces of Augustinianism in the reliance on the direct agency of the soul in vital operations. If not only the material species but also the (animate) power are parts of the integral instrumental cause of the sensory act (a claim that perfectly coheres with Suárez's view that the *species sensibilis* is not received in the organ but in the power), then the instrumental cause can be considered to be more perfect than a perceptual act, or at least equally perfect. Accordingly, this instrumental cause will comprise both indispensable components. One is the more perfect part, namely the animate power, which is not accountable for the representation of the sensible thing; the other is the less perfect one, sc. the sensible species, which, on the contrary, provides the cognitive act with perceptual content. While the first element provides the cognitive act with entitative perfection, the second one gives it representational content. While the sensible species does not attain the perfection of a cognitive act,

[91]On the activist *Zeitgeist* in Renaissance philosophy of perception, see Spruit (2008).

[92]For Suárez's remark about Henry's view, see *CDA* disp. 5, q. 4, n. 8, t. 5, 358. For Henry's Augustinianism in his theory of perception, see Silva (2018).

the animate power does.[93] To sum up, Suárez's abovementioned conciliatory conclusion can be more precisely stated in this way: the integral principle producing sensory cognition is the power informed by the species, and *the soul* of the percipient.[94]

How would this conclusion be spelled out if the technical distinctions related to the various kinds of efficient causes introduced in the later *DM* were again taken into account? In *DM* disp. 18, s. 7 Suárez clarifies this view of the principles of a perceptual act by bringing in the various kinds of principal and instrumental causes. He says that the soul is not the per se principal or substantival principle of acting *quod*, that is, it is not that which operates but only the substantival principle *quo* of acting, that is, the principle through which an agent operates. As a true Aristotelian he notes that the main principle *quod* of acting is a supposit, that is, the composite of body and soul. It is a man who walks, sees and understands.[95] In line with this claim Suárez (at least verbally) avoids the view of Neoplatonists according to which the soul is an operating supposit and the body is a moved instrument. (We have seen that some of Suárez's formulations suggest this view, however). Regarding the principal principle *quo* of receiving, Suárez distinguishes between sensation and intellection. While in perception this principle is the body or a composite by means of the body, in intellection it is the simple substance of the soul. From this point of view, intellection can be considered to be more immanent than perceptual acts since the principal principle *quo* of acting and the principal principle *quo* of receiving in the case of intellection are the same.[96] How is it with the total *instrumental* cause that comprises the power and the species? Suárez says that the instrumental cause is the proximate or accidental principle *quo* of acting. How is this accidental principle *quo* of acting related to the accidental principle *quo* of receiving? Going through the various kinds of cognition, Suárez claims that cognitive acts are of assimilative nature. They need the species causally concurring in the production of cognitive acts.[97] Since it is only the power *without* the species

[93] *CDA* disp. 5, q. 4, n. 16, t. 2, 366–8.

[94] I cannot agree with Cees Leijenhorst's view that Suárez's model of sense perception is fully identical with that of Scotus. For this claim, see Leijenhorst (2014, 172). When laying out Scotus's theory Leijenhorst conceives the cognitive power and the sensible species as two concurring partial causes constituting one total or integral cause. Even though this statement corresponds to what Suárez literally says in *CDA* disp. 5, q. 4, n. 15, it does not square entirely with the qualifications adduced in the paragraph below (*ibid.*, n. 16). There, Suárez explicitly states that these two concurring partial causes constitute the total instrumental cause, and not the total cause *simpliciter*. I have reformulated Suárez's conclusion accordingly.

[95] *DM* disp. 17, s. 2, n. 7, t. 1, 586.

[96] *DM* disp. 18, s. 7, n. 46, t. 1, 645.

[97] There is a difference in the causal concurrence of an object of an appetitive power, such as the will, and that of a cognitive power. While the cognized object through its intentional species concurs *efficiently* in the production of its own act, the intended object as a volitional agent constitutes the final cause. The final cause or the goal does not concur efficiently in the production of its own action but it concurs only *metaphorically*. It only attracts and inclines an appetent through presenting a desirable intentional object.

that receives the species, the accidental principle *quo* of acting and the accidental principle *quo* of receiving will be different. They will differ as that which includes and that which is included. While the power with the species (the accidental principle *quo* of acting) is that which includes, the power alone (the accidental principle *quo* of receiving) is that which is included.[98] However, despite this difference Suárez underlines that it is the *same cognitive power* which is both the proximate principle *quo* of acting and the proximate principle *quo* of receiving:

> It is certain that the same cognitive faculty in which a cognitive act is proximately received is simultaneously the proximate principle that effects that same act.[99]

As we have seen, this partial identity between the agent and the subject of reception makes no problem for Suárez. On the contrary, it expresses the quintessence of Suárez's approach. As said in Sect. 2.7, he endorses Scotus's axiom according to which the same faculty can be in virtual or eminent act and in formal potency at the same time with regard to the same thing.[100] It is above all a vital power that can be both in virtual act and formal potency at the same time with regard to the same act. As said above, the motivation for this claim is to be searched for (as it was in Olivi[101]) in Suárez's libertarian concept of free will.[102] His libertarian theory of free will rules out an external efficient cause, whether divine physical premotion, intellectual judgment, movement of the stars, or of bodily humours.

The metaphysical root of Suárez's embracement of Scotus's axiom and his emphasis on the soul's immediate causal efficiency within the acts themselves can be observed in his views of the soul's causal efficacy in *DM* disp. 18, s. 5. Here Suárez envisages the question of a precise causal *influxus* or the communication of *esse* (this expresses the ontological core of his causal theory) in the production of cognitive acts. He introduces his own theory (see also Sect. 2.2) stating proximate causal influence in vital acts against the background of a theory denying the soul's proximate causal influence, which stands behind the passivist account of perception. It shows that Suárez clearly perceives the doctrinal difference between the soul's proximate causal influence in the acts and its position of being only a remote cause. Albeit he does not explicitly mention the names of the advocates of this latter view, we may be sure that they would have been recruited from among the Thomists. For

---

[98] *DM* disp. 18, s. 7, n. 47, t. 1, 645–6.

[99] Cf. *DM* disp. 18, s. 7, n. 47, 645: 'Certum ergo est eamdem cognoscitivam potentiam, in qua proxime recipitur actus cognoscendi, esse simul principium proximum efficiendi eumdem actum'. The translation is taken from Francisco Suárez (1994, 167).

[100] See *DM* disp. 18, s. 7, n. 4, t. 1, 632.

[101] Olivi, *Quaestiones In secundum librum Sententiarum*, vol. II, q. 58, 461 : 'Quia autem illud quod meo iudicio super omnia movit multos ad credendum quod voluntas nostra sit totaliter passiva fuit et est hoc quod pro firmo tenent omnes alias potentias esse passivas [. . .]'; Olivi, *Quaestiones In secundum librum Sententiarum*, vol. III, q. 74, 124: 'Nam et principales rationes, quibus philosophantes conantur probare potentias cognitivas non esse activas sed passivas, non minus probant hoc de voluntate'. For the affinity between the passivity/activity of the cognitive powers and the will, see Pasnau (1999a, 1999b, 25); Toivanen (2013b, 119, 126, 144, and elsewhere).

[102] For this connection, see Burillo (1992).

them, a substantial form or the soul seems to be only the immediate principle of accidental faculties. The soul's causal influx, so to say, does not attain cognitive acts since the soul is related to them only as a *remote* principle. A proximate causal influence, as Suárez states, is impossible for them for two reasons. These reasons are similar to those that were mentioned in Sects. 2.6 and 2.7. First, created substances cannot be the immediate and proximate principles of their accidents. If they were, the instrumental causes, that is, the accidental powers, would become redundant and finite substances would be unduly divinized. Second, the proximate efficiency of substantial forms goes against ontological parsimony. The quality of heat in water is by itself a sufficient cause for heating air. It is redundant to posit the substantial form of fire as a proximate efficient cause. Most importantly, the same holds also for immanent acts.[103] Needless to say, both reasons go against the core of Suárez's philosophical psychology.

3.5 Endpoints of Perceptual Acts and the 'Representationalism vs. Direct Realism' Issue

Although so far perceptual states have been mostly referred to as perceptual *acts*, in Sect. 3.3 it has been indicated that they are also *qualities*. It is in *CDA* disp. 5, q. 5, entitled 'Whether a *terminus* is produced through an act of a cognitive power; and on the mental word', where Suárez spells out more clearly the nexus between these two aspects and treats *ex professo* the issue of the nature of the cognitive act. In this question he substantiates the claim that this cognitive quality is a *produced* quality, that is, a quality caused by essentially ordered causes, which are operative through *action*, and not through a noncausal natural resulting or some other mysterious 'inference' supervening on the reception of a sensible form from a cognitive subject. Only such an investigation can provide evidence that the production of the perceptual process should be investigated above all in the metaphysical framework of efficient causality.[104] For this reason I will have to include in this section a digression to the issue of efficient causality, as expounded especially in *DM* 48 in the context of the category of action, to clarify Suárez's claim about the modal

[103] *DM* disp. 18, s. 5, n. 1, t. 1, 627–8.

[104] For this interpretation, see Heider (2019c). Notwithstanding the soul's a-causal mediation of the soul in the production of the intentional species of the interior sense and the intelligible species, the topics of Sect. 5.1, the *causa efficiens* with its *extrinsic* communication of *esse* to the effect represents the central model of causality for Suárez in *DM*. In his recent study Perler (2018, 382–3) confirms this view by saying that '[...] Suárez explains the activity of all these modules [the powers; D.H], habits in particular. He describes them as being part of a chain of efficient causes: they are efficiently caused by acts, and in turn efficiently cause other acts in cooperation with the rational faculty. In fact, all the items at the sub-personal level are efficient causes'. For efficient causality as a dominant type of causation in Suárez, see Carraud (2002, 163–6), Schmaltz (2008, 24–44), Freddoso in Suárez (2002, xi–cxxiii) and Schmid (2015a, 84–121).

distinction between perceptual act *qua* quality and perceptual act *qua* action. This metaphysical framework is employed also in *CDA* where Suárez suggests that the production of the cognitive act is to be analyzed in the context of efficient causation because it pertains to the category of action producing a quality.[105] This fundamentally productive character, connected with Suárez's general emphasis on the vitality and the immanency of cognitive operations, holds for *all* cognitive powers including the external senses. Importantly, this productive character and the ontological modelling of cognitive act in the manner of the produced quality give rise to a doubt concerning the immediate object of perception: what do we immediately perceive? Do we perceive a produced intramental quality, or an object in this quality, or do we apprehend an extramental sensible object directly? Accordingly, after the treatise of the ontology of perceptual (cognitive) act I will treat this 'representationalism vs. direct realism' issue too.

Before presenting Suárez's conclusions about the nature of the endpoint or terminus of the perceptual act, the important distinction between intuitive and abstractive cognition needs to be briefly presented.[106] While intuitive cognition apprehends objects that exist *hic et nunc* with respect to the percipient, abstractive cognition cognizes them *in absentia* or in abstraction from existence. Rather than from universality abstractive cognition abstracts from the presence of an object. Traditionally, there are two main theories related to the issue of the production or non-production of cognitive endpoints in these two kinds of cognition. Both constitute an important background of Suárez's reasoning. In conformity with the first theory, advocated especially by Aquinas and his followers, the *idolum* – the terminus or expressed species in perception – and the mental word (*verbum mentis*) in intellection are formed *only* in abstractive cognition, that is, in acts of the interior senses and the intellect.[107] Only these two powers apprehend their objects in their absence or in a manner that differs from the way in which they extramentally exist. Cognition of the external senses, the *visio beatifica* of the blessed in the heaven and angelic self-awareness are all examples of intuitive cognition in which no terminus is produced. They do not bring it about since in them a cognizant gets *immediately* identified and assimilated to an intelligible object known which is intimately present to the percipient's power.[108] According to this approach the *idolum* cannot be regarded as a necessary item since it is produced only to get an intramental substitute or expressed representation for an

[105] *CDA* disp. 5, q. 5, n. 4, t. 2, 374: '[...] actio cognoscitiva est vera actio de praedicamento actionis, quia vere producit aliquam qualitatem de novo'.

[106] For a detailed historical presentation of this crucial epistemological distinction in late medieval philosophy, see Sect. 3.9.

[107] A clear statement of this view can be found in Cajetan, *In STh.* I, q. 27, art. 1, XVI, 308.

[108] In *Quodl.* 5, q. 5, art. 2, ad 2 Aquinas explicitly rejects such production in the external senses: '[...] cognitio sensus exterioris perficitur per solam immutationem sensus a sensibili: unde per formam quae sibi a sensibili imprimitur, sentit: non autem ipse sensus exterior format sibi aliquam formam sensibilem: hoc autem facit vis imaginativa, cuius formae quodammodo simile est verbum intellectus'.

absent (nonexistent) object or for an object such a material quiddity, which (in line with Aquinas's moderate realism) does not exist extramentally in the way it exists in the intellect, that is, as a universal. This character of the *idolum* and of the mental word is a reason why the expressed species is to be conceived as that in which (*id in quo*) as if in a mirror (see the mirror metaphor above) objects are cognized, rather than as that by which (*id quo*) they are apprehended. Accordingly, the expressed species is to be construed as something really distinct from cognitive acts, as a *res* differing from another *res*.[109] Beside this view Suárez mentions another opinion, which is advocated especially by Scotus. As far as the production of the expressed species is concerned this view assumes a fundamental parity between intuitive and abstractive cognition. Although the *verbum* in the intellect and the *idolum* in the senses are not produced through a cognitive act, there are items that are appropriately indicated as the expressed species in all the powers. These items are nothing else than occurrent cognitive acts themselves.[110] As Scotus suggests, these cognitive operations such as *intelligere* and *sentire* are produced not by an act of cognition but by a really distinct act that is called *diction*. It would be a category mistake to say that cognitive operations can physically produce anything. Accordingly, he makes it clear that the productive *dicere*, which pertains to the category of action, precedes by nature the cognitive act that falls into the category of quality.[111] This expressed species or quality is an integral part of all cognitive operations, since all cognitive powers are in an important aspect active and productive. Since the formation of a terminus is not primarily posited for the sake of having an intramental representation or substitute of an absent, nonexistent or universal object, the endpoint must be construed differently than as 'that in which'. It should be conceived as 'that through which' (*id quo*) an extramental object is apprehended.[112]

As suggested, Suárez is intent to show that cognition as such exhibits not only the assimilative trait, but also an expressive/articulative aspect. In his first conclusion he asserts that only the second of the two abovementioned tenets can be true: an intrinsic endpoint is formed in *every* cognitive act regardless of whether it is an act

---

[109] *CDA* disp. 5, q. 5, n. 2, t. 2, 370–2. For this exposition of Aquinas's theory of the *verbum mentis*, see also Pasnau (1997, 256–71) and Perler (2004, 89–100).

[110] Scotus, *Ord.* I, d. 27, q. 1–3, n. 59 (ed. Vat. 6: 87): 'Sequitur ergo, per viam divisionis, quod verbum est actualis intellectio [. . .]'. For Scotus's theory of the mental word, see also Cross (2009, 320–32; 2014, 117–21, 171–3).

[111] Scotus, *Ord.* I, d. 27, q. 1–3, n. 84 (ed. Vat. 97–8): 'Ad tertium concedo quod notitia est proles et vere genita, scilicet actualis intellectio, sed illa non est actio de genere actionis (quia, ut dictum est supra, actualis intellectio non est actio de genere actionis), sed est qualitas nata terminare talem actionem, quae signatur per hoc quod est 'dicere' et - in communi - per hoc quod est 'elicere'. Non ergo verbum est aliquid productum actione quae est intellectio, quia ipsa intellectio non est productiva alicuius, sed ipsa est producta actione quae est de genere actionis, sicut dictum est supra'.

[112] *CDA* disp. 5, q. 5, n. 3, t. 2, 372.

of intuitive cognition or an operation of abstractive cognition. In a reference to his unpublished *De sex ultimis praedicamentis*, a text written in Segovia before the composition of *CDA*, Suárez distinguishes two aspects in all acts, including transitive and immanent operations. The first aspect is the active aspect of producing or causing. This aspect is the way (*via*) or the becoming (*fieri*) that leads to its terminus or effect. The second 'ratio' is the character of a produced quality (*ratio qualitatis productae*) or the cognitive act in its *factum esse*. This feature, on the other hand, is the cognitive act's *terminus*.[113] Having denied the Thomist notion of the *verbum mentis* and the expressed species, in his second conclusion Suárez says that these termini are not really but only *formally* distinct from cognitive acts.[114] The cognitive act *qua* quality can function as a representing vehicle, namely as a second-act formal likeness or mental sign of extramental objects. Suárez denies Scotus's view according to which *dicere* and *intelligere* (*sentire*) are to be conceived as two really distinct acts. Both aspects can be part of the same act since one and the same act can be related to itself as a production to its terminus.[115] In the third upshot, he adds that beside these termini, which are *realiter* identical to cognitive acts *qua* qualities and formally distinct from cognitive acts *qua* becoming, no other entity, whether really or formally distinct, is produced by a cognitive act. Of course, we know that a habitus can be produced. But this is produced by a cognitive act *qua* quality and not by a cognitive act in its *fieri*.[116] In a structural analogy to the distinct production of the powers from the soul, Suárez highlights that a single action can produce only a single effect.[117] An additional representational entity is redundant and theoretically implausible. Moreover, as such, as we shall see below, it would impede the direct and immediate apprehension of the extramental objects. It would be an unwelcome *tertium quid* or intramental object, which would have to be somehow cognized before the apprehension of the extramental object.[118]

In his last conclusion, Suárez summarizes all the doctrinal points of his theory of the expressed species and brings up an important qualification:

> A [mental] word or something proportional [*idolum* in the case of senses] is produced by every cognitive action; this word is formally and really nothing else than the act of cognition itself in so far as it is a (categorial) quality; nevertheless, it is *modally* distinguished from the cognitive act in so far as it is a production. This is why a [mental] word is not posited from the lack of an object but from the very nature of cognition.[119]

[113] *Ibid.*, n. 4, t. 2, 374.

[114] *CDA* disp. 5, q. 5, n. 5, t. 2, 374.

[115] *Ibid.*, n. 26, 400.

[116] For the efficient cause of habits, see *DM* disp. 44, s. 8, n. 13, t. 2, 684. See also Perler (2018, 372–5).

[117] *CDA* disp. 5, q. 5, n. 6, t. 2, 376.

[118] *Ibid.*, n. 8, 378.

[119] My italics. *CDA* disp. 5, q. 5, n. 9, t. 2, 380: 'Per omnem actionem cognoscendi producitur verbum, vel aliquid illi proportionale, quod realiter et formaliter non est aliud quam ipse actus cognoscendi, ut est qualitas, tamen distinguitur [modaliter] ab illo, ut est productio. Quare verbum non ponitur ex indigentia obiecti, sed ex vi et natura cognitionis'.

Whatever this formal distinction in *CDA* precisely may mean, Suárez rejects the view according to which the expressed species is a thing distinct from the act, or that it is that in which an object is cognized. The existence of such a pictorial or mirror-like terminus is not necessary to explain mental representation.[120] But it is especially his remark about the modal distinction between the cognitive act *qua* quality and the cognitive act *qua* production that is rich in his metaphysics of efficient causality. Considering this remark about the modal distinction I cannot help thinking that already by the early 1570s Suárez had formulated the basics of his metaphysics of efficient causality, which he at length analyzed in *DM*. Remaining within the framework of *CDA*, this formulation about the modal distinction between the cognitive act *qua* production and the cognitive act *qua* quality is not clear. This is why we have to turn now to *DM*.

The distinction between the cognitive act *qua* production and the cognitive act *qua* quality is related to what has recently been called Suárez's nonreductive theory of efficient causality.[121] As it has been said above, it is mediation through *action* that distinguishes efficient causality not only from internal causes, in which the effect emanates from the cause through the intrinsic unification, but also from the final cause. This theory of efficient cause is based on the irreducibility of the category of action. For Suárez, action cannot be reduced to any other category, such as relation, substance, or quality.[122] Action is the 'nature of causing' (*ratio causandi*) in the efficient cause.[123] Action is the medium between the cause and its effect, the way (*via*) to its endpoint, the becoming (*fieri*) of the effect, or the tendency to its term. Most importantly, action is metaphysically construed as an ontological *mode* of the effect. Albeit contained in the scope of being and therefore part of the metaphysical investigation,[124] in order to be produced this mode does not require another action through which it would have to be caused. If it did, an infinite regress of causations would result.[125] Moreover, action cannot be a free-floating thing which exists or can exist without an endpoint. As an essentially relational entity (it is not only the tendency to its effect but also extrinsically denominates the agent as 'acting'[126]) it can be neither a substance nor an accident considered as *res*. Based on the theological lesson from the Eucharist, Suárez distinguishes two basic kinds of accidents, which are significantly operative in his metaphysics. There is the accident of quantity which can exist without actually inhering in its subject. Its essence is only that of

[120]This non-pictorial character of the expressed species goes well together with the non-pictorial character of the impressed species. For Suárez's rejection of the conception of the impressed sensible species as *figura*, see *CDA* disp. 5, q. 2, n. 23, t. 2, 324.

[121]For this, see Tuttle (2016).

[122]For a nominalist reduction of an action to an effect (creatures) in the divine creative action, see *DM* disp. 20, s. 4, n. 6, t. 1, 770.

[123]*DM* disp. 18, q. 10, n. 1, t. 1, 680.

[124]*DM* disp. 48, Prologue, n. 2, t. 2, 867.

[125]See *DM* disp. 18, s. 10, n. 8, t. 1, 682.

[126]*DM* disp. 48, s. 1, nn. 17–20, t. 2, 872–3.

aptitudinal inhering in the subject. However, action is not of this kind. It can be only a modal accident or entity that exists only as *actually* modifying a *res*, that is, its endpoint, just as the position of sitting cannot exist independently of a *res* such as Peter.[127] Applying this distinction to the topic of the metaphysics of cognition, it may be said that the cognitive powers informed by the species and operated by the principal cause, namely the soul, constitute the total efficient cause, which through a mediating action necessarily produces the endpoint in all kinds of cognition, that is, a cognitive act conceived categorially as quality.

Suárez has at least two reasons to conceive action as a mode of the effect, rather than of the agent.[128] First, although his view of efficient causality can be regarded as 'a conservative extension of Aristotelianism',[129] he follows the main thrust of Aristotle's analysis of change in his *Physics*. In *Physics* III.3 Aristotle notes that actual movement is realized in that which is in potentiality capable of motion. An act of moving takes places in a patient.[130] An act of heating by fire takes place in a *passum*, for example, in a piece of wood. The second reason is theological. Since God is perfect, He is immutable.[131] He has no unactualized passive potentialities. He cannot be changed by creating the world. However, if He is immutable, His creative action cannot inhere in Him. If it inhered in Him, by creating the world He would have to have been changed. The creative action emanating from God can only be a mode of His effect, that is, the created universe. At the same time, since creation does not operate on a patient, it cannot, properly speaking, inhere in the subject. Therefore, in order to accommodate *all* the causal instances of efficient causality including creation and immanent acts, Suárez makes it clear that action as a mode of the endpoint does not inhere in its effect but only *adheres* to it, or, in other words, it is terminated in its effect.[132] Action can inhere in the subject only by virtue of its formal endpoint.[133]

Suárez is clear that for him the production of a cognitive act proceeds according to the principles of efficient causality. The production of the cognitive act as quality is a *variable* and *contingent* item. If God as the primary cause can substitute for *all* secondary causes, He can also substitute for the active concurrence of a cognitive power. He alone can elicit these acts in the power without its concurrence. However, in order to avoid a tension with the abovementioned axiom about the essential

[127]For the distinction between these two types of accidents, see *DM* disp. 37, s. 2, nn. 9–10, t. 2, 495.

[128]For these reasons, see Schmid (2015a, 98ff).

[129]*Ibid.*, 101.

[130]Aristotle, *Physics* III.3, 202a13ff, 207ff.

[131]For the divine attribute of immutability, see *DM* disp. 30, s. 8, n. 2, t. 2, 113.

[132]For this adherent termination of an action in the endpoint, see also Tuttle (2016, 153–5).

[133]*DM* disp. 48, s. 4, n. 13, t. 2, 891–2.

dependence of the powers on the soul's and the powers' efficiency,[134] our author is quick to introduce another example. The same cognitive act *qua* quality could have been produced differently from the way in which it was caused since it could have been produced with or without the causal concurrence of a habitus; it could have been caused by a less perfect sensible species or by a more perfect species; it could have been produced with or without divine assistance etc.[135] The same holds also for divine creative and conservational action. It is not part of the Catholic faith to assert that the universe could have been created and conserved only through the particular action by which it was created. The creational causal action could have been different. God could have created the world differently from the way He did. He could have employed angels as instruments of creation. The divine conservation of the world in *esse* could have been replaced by a different act of God.[136]

What are for Suárez the competing positions denying that the cognitive act is a produced quality? In *DM* disp. 48, q. 2, called 'Whether action, as action, essentially respects some terminus—even if [the action] is immanent—and therefore is indeed included in this category', Suárez presents two theories, both advocated by Thomists and both disclaiming the necessary production of an endpoint in immanent operations. According to the first theory, defended by Paul of Soncino (ca. 1458–1495),[137] we need to distinguish between transitive and immanent actions. Transitive acts, whose effect passes to another subject, are part of the category of action and produce their own endpoint that remains in the subject even after the act has ceased. Immanent actions do not pertain to the category of action. They are part of the category of quality. Suárez presents four arguments for this view, all of which are taken from Aristotle.[138] First, since no result comes from immanent acts except for the operation itself, these operations cannot be properly speaking actions.[139] Second, since every action is a motion and motion is a successive action, every action must be a successive action. However, immanent acts are not successive. They are instantaneous events.[140] Third, immanent actions do not inflict a passion on that toward which they operate. Vision does not transmute the visible object it sees. But if it does

---

[134]As we have seen in Sect. 2.2, not even God can cause vital operations *qua* vital to occur without the proximate and internal causal influence of the soul.

[135]For the possibility of a divine insertion of a sensible species representing an absent sensible and thus for the (metaphysical) possibility of the external senses' abstractive cognition of absent or nonexistent objects, see the Sect. 3.9.

[136]*DM* disp. 48, s. 2, nn. 12–3, t. 2, 876–7. For an analysis of this Separability Argument, see Tuttle (2016, 145–50).

[137]For his view, see Paulus Soncinas, *Quaestiones metaphysicales acutissimae*, *In IX Met.*, q. 21, 227.

[138]For the first argument, see Aristotle, *Metaphysics* IX.8, 1050 a23–b1, 459–61; for Aristotle's second claim, viz. that action and passion are movements and thus successive *doings* and successive *sufferings*, see Aristotle, *Physics* III.3, 202a22ff, 209ff; the third and fourth arguments are obvious from the overall thrust of *Physics* III.3.

[139]*DM* disp. 48, s. 2, n. 2, t. 2, 874.

[140]*Ibid.*, n. 3, 874.

not affect the patient, it cannot be thought of as producing an endpoint. Fourth, no categorial action perfects its agent. Fire is not perfected by heating water. But immanent acts perfect their agents. Therefore, they cannot belong to the category of action.[141] The second position, advocated by some other Thomists, agrees with the first one in the crucial point: properly speaking, immanent actions do not produce endpoints. But it substantially differs in how it explains the absence of this production. On this view immanent actions are part of the category of action. Cognitive acts of the external senses or of the common sense are 'pure actions', that is, they attain *immediately* their sensible objects. As such they do not need to produce a mediating endpoint.[142] Their operations belong to the external sensorium. As critics of Suárez's theory say, if these acts did not attain the sensibles immediately and if they produced an endpoint, direct perceptual realism would be jeopardized. The theory of direct realism would have to be replaced by a representational view of perception, which ushers in the early modern philosophy represented by Descartes, Locke, and others. According to some interpreters, just this happened to Suárez's theory of perception.[143] The reasons for this second position are again taken from Aristotle. Referring to the above-quoted passage from Aristotle's *Metaphysics* IX.8, the first reason claims that there are actions that exist for the sake of themselves. These actions do not produce an endpoint. Second, corporeal vision produces nothing, whether in the visible or in the percipient. Since the multiplication of entities beyond the visual power and the object is entirely redundant, these perceptual acts cannot produce anything. Third (and this is highly relevant for our topic), these actions are not instituted for the sake of production or for the agent's communication of *esse*, but for *uniting* the power from which these acts 'emanate' with its objects. The production of an endpoint would be not only redundant but also detrimental to this intentional identification and assimilation. It would prevent the possibility of perceptual direct realism, which is ultimately based on immediate experiential contact with the external word.[144]

In his replies to the first arguments for the first and second view, Suárez proposes his own reading of Aristotle. True, immanent acts do not produce effects that remain outside the operating agent, nor do they produce anything that remains when the operation has passed. However, this does not mean that Aristotle aimed to show that immanent acts produce nothing. Second, it is not true that every action must be successive. Leaving aside creation, there are instantaneous actions, such as illumination, that belong to the category of action.[145] Third, it is not necessary for *X* to be

---

[141] *Ibid.*, n. 4, 874.

[142] *Ibid.*, n. 5, 874–5.

[143] For a substantial emphasis on an immediate and experiential cognitive contact with extramental reality and for a critique of the productive character of Suárez's theory of the external senses' aprehension, see especially Lisska (2016), 235 and *passim*. For my reply to Lisska, see Heider (2020).

[144] *DM* disp. 48, s. 2, n. 6, t. 2, 875.

[145] *Ibid.*, nn. 22–3, 880.

an act to cause a passion in an object. Fourth, if considered in its becoming, it is true that a cognitive act does not perfect the cognitive subject. However, if considered in its *factum esse*, it is clear that it does perfect her. In a reply to the second argument of the second view Suárez notes that it is not true that a cognitive act does not imply a passion in the subject in which it exists. Vision does inflict a passion on the cognitive agent in which it exists. As we have seen, the perceptual process is a three-phase process. It is in the last phase, in which a cognitive act is received in the power through which it was produced.[146] In a response to the third argument of the second view, based on the fact of immediate cognitive unity with the object, Suárez is more than clear: this cognitive unity may be called a unity only *metaphorically*. In line with the overall thrust of the cognitive segment of *CDA* he refuses to conceive this intentional identification and intentional unity in the literal sense. It must be said that this refusal is in line with his deliberate omission of Aristotle's comparison of perception to a signet ring received in a piece of wax.[147] Intentional unification can proceed only through a cognitive act conceived as quality, also called formal concept by Suárez, which can be called quasi-formal concept in the case of perception, in which a cognized object (a quasi-objective concept) is *vitally* formed.[148] In line with Augustine, only in this cognitive act *qua* quality, conceived as an 'offspring' of the 'parental' activity of the cognitive powers and the species an intentional object exists and an extramental object (if it exists) can be cognitively 'intended'.[149]

In his replies to the objections to his act-theory of the external senses' expressed species Suárez further clarifies his direct-realist concept of the *idolum*. Someone may insist that *idola* must be posited as entities that are really distinct from the operations and conceive them as 'that in which' an object is apprehended, since only if they are conceived in this way can the intentionality of the cognitive acts *of* absent objects and objects that exist differently from the way they are apprehended be explained at all.[150] These cognitive phenomena show that perceptual intentionality can be explained only if we assume that perceptual acts are epistemically related to intramental images conceived as that in which as if in a mirror an object is cognized

[146] *Ibid.* n. 24, 880–1.

[147] For this see Sect. 3.2. For Suárez's rejection of this literal interpretation of the *dictum* 'ipsa potentia per speciem fit ipsum cognoscibile', see also *CDA* disp. 5, q. 4, nn. 5–6, t. 2, 354–6.

[148] *DM* disp. 48, s. 2, n. 26, t. 2, 881.

[149] *CDA* disp. 5, q. 5, n. 17, t. 2, 388: '[...] nam ut res possit intelligi necesse est, ut in intellectu [vitaliter] formetur [...]'; *Ibid.*, n. 8, 378: '[C]ognoscere nihil aliud est, nisi vitaliter operari formando in se rem ipsam quam cognoscit'.

[150] As is well-known, visual illusions and visual error motivated similarly Peter Auriol to articulate his concept of *esse apparens* (which can be called a quasi-objective concept) produced by every perceptual act. In his *Scriptum* on the first book of the *Sentences*, d. 3, Auriol adduces eight (visual) experiences to argue for the conclusion that every act of the external senses puts the perceived things into intentional being, which he calls ' apparent being'. For this see Petrus Aureolus, *Scriptum super primum Sententiarum*, vol. II, d. 3, s. 14, art. 1, n. 31, 696–8. For Suárez's analysis of perceptual error and scepticism, see Sects. 3.8 and 3.9.

or in which an object is appearing. If *idola* did not take on this object-like form, perceptual intentionality would not be explained since in the case of nonexistent objects no relation between cognitive acts and extramental objects is conceivable. This objection can be reinforced by reference to the theory of ideas or exemplary causes regulating an artist's production. An artist creates an artefact according to a pattern she has in mind. But this pattern or exemplary cause can be only an object-like *idolum*, in which we see the object which is to be made. Only if known in this way can it regulate the artist's operation.[151] Suárez's reply is clear: it is far from necessary to assume the abovementioned kind of *idola* as a necessary termination of cognitive acts since this termination is not to be construed *materialiter* in the manner of a line terminating in a point, but only *spiritually* or *intentionally*.[152] Intentional termination or reaching out does not require the existence of things that are *realiter* distinct from the cognitive acts. To explain the intentionality of cognition involving nonexistent (absent) objects or objects that exist in a way different from the way they are apprehended we do not have to posit an object-like *idolum* in the mind. Love can terminate in a beloved dead person even though no image of the deceased person is formed. Pictorial depiction is not necessary even in the cognition of beings of reason. What *is* necessary is the existence of objects vitally formed by cognitive acts in intentional (objective) being. Nevertheless, it is just this theory of intentional objects that can give rise to a new objection to Suárez's cognitive theory, which again accuses him of perceptual representationalism. According to this objection, the directly perceived object seems to be a *res* in intentional (the quasi-objective) being vitally formed in every perceptual act.[153] If not a quasi-formal concept in the manner of an *idolum* conceived as that in which a sensible object is conceived, the objection goes, so this quasi-objective concept, or objective correlate of perceptual acts-qualities, will be a *tertium quid* wedged between our *idola* and the external sensibles.[154]

Considering this objection, not presented in *CDA*, it is no wonder that not a few scholars formulated their conclusions about Suárez's perceptual representationalism by referring to this quasi-objective concept.[155] So what does Suárez mean by this

---

[151] *CDA* disp. 5, q. 5, n. 18, t. 2, 390.

[152] For an analysis of this spiritual termination in Suárez, see also Hernandéz (1948, 767–91).

[153] *CDA* disp. 5, q. 5, n. 23, t. 2, 394–6.

[154] For Suárez's remark about the objective concept conceived as cognized thing in *CDA* see disp. 5, q. 5, n. 17, t. 2, 388.

[155] The secondary literature on Suárez's theory of objective concept and objective being is huge. In particular, it bears on the debate regarding Suárez's view of metaphysics (the object of his metaphysics conceived as the objective concept of real being) and with his analysis of divine knowledge of creatable beings. Since the subject matter of the present analysis is perception I cannot delve into these debates here. Elsewhere (Heider 2011a, 61–7), however, I argued against the mentalist and representationalist interpretation which conceives the objective concept as a *tertium quid* interposed between cognitive acts and the extramental reality. My reading was based on Suárez's crucial passage from *DM* disp. 2, q. 1. n. 1, t. 1, 65 where the objective concept is defined by means of an extrinsic denomination from the formal concept: 'Conceptus objectivus dicitur res illa, vel ratio, quae proprie et immediate per conceptum formalem cognoscitur seu

thing formed within the cognitive powers, including (due to his cognitive parallelism) also the external senses? Interestingly, the young Suárez says almost nothing about this issue in *CDA*. Perhaps he did not see any problem in it. Clearly, when he speaks about a thing in intentional being to which no extramental object corresponds, this quasi-objective concept can be neither an external thing, nor an impressed species. While the impressed species is received in the power, the *res* in quasi-objective being is actively and vitally formed. Whereas the power is related passively to the former, it is active in relation to the latter. But could it not be that this vitally formed thing possesses a kind of special diminished being (*esse deminutum*)? Could it not be a thing with a special kind of shadow real being that differs from the being of cognitive acts, extramental objects and beings of reason?[156] Albeit much has been written on this topic in the context of intellectual knowledge and much of this is discussed in *DM*, it seems that a reductionist account of objective being is available and is in line with the overall thrust of Suárez's *CDA*.

As the history of early Scotistic interpretation of Scotus's theory of *esse obiective* shows, similar problems connected with the ambiguity of *esse obiective* plagued the Subtle Doctor's thought.[157] Richard Cross has recently proposed an interesting disambiguation of Scotus's nomenclature of 'real' and 'intentional' that, I think, can be useful also for interpreting Suárez's theory.[158] Cross states that when talking about the intentionality of sensation Scotus has two kinds of meaning of 'real' and 'intentional' in mind. Regarding the meaning of 'real', Cross calls one $real_R$, the other one is indicated as $real_I$. The reception of a form $real_R$ equals the natural reception of a form with its matter. In line with what has been said in Sect. 3.2, upon this reception the recipient is made to be an instance of the received form's kind. It turns red when the form of red colour is received, as it happens in a wall painted with red paint. The reception of a form $real_I$, however, is the reception of a

---

repraesentatur; [. . .] homo autem cognitus et repraesentatus illo actu dicitur conceptus objectivus, conceptus quidem per denominationem extrinsecam a conceptu formali, per quem objectum ejus concipi dicitur, et ideo recte dicitur objectivus, quia non est conceptus ut forma intrinsece terminans conceptionem, sed ut objectum et materia circa quam versatur formalis conceptio, et ad quam mentis acies directe tendit, propter quod ab aliquibus ex Averroe, *intentio intellecta* appellatur; et ab aliis dicitur *ratio objectiva*'. For the detailed defence of this realist exposition, see Gracia (1991, 301–7). For a 'mentalist' reading of Suárez's *conceptus objectivus* see Cronin (1966a); Cronin (1966b, 79); Courtine (1990, 169–82); Wells (1993); Doyle (1967). For putative Suárez's representationalist anticipation of Descartes's *realitas obiectiva* from the third Meditation of his *Meditationes de prima philosophia*, see Rabade Romeo (2003, 147–8).

[156]For Suárez's intellectual formation of beings of reason and logical intentions (these are formed only in a second step, i.e., only if an abstracted and univerzalized nature that is cognized directly is conceived *as if* it were a real being), see Novotný (2013, 72–4) and Heider (2014, 80–4).

[157]In early Scotism two opposing interpretations of Scotus's *esse deminutum* became established. While for Jacob of Aesculo (fl. early fourteenth century) intentional being constitutes a third realm of being distinct from the subjective being of cognitive acts and from the extramental reality, for William of Alnwick (d. 1333) they are equivalent to a mental content that is identical with a representing form. For these two different interpretations of Scotus's theory of objective being, see Perler (2004, 230–245) and Perler (2011).

[158]Cross (2014, 35–9).

representational form, for example, of the visual species of red colour. Obviously, affection by this form does not result in such an alteration. Cross at the same time describes this second form as intentional$_R$. Accordingly, he identifies intentional$_R$ with real$_I$ in Scotus's theory.[159] This is in consonance with Suárez's account. Besides these two meanings he mentions another one that he calls intentional$_I$, which he takes to be identical with a pure representational *content* that is inherent in the real$_I$ or intentional$_R$ item. A reductionist step in Cross's account of Scotus's view of the cognitive act lies in the reduction of Scotus's talk about the intentional object possessing objective being to a 'discourse' about the content of perceptual acts which is originally encoded in the impressed sensible species and transmitted to the power. Just this reductive account or this reductive 'translation' from the language of objects to the language of contents seems to be transferable to Suárez's *CDA* too. Although Suárez explicitly speaks about 'the formation of (intramental) *things* in the power', these things can be interpreted as nothing else than mental contents of the quasi-formal concepts. As such they can be taken as equivalent to the content of mental acts. This interpretation seems to be implied also by Suárez's abovementioned third conclusion related to the topic of the terminus of cognitive act, according to which it is stressed that *nothing else* beside the above-characterized terminus conceived as 'id quo', whether really or formally distinct (no doubt, objective being in its reified interpretation can be conceived as formally distinct from acts), is produced by a cognitive act. In other words, this third conclusion can be read as follows: what is produced is only a cognitive act which is structured in this or that way depending on the reception of this or that likeness coming in this or that way through this or that medium from a sensible object, the result of which is this or that appearing of the object to a given power. Although in nonexistent objects this mental content will exist without external objects, in the case of veridical perception it will be identical with the aspects of the external sensible objects themselves.

3.6 Attention and How It Is Divided

As I have pointed out above, for Suárez the sensible species is not the principal cause, let alone the only cause of perception; that is the soul. Through the soul's substantial causal influence Suárez brings to the fore the phenomenon of attention. The emphasis on attention makes Suárez a follower of Augustine's and late medieval Augustinian theories of perception.[160] At the same time, it also adverts to Suárez's affinity with early modern subjectivity.[161] Today the phenomenon of attention is a

---

[159] *Ibid.*, 171–3.

[160] For the emphasis on attention in Augustine and in late medieval Augustianianism, see Silva (2014a) and (2014b).

[161] Since early Enlightenment, attention became one of the major topics in the emerging experimental psychology. Gary Hatfield (1998, 5–7) shows that the introduction of attention into

commonplace in philosophical and psychological analysis of mental states such as perception, thought, desire, introspection and agency.[162] It might be said that the notion of attention is first of all correlated with the notion of consciousness, in contemporary philosophy of mind an umbrella term covering a variety of mental phenomena.[163] If compared with the contemporary classification, Suárez's application of the terms 'attendere' and 'attentio' is not systematic. In fact, he uses these terms in different meanings without providing a clear differentiation. This is why before presenting his theory of the division of attention, several meanings of the notion of consciousness must be disentangled. Such differentiation will help us to see what Suárez means by the notion of attention too. At first Suárez underlines the connection of attention with the transitive and the access consciousness by referring to *conscious attending* to an object. As said, without this attending to an object the receptions of stimuli or species are 'dead'. Without attending to visual information encoded in the species the information cannot be available to percipients. Two people receiving species representing the same object often have different perceptions of the object since one attends better than the other. Moreover, we themselves introspect this variability in the 'quality' of our percepts too. When we are not attentive, we do not apprehend objects in the way we do when we are attentive and after rest.[164] Besides, this access consciousness is necessary not only for the production of cognitive acts attaining the external objects but (as we shall see more below in Sect. 5.1) also for the production of intentional species in the higher powers. The internal sense does not produce sensible species unless the external senses are attending to the sensibles and unless they are moved to second act.

Suárez does not refer only to a psychic facet of attention related to the access consciousness. He also refers to a physiological aspect, which bears upon the activation and spreading of the animal spirits from the brain, which for Suárez is the principle and root of sentience, to the sensory organs. Without 'irrigation' through these spirits the pertinent organ would not be disposed for its operation. As such these spirits form the necessary material condition for the exercise of the sensory powers. As Suárez stresses, in respect to the soul's causal influence they are not its cause but rather its effect.[165]

psychology as a major topic was marked especially by its discussion by Christian Wolff (1679–1754).

[162]For an excellent treatment of attention in contemporary philosophy, see Wu (2014).

[163]For a disambiguation of this notion and its various divisions applied, see van Gulick (2018). For a close affinity of the concepts of attention and consciousness, see also Wu (2014, 146–208) and Mole (2017, esp. 3.1 'Attention and Consciousness').

[164]For this, see *CDA* disp. 5, q. 4, n. 5, t. 2, 354.

[165]In *CDA* disp. 6, q. 6, called 'Utrum principium et radix sentiendi sit in cerebro vel in corde', Suárez distinguishes two kinds of spirits, which both are corporeal vehicles through which bodily heat in the nerves is conveyed to all parts of the body. While the so-called vital spirits (*spiritus vitales*) have their origin in the heart, the animal spirits (*spiritus animales*) come from the brain's cells. Just these latter *spiritus* are activated through the soul's attention and carried to the sense organs heated by them. Through this heating they get materially disposed for sensation and are

It was William James (1842–1910) who described attention also in a different sense, namely in the terms of selective attention which could be imputed also to Suárez:[166]

> Focalization, concentration, of consciousness are of its [attention's] essence. It [attention] implies withdrawal from some things in order to deal effectively with others, and is a condition which has a real opposite in the confused, dazed, scatterbrained state which in French is called *distraction*, and *Zerstreutheit* in German.[167]

Without selective attention filtering the chaos of stimuli emitted from the sensibles it is impossible to process information and focus on a task. As contemporary psychologists and cognitive scientists say, attention is an irreplaceable mental process and a unique tool that prevents the processing mechanism of the brain from overloading through an enormous amount of information, which is a constant in our world. Given our limited attentional reservoir attention functions as 'a bottleneck' capable of covering only a limited amount of information.[168] Suárez is alive to this *filtering* and *selective* role of attention. He is explicit that the air is full of sensible species multiplied from the objects and they need to be filtered if anything is to be perceived at all.[169]

Furthermore, as we have seen, Suárez is sensitive also to the phenomenon of focalization which functions as a flashlight *preceding* the reception of species: 'The soul can attend even though it has not received a species yet, as it happens when someone focuses to be able to hear something'.[170] In this attentional anticipation or expectation, which differs from the selective attention, we can wonder what this or that object might be or where it might be. We strain our attention to a yet unknown object which we are impatient to see or hear as well as when we look forward to a pleasant visit. Last but not least, Suárez stresses also the interconnection between attention and the introspective consciousness: not only that attention is an essential

---

impressed with sensory stimuli. This is often the reason why when the brain gets injured the sense organs cease to function too. For this, see *CDA* disp. 6, q. 6, n. 10–11, t. 2, 540–4. For a readable exposition of the theory of animal spirits in connection with anatomy of the brain and 'cellular psychology' in the Middle Ages and the Renaissance, see Smith, Frixione, Finger and Clower (2012, 71–94).

[166] Although William James played down the perceptual aspect of attention and attributed an important explanatory role in the behavioral sense to attention, his definition can be applied to cognition and perception as well. For this, see Mole (2017, '1.5 William James and His Contemporaries: Deflationary Theories').

[167] James (1890, 403).

[168] For this metaphor, see Wu (2014, 15).

[169] *CDA* disp. 5, q. 7, n. 10, t. 2, 450.

[170] For this, see *CDA* disp. 5, q. 3, n. 4, t. 2, 346: 'Attendere enim [potest] anima etiam si nullam speciem recipiat, ut quando attente specto an aliquid possim audire'. As it has been said above, such attention can be exerted by a blind person whose organ has been fatally damaged.

factor already at the level of external sensation of external sensibles, at this level it is also important introspectively; while perceiving external sensibles we are at the same time aware of our perceptual operations too.[171]

Having stated these kinds of consciousness correlated to the notion of attention, this does not mean that for Suárez attention is only an all-or-nothing phenomenon. It is not only a selective tool that adverts to one task and neglects entirely another goal, or pays attention to one object without considering another. Although he agrees that a stronger stimulus often removes a weaker one (e.g. a loud blow near our ears often takes our attention away from a person standing in front of us), our attention often comes to be *divided* and related to *multiple* objects and tasks at a time.[172] Moreover, it can be divided so that in different channels it achieves different degrees of intensity and different degrees of clarity. I can be attentively and distinctly conscious of an object *X* and simultaneously be less attentively and indistinctly conscious of an object *Y*. I can be sensorily attending to an apple while being only marginally but not unconsciously of the tree the apple is hanging from on the edge of my visual field. In scholastic treatises, including Suárez's *CDA* disp. 5, q. 7, called 'Whether one and the same cognitive power can produce multiple acts at a time', this phenomenon is analyzed through the following main question: can the human/angelic intellect cognize many things (*plura*) at a time? In Suárez's treatment 'in communi' in disp. 5, in which all kinds of cognition are treated uniformly, it is broached as follows: can a cognitive power simultaneously cognize many things?[173]

As elsewhere, the doctrinal background of Suárez's answer to this query is represented by Aquinas's theory. Aquinas was also aware of the philosophical brisance of this phenomenon of attention and how it is divided. In fact, both Aquinas and Suárez agree *that* multiple objects can be cognized at a time. Nevertheless, both substantially differ in the manner of *how* these multiple objects are apprehended. Suárez allows that cognitive powers including the external senses exercise multiple acts cognizing the objects as multiple objects (*per modum plurium*). His statement is based on the theory affirming the possibility of the reception of *multiple species* in a single power at a time. Aquinas denies this view. He rejects the view that two species

---

[171] For this introspective perceptual self-awareness, see the following section.

[172] A popular psychological experiment concerning the phenomenon of divided attention is called 'dichotic listening task'. A subject is instructed to listen to two different conversations at a time. She is asked to 'shadow' only one of them. The experiment shows that the focus on the attended channel significantly filters out the information from the unattended channel. On the other hand, however, it also shows that the percipient can also recall something from the unattended channel. Not only can the subject detect the timbre, intonation, pitch and other, so to say, 'material' qualities of the voice, as the so-called cocktail party phenomenon shows, she can also hear the articulation of her name from the untracked conversation. Moreover, an auditor can improve her understanding of the parallel conversation if she (as a spy or a private eye) receives a training and schooling in such parallel listening. For dichotic listening and the coctail-party phenomenon, see Wu (2014, 16–9).

[173] Aristotle was already aware of this phenomenon. In *On Sense and Sensible Objects*, ch. 7, 447a13–5, 269, he says the following: 'There is a further question about sensation, whether it is possible to perceive two things in one and the same indivisible time or not, if we assume that the stronger always overrides the weaker stimulus [. . .]'

can co-cause two distinct cognitive acts of the same power at once. Multiple objects can be cognized only as a single object (*per modum unius*).[174]

Before presenting Aquinas's view, Suárez lays out three arguments for the least liberal and thoroughly negative opinion according to which cognition of multiple objects at a time is to be entirely dismissed. First, Aristotle himself regarded such production of multiple acts at a time as impossible: '[...] for it is possible to know many things and not to be thinking of them'.[175] It is possible to have habitually a knowledge of a plurality of things but it is not possible to think about them at once. Second, the simultaneous production of two acts at a time is not in line with the finite and limited attentional reservoir of our capacities. Third, a mental event with concurrent acts is as such contradictory. We can see multiple persons only through a single species. However, if we saw them through multiple species, we would have to see them through numerically distinct species since the people represented by them are (merely) numerically distinct too. However, considering the (Thomist) relational principle of the individuation of accidents, sc. substance, this is not possible. There cannot be two (merely) numerically distinct species inhering in a single subject.[176] Suárez leaves no doubt that this general view that *tout court* denies the simultaneous cognition of multiple objects cannot be attributed to Aquinas either. Since his presentation of Aquinas's view is fragmentary, it will be useful to focus on his tenet in greater detail.

Aquinas shows that his denial of the thesis that two distinct species can co-cause two distinct acts at once applies only to intentional species representing sensible objects of the same kind, such as colour. Specifically differing qualities, such as colour and sapour, can well be co-exemplified, for example, by milk, and the species representing them can be received in two distinct capacities and co-cause two distinct perceptual acts in two distinct powers at the same time. As said, the *status quaestionis* of *CDA* disp. 5, q.7 is related to a *single* cognitive power. Why one and the same cognitive power, such as the intellect and the external senses, cannot apprehend (according to Aquinas) multiple objects as the many things?[177] Speaking

---

[174]Speaking about the human intellect in *STh*. I, q. 85, art. 4, c., 339, Aquinas says: 'Respondeo dicendum quod intellectus quidem potest multa intelligere per modum unius, non autem multa per modum multorum: dico autem per modum unius vel multorum, per unam vel plures species intelligibiles. Nam modus cuiusque actionis consequitur formam quae est actionis principium. Quaecumque ergo intellectus potest intelligere sub una specie, simul intelligere potest: et inde est quod Deus omnia simul videt, quia omnia videt per unum, quod est essentia sua'.

[175]Aristotle, *Topics* II.10, 114b34–5, 373.

[176]*CDA* disp. 5, q. 7, n. 3, t. 2, 440.

[177]Thomas de Aquino, *STh*. I, q. 85, art. 4, c., 339: '[...] impossibile est idem subiectum perfici simul pluribus formis unius generis et diversarum specierum, sicut impossibile est quod idem corpus secundum idem simul coloretur diversis coloribus, vel figuretur diversis figuris. Omnes autem species intelligibiles sunt unius generis, quia sunt perfectiones unius intellectivae potentiae; licet res quarum sunt species, sint diversorum generum'. In *De veritate*, q. 8, art. 14, c. he develops this claim by saying that the denial of the simultaneous co-existence of diverse forms is not related to forms that are *of different genus* and related to different powers. He adds that under certain conditions also forms of the same genus, such as whiteness and blackness, can be co-instantiated in

about the intellect (this can be extended also to perception), in *Quodlibet* VII, q. 1, art. 2, c. Aquinas formulates this reason:

> But it is primarily and in principle impossible for the intellect to understand the many intelligibles simultaneously. The reason is that *the intellect as it is in its actuality is entirely, i.e., perfectly, a cognized thing*, as it is said in III De anima. This is not to be understood in the sense that the essence of the intellect becomes a thing known or its species but *that it is completely informed by the species of a thing when it understands it*. Therefore, for the intellect to cognize many things at the same time is the same as if one thing were simultaneously many.[178]

The reason for this is that the intellect and the other cognitive powers can receive only *one* intentional species at a time because *in actu* they are fully identified with a cognized thing represented by the species. If informed by the intentional species the intellect's and other powers' capacity gets entirely *filled up* with the informing or impressed species in such a way that they cease to be open to the reception of another one.

Accordingly, they can cognize multiple objects only as a single object. Therese Cory pointedly calls such apprehension 'the participated attention'.[179] This attention is related to the part-whole relation. Cognition of the many in the manner of a single thing equals the apprehension of the parts of a single whole, which only is the primary focus of attention. A single thing is a whole, whether intelligible or perceptual one, to which attention is primarily applied. The many objects *alias* the parts are cognized only as *following upon* (*ex consequenti*) this primary cognition of the whole. These parts participate in a cognized whole. As such the parts/the many are participating entities. Aquinas illustrates this 'following' with three examples:

> For instance, when the intellect understands a line, it simultaneously understands the parts of this line, as it is said in III De anima; and likewise when it understands a proposition, it

---

the same object. These conditions are twofold. (a) If they are considered as *potentially* existing, and (b) if they exist *in imperfect act* or in becoming (*in fieri*). Clearly, one subject can potentially be receptive of diverse forms belonging to the same genus. It can, diachronically, be informed by contrary forms as a persistent and common subject. Equally, if these qualities are in imperfect act, they can co-exist in the same subject at a time as well. An object, say, a wall which is being whitened, can in a certain moment be at once white and black. Aquinas is nevertheless sure that if these forms are considered in *perfect* act, it is impossible for them to inhere at once in the same subject. He points out that all intelligible species, albeit representing things of various kinds, are of *one* genus. They can co-exist in the same subject only in their potential or incomplete state. While the potential co-existence (a) is manifested in the fact that the (potential) intellect can become everything, the other co-inherence (b) is connected with the instance of a *habitual* possession of the species.

[178]My italics and translation. Thomas de Aquino, *Quodlibet* VII, q. 1, art. 2, c.: 'Sed quod intellectus simul intelligat plura intelligibilia, est primo et principaliter impossibile. Cuius ratio est, quia intellectus secundum actum est omnino, id est perfecte, res intellecta, ut dicitur in III de anima. Quod quidem intelligendum est, non quod essentia intellectus fiat res intellecta, vel species eius, sed quia complete informatur per speciem rei intellectae, dum eam actu intelligit. Unde intellectum simul plura actu intelligere primo, idem est ac si res una simul esset plura'.

[179]Cory(2014, 138–40).

simultaneously understands the subject and the predicate; and when it understands the similarity and the difference of various things, it simultaneously understands those things that are similar and different.[180]

Importantly, he applies this principle also to sensory cognition:

[T]he same should be said about the sense power, i.e., it cannot simultaneously sense many [objects] but only as a consequence, in so far as they are conceived as one [. . .] the many sensibles that are parts are unified in a whole; therefore, when the whole is sensed, at the same time the parts are sensed as a consequence: and thus the intention of the senses is not carried principally to the parts, but to the whole; if it were carried to some of these parts as to its principal sensible, [the sense] would not apprehend the other parts at the same time.[181]

This participated attention leads Aquinas to the affirmative conclusion according to which the cognitive power can 'touch' the different realities at once:

[. . .] just as a body can touch diverse things with its diverse parts due to its dimensive quantity, so the power can be applied in diverse ways to diverse things according to diverse relations when it is a power perfected in act, just as fire can simultaneously heat diverse bodies from all sides. And so too the intellect perfected by a form can be carried simultaneously to diverse things to which the representation of that form extends; and there will be many intentions with respect to which the intellect is carried, but only one with respect to the unity of the intellect and the form.[182]

The reception of a single species does not imply for Aquinas that the power cannot be intentionally related to distinct realities represented by the species. Within the representation of the single species the intellect (and other powers) can be flexible enough to direct attention to various aspects of the represented thing (whole). Following Cory, the flexibility of this directing can be explained through the distinction between two kinds of intentionality. While one kind can be called static representation of intentional species (it is a representation fixed by the received

[180] My translation. Thomas de Aquino, *Quodlibet* VII, q. 1, art. 2, c.: '[S]icut cum intellectus intelligit lineam, simul intelligit partes lineae, ut dicitur in III de anima; et similiter cum intelligit propositionem intelligit simul subiectum et praedicatum; et cum intelligit similitudinem vel differentiam aliquorum, simul intelligit ea quorum est similitudo vel differentia'.

[181] My translation. *Ibid.*, ad 1: '[S]imiliter dicendum est de potentia sensitiva, id est, quod non potest plura simul sentire, sed ex consequenti, in quantum plura accipiuntur ut unum [. . .] plura sensibilia quae sunt partes uniuntur in uno toto; unde quando sentitur totum, sentiuntur simul plures partes ex consequenti: et tunc intentio sensus non fertur ad aliquam partium principaliter, sed ad totum: quia si ad aliquam partium ferretur ut ad sensibile principale, non simul sentiretur alia'.

[182] My translation. Thomas de Aquino, *De veritate*, q. 8, art. 14, ad s. c. 2, 267: '[. . .] sicut corpus per quantitatem dimensivam potest ex diversis suis partibus diversa tangere, ita et virtus potest diversis applicari secundum diversas comparationes ad diversa dummodo sit virtus perfecta in actu, sicut ignis simul undique calefacit diversa corpora. Et ita intellectus perfectus per formam potest simul ferri in diversa ad quae se extendit repraesentatio illius formae; et erunt multae intentiones ex parte eius in quod fertur intellectus, sed una ex unitate intellectus et formae'.

species), the other can be indicated as dynamic representation (corresponding to the volitional *use* of the species).[183]

Typically, Suárez holds that this Aquinas's doctrine 'stands in need of interpretation' ('haec doctrina expositione indiget').[184] Before formulating his four conclusions he makes two preliminary remarks that are necessary for the understanding of his theory. (1) The first concerns the following 'mereological axiom': the soul's power is finite and limited. If its attentional reservoir is applied to many objects, it will get necessarily diminished in respect to each of them. Clearly, a part of an attentional whole will be lesser than the whole itself. It makes no sense to claim that one and the same capacity can be applied as a whole in full-size and equally to its two objects. If that were possible, one would have to say that the attentional resource is twice as big as it really is. (2) The second remark is concerned with the definition of perfect and imperfect cognition. Perfect cognition can be seen from two different sides: (i) from the side of the power (*ex parte potentiae*) and (ii) from the side of the object (*ex parte obiecti*). Regarding the first kind of perfect cognition, a cognitive act is perfect if it is adequate to the attentional resource of a power (*virtuti potentiae*). It is perfect if cognition explicates the whole attentional reservoir of the power. It is imperfect if it is not adequate to it or, more precisely, if the power or the soul applies only a part of its attentional resource. The second kind of perfect and imperfect cognition is related to the object of cognition. While perfect cognition is a cognition that apprehends an object in its whole cognoscibility, imperfect apprehension is a cognition that does not pick out an object's total cognoscibility.[185]

After these remarks Suárez formulates his first conclusion: 'It is impossible for a power to have two acts at the same time which are of the same perfection as far as the power is concerned'.[186] This conclusion, advocated by both Aristotle and Aquinas, is clear from what has been said above. The application of two equally perfect acts from the side of the power would have to result in the duplication of the single attentional reservoir. In the second conclusion, however, he departs from Aquinas:

> There is no contradiction in a state of affairs in which a cognitive power generates two acts at once, whether the objects are apprehended as one object or as *the many objects* [my italics].[187]

Suárez proves this conclusion by referring to common experience: we simultaneously see light, white and black colour; we simultaneously hear two sounds together with their differences. The intellect considers two things with their mutual

[183] For this distinction between these two kinds of intentionality in Aquinas, see Cory (2015, 177–83). It must be said, however, that Suárez does not refer to Aquinas's *Quaestiones disputatae de malo*, q. 16, art. 8, which is the text analyzed by Cory.

[184] *CDA* disp. 5, q. 7, n. 3, t. 2, 440.

[185] *Ibid.*, n. 4, 440–2.

[186] *Ibid.*, n. 5, 442: 'Impossibile est quod una potentia habeat duos actus simul perfectos ex parte potentiae'.

[187] *Ibid.*, n. 6, 442: 'Non repugnat unam potentiam cognoscitivam habere simul duos actus cognoscendi, sive res cognitae cognoscantur per modum unius, vel plurium'.

agreements and differences. But how does he argue for the conclusion that we can produce two or more acts apprehending objects as the many things and not in the aforesaid manner of 'participated attention'? The main argument can be called 'Argument from a more complicated case to a less complicated one'. If the intellect can consider two things at once with their comparison or ordering, it can *ipso facto* consider them *without* this comparison and ordering. If the intellect and the senses can comparatively apprehend multiple objects in their similarity and difference, they must be *a fortiori* able to apprehend them absolutely, that is, without this comparison and ordering. Clearly, absolute cognition of the *relata* must be prior to the comparative and relational cognition of their similarity and difference. Suárez shows that attention is flexible and has a latitude (*latitudo*). This or that cognitive power can apply its attentional reservoir either totally (*totaliter*), or partially (*ex parte*) to a particular cognitive act. It is not determined by the object so that it can receive only a single species representing it. A particular faculty can cognize the object *X* with one part of attention and the object *Y* with another.[188] It is in the third conclusion where Suárez formulates his negative attitude to Aquinas most clearly: 'One and the same cognitive power can be operative through different intentional species at a time'.[189] We distinctly see white and black colour. However, we can sense them *only* through the different sensible species. In *CDA* disp. 5, q. 2, Suárez espouses this view by saying that two distinct and most contrary things, such as white and black colour, cannot be represented by a single species. The vestigial sensible species of colour cannot eminently represent contrary qualities.[190]

In his last conclusion Suárez refers to the second meaning of perfect and imperfect cognition which has to do with perfect and imperfect cognition from the side of the cognized objects:

> It can happen that a cognitive power at once perfectly cognizes the multiple objects from the side of objects, that is, it can happen that by cognizing them it apprehends them in the whole field of their cognoscibility in so far as these are attainable by such a power.[191]

Although a cognitive power cannot elicit two perfect operations *ex parte potentiae* at a time, if it is perfect enough it can produce multiple cognitive acts that are perfect in relation to the overall cognoscibility of the objects. Suárez admits that this occurs especially in angelic cognition. Its *latitudo* is wider than that of humans. Although divided between multiple objects, the angelic intellect can do epistemic justice to each of them at the same time. Referring to *De angelis*[192] he states that angels can immediately, that is, without intentional species (one of the few cases in Suárez's

---

[188] *Ibid.*, 442–4.

[189] *Ibid.*, n. 7, 444: 'Eadem potentia cognoscitiva potest simul operari per diversas species intentionales'.

[190] *CDA* disp. 5, q. 2, n. 9, t. 2, 308.

[191] *Ibid.*, n. 8, 446: 'Contingere potest, ut potentia simul plura cognoscat perfecte ex parte obiectorum, id est, cognoscendo illa quantum cognoscibilia sunt a tali potentia'.

[192] Francisco Suárez, *De angelis*, book 2, ch. 37, n. 2, 350 (for this reference, see *TDA* book 3, ch. 7, n. 2, 641).

philosophical system where the intentional species is dispensable) cognize themselves while at the same time being able to perfectly cognize natural things through species innate to them. Due to the significant limitation of the human attentional reservoir this occurs rarely or never in our intellects. Dividing our attention results not only in the imperfection of cognition from the side of power, but almost always also in imperfection from the side of objects. When divided our attentional resources can hardly be fully adequate to the multiple objects under consideration.[193]

All these conclusions make *a fortiori* clear that the abovementioned two arguments for the overall denial of the concurrent production of acts apprehending multiple objects are false. Suárez also rejects the third argument 'from the impossibility of the co-instantiation of numerically distinct species in the same subject'. He rejects Cajetan's reply distinguishing between the real and the representative being of the species. According to Cajetan, while two merely numerically distinct species in their real being cannot inhere in the same subject, they can do so in their representative being. As stressed in Sect. 3.2, the distinction between these two aspects is not applicable for Suárez. He is sure that there is no inconvenience related to the co-inherence of two merely numerically distinct species in the same subject.[194] Contrary to Aquinas and Thomism, for him it is not the subject that individuates accidents but their whole entity (*entitas tota*).[195]

It may be concluded that – contrary to Aquinas – for Suárez the volitional use of intentional species need not occur only at a later stage, that is, after the impressed species is imprinted in the power conceived as a whole within which attention can be shifted. Importantly, it can come even at an *early* stage, that is, before this reception. A cognitive power is not obliged to undergo the *full* imprint and is not forced to operate only within the borders of the representation given by the informing species. Its attention can be applied at an earlier stage too.[196] Through its attention a cognitive power can use the species (generating efficient causality) inadequately so that it can leave a 'space' for an epistemically relevant imprint coming from another species. Accordingly, we can say that Suárez in a way subordinates the species to the attentive power. This is also why (as said in Sect. 3.4) only the power and the species together constitute the whole instrumental cause of a perceptual act for

---

[193] *CDA* disp. 5, q. 7, n. 8, t. 2, 446.

[194] *Ibid.*, n. 10, 448–50.

[195] For the principle of individuation of accidents, see *DM* disp. 5, s. 8, n. 15, t. 1, 194.

[196] In *De angelis*, book 2, ch. 37, nn. 20–23, 355–6 Suárez strives to harmonize his theory with Aquinas's view by distinguishing the various meanings of what Aquinas could have meant when he endorsed the impossibility of cognizing multiple objects in the manner of many objects. As the last interpretation, which coincides with Suárez's view, he mentions one according to which this phrase means 'the production of more acts of which a single one is adequate to the attentional resource of a given power'. Clearly, as stated in the first conclusion, Suárez would not admit that such cognition of multiple objects in the manner of many objects is possible. Nevertheless, despite this he is quite pessimistic about the possibility of ascribing this to Aquinas. In *De angelis*, book 2, ch. 37, n. 20, 355 he says: 'I do not dare to say that Saint Thomas spoke in this sense' ('Sed non audeo dicere, divum Thomam in hoc sensu fuisse locutum'.).

Suárez. From this point of view, this claim about the two partial instrumental causes composing the whole instrument makes perfect sense. When compared to Aquinas, Suárez's overall emphasis on substantial attention can be regarded as a manifestation of the emphasis on the human subjectivity and a voluntarist feature of his thought. Contrary to Aquinas, who stands close to what is today called *Late Selection Theory*, Suárez's view can be assessed as an example of *Early Selection Theory*.[197]

3.7 Perceptual Self-Awareness

Suárez's view of the soul's attention gives rise to his teaching about perceptual self-awareness or, more precisely, about the external senses' perception of their operations. Not only that I can attentively perceive an apple and less attentively a tree from which the apple is hanging, I can also attentively perceive an apple and less attentively be aware of the act through which this apple is sensed. This kind of less attentive self-perception or apprehension of one's own operations can be regarded as one of the most representative manifestations of Suárez's theory of the soul's substantial attention. It would not make sense to speak about the attentional sensorial focusing on the apple without being at the same time aware of this directing. The access consciousness is closely related to the introspective consciousness. Suárez's theory of thing-consciousness can hardly be separated from his theory of act-consciousness. Cognitive operations including those of the external senses are *bipolar*. They are conscious both of the external things and of their acts through which the sensibles are apprehended.

Since a cognitive act is the most accessible intramental item and provided that all other entities such as habits, powers, intentional species and the soul can be apprehended only discursively through the *intellect* on the basis of a knowledge of these acts, it is only a perceptual operation, whether of the external or internal senses, that, properly speaking, constitutes the object of *CDA* disp. 6, q. 4, in which the topic of sense perception is treated.[198] Beside the ontological query about the object of

[197] Of course, this distinction between early and late selection theories in contemporary psychology is not the distinction between the stages of which one precedes and the other one follows the reception of sensible species. It concerns the issue of whether selection filters basic (timber, pitch, intonation of voice) or categorial (meaning or identity of voice) features. However, employing a structural analogy, this doctrinal distinction can be *cum grano salis* employed also in the description of the doctrinal difference between Aquinas's and Suárez's theories of attention. For these two theories, see Wu (2014, 14–21).

[198] For Suárez's indirect intellectual knowledge of habits, powers, intelligible species and soul, see *CDA* disp. 9, q. 5, t. 3, 169–78. For an analysis of intellectual self-awareness with an emphasis on his theory of cognition of the acts through intelligible species, see Perler (2014a).

self-perception, an important question is also related to *how* perceptual acts are cognized.[199] Are they grasped by an external sense, or only by an internal sense or by both in a different way? If the last holds, is there any difference in the perfection of their self-awareness? Moreover, are sensorial acts apprehended by their proper species, or not?

Before presenting two theories that substantially differ in how they distribute perceptual self-awareness among the powers, Suárez premises two remarks. At first, he formulates a distinction between two kinds of self-awareness that are central both in early modern and in present-day philosophy. (a) An act can be cognized *properly* or, as he says in *TDA*, through formal reflection (*reflexio formalis*), that is, intentionally as the object (*quod*) of a higher-order act. A first-order act representing an apple can be intentionally and deliberately apprehended through a distinct second-order act that reflexively turns to the lower-order act. If the first-order act is apprehended in this way, it is known *expressly* and *directly* in a designated act (*in actu signato*).[200] The first-order act gets cognitively 'stamped' by the second-order act in the same way as my joy is 'stamped' by the proposition 'I am happy'. In contemporary philosophy of mind, this kind of self-awareness as an instance of proper reflexivity is called 'Higher-Order Self-Awareness'. (b) A perceptual act that objectively terminates in an apple can be cognized also *improperly*, *implicitly*, and *indirectly* in an exercised act (*in actu exercito*). It can be immediately apprehended by the same first-order act, namely as a 'quo', through which an external object is apprehended. When I perceive an apple I am indirectly and experientially aware also of this act through which I see the apple. This kind of cognition is similar to my expression of joy through the spontaneous outcry 'Yipee!'. This expression of joy is not expressed propositionally. In contemporary philosophy, this manner of self-awareness that takes place automatically and spontaneously is called also 'Same-Order Consciousness'.[201] In his later *DM* and *TDA*, Suárez introduces another specification of this improper reflection that was to become a famous 'label' in seventeenth-century scholastic and early modern philosophy in general. In those treatises he refers to this distinction as to a *virtual* reflection (*reflexio virtualis*).[202]

[199]For these issues connected with self-awareness supplemented by an epistemological enquiry into the certainty of self-knowledge, see Perler and Schierbaum (2014, 14–7).

[200]For this term 'designated act', see *CDA* disp. 6, q. 4, n. 12, t. 2, 516.

[201]For this distinction in contemporary philosophy of mind as connected with two rival theories of introspective consciousness, see Kriegel (2007), who is regarded as one of the foremost advocates of Same-Order (Monitoring) Theory of Consciousness. To quote Kriegel, 'When I have a conscious experience of blue, I am aware of my conscious experience. But the awareness is not an extra mental act, which occurs *in addition to* the experience. Rather, the awareness is *inherent in* – it is built into – the experience' (*ibid*., 365). Higher-Order Theory of Consciousness, on the contrary, claims that 'a mental state M of a subject S is conscious iff S has another mental state M*, such that M* is an appropriate representation of M' (*ibid*., 361). For a defence of this theory, see, among others, Armstrong (1981). For further references to representative scholars and literature related to both theories, see Rode (2015, 13–6).

[202]See *TDA* book 3, ch. 11, n. 1, 652–3; *DM* disp. 25, s. 1, n. 39, t. 1, 909–10. For the denomination 'virtual reflection' as intrinsic to the will's elicited act, cf. also Suárez, *De voluntario et*

Beside this first preliminary distinction, the second introductory remark concerns the distinction between the apprehension of an act's existence and the cognition of its quiddity. While by means of the former we can know only that something (an act) exists, according to the latter we know its essence. Suárez is clear that perceptual self-awareness cannot be the cognition of an act's quiddity; due to its imperfection it can involve at most its existence.[203]

Suárez introduces two doctrines copying conflicting statements from Aristotle's *On the Soul* and *On Sleep and Waking*. In *On the Soul* III.2 the Stagirite says:

> Since we can perceive that we see and hear, it must be either by sight itself, or by some other sense. But then the same sense must perceive both sight and colour, the object of sight. So that either two senses perceive the same object, or sight perceives itself. Again, if there is a separate sense perceiving sight, either the process will go on *ad infinitum*, or a sense must perceive itself. So we may assume that this occurs with the first sense.[204]

Aristotle starts with evidence that there is a self-awareness in our visual and auditive experiences. However, there are two ways to explain this. One can say either that there is a higher sense, sc. the common sense, that cognizes the first-order act or that the external senses themselves exercise this self-cognition. Since the first option leads to an infinite regress,[205] in *On the Soul* Aristotle seems to opt for the view

---

involuntario, disp. 1, s. 1, nn. 5–6, 160–1. As Knebel (2016, 80) states, Suárez was the first to coin the notion of a 'virtual reflection'. However, even though this notion is regarded as 'an index fossil of Suarezianism' (Knebel 2016, 81) and regardless of how it spread to all religious schools in Baroque scholasticism and influenced Cartesian thinkers headed by Antoine Arnauld (1612–1694) and expressed Descartes's mind, Suárez does not seem to place a special emphasis on this term. Recently it has been argued in Schmal (2020, 724–5) that the expression 'reflexio virtualis' is to be understood primarily in connection with the notion of the virtual distinction, which is the kind of distinction that exists between two or more perfections (grades) within a single thing. It is by virtue of virtually distinct perfections that a thing which is not structured *realiter*, such as the Sun, is capable of producing really distinct effects, such as heating, illuminating, and drying. Accordingly, it is due to the virtually distinct perfections inherent in the first-order act that this act is capable of delivering information not only about the external objects but also about itself. As Schmal (2020, 725) shows, this act is 'capable of producing two actually distinct effects through its virtually distinct perfections. These two effects – the knowledge of the primary object and that of the subject performing the act – have common roots in the cognitive act itself which is able, by its nature, to produce them'. I agree that this interpretation makes good sense, especially if we take into account the reified character of cognitive acts-qualities in Suárez's view. Nevertheless, in this book I defend the claim (which, after all, might well be compatible with that of Schmal) that this concomitant and spontanenous self-awareness is, at its ontological level, made possible by the soul's causal proximate influence (through the powers) in the operations. At any rate, this indirect self-consciousness is one of the main reasons why I interpret Suárez's philosophy of perception as Aristotelian subjectivism (see further Sect. 6.3). For this implicit reflection conceived as consciousness with its medieval roots in Durandus of Saint-Pourçain, cf. also Knebel (2014).

[203] *CDA* disp. 6, q. 2, n. 2, t. 2, 502–4.

[204] Aristotle, *DA* 425b13–7, 145–7.

[205] For the same objection from infinite regress against Ockham's higher-order theory of consciousness, similar to the one mentioned by Aristotle, which was formulated by Walter Chatton (ca. 1290–1343), see Brower-Toland (2014, 204–8).

according to which sight and hearing themselves exercise self-cognition. In *On Sleep and Waking* Aristotle formulates a different position, though:

> Now every sense has both a special function of its own and something shared with the rest. The special function, e.g., of the visual sense is seeing, that of the auditory, hearing, and similarly with the rest; but there is a common faculty [sc. the common sense] associated with them all, whereby one is conscious that one sees and hears (for it is not by sight that one is aware that one sees [. . .])[206]

In this passage it is the common sense that perceives the acts of the (lower) external senses. The only function that remains to the external senses is to apprehend the proper and the common sensibles. As such these seem to be unaware of their operations. Sight sees only colours and coloured extended objects having shape, hearing hears sounds and motion. Perceptual acts stand outside the ambit of their proper sensibles and of the common sensibles and, therefore, can be apprehended only by a distinct sensorial capacity, namely by the common sense.[207]

In line with the distinction between the abovementioned two kinds of self-awareness, Suárez formulates the first conclusion: 'No sense power can properly and expressly cognize its own act'.[208] This perfect and reflexive self-cognition exceeds the capacities of the material powers. In his argument Suárez refers to the Neoplatonic justification advocated by Proclus,[209] namely that no extended entity can turn to itself since no part of an extended whole can touch or be applied to itself.

---

[206] Aristotle (2000), *On Sleep and Waking*, 455a13–8, 327.

[207] These two opposing views of Aristotle's are mirrored in the views of Aquinas who oscillated between these two positions at various points of his carrier – for this see Cory (2021). In the early III *In Super Sent.*, d. 23, q. 1, art. 2, ad 3 Aquinas says: 'Hoc autem non potest esse ita quod aliqua potentia utens organo corporali reflectatur super proprium actum, quia oportet quod instrumentum quo cognoscit se cognoscere, caderet medium inter ipsam potentiam et instrumentum quo primo cognoscebat. Sed una potentia utens organo corporali potest cognoscere actum alterius potentiae, inquantum impressio inferioris potentiae redundat in superiorem, sicut sensu communi cognoscimus visum videre'. In *STh.* I, q. 78, art. 4, ad 2, 256 a similar view is advocated: '[. . .] a quo [the common sense] etiam percipiantur intentiones sensuum, sicut cum aliquis videt se videre. Hoc enim non potest fieri per sensum proprium, qui non cognoscit nisi formam sensibilis a quo immutatur; in qua immutatione perficitur visio, et ex qua immutatione sequitur alia immutatio in sensu communi, qui visionem percipit'. However, for the opposite view, see *De veritate*, q. 1, art. 9, 29: '[. . .] quamvis enim sensus cognoscat se sentire, non tamen cognoscit naturam suam, et per consequens nec naturam sui actus, nec proportionem eius ad res, et ita nec veritatem eius'.; *De veritate*, q. 10, art. 9, 328: '[. . .] sicut videmus quod visus primo dirigitur in colorem, sed in actum visionis suae non dirigitur nisi per quandam reditionem dum videndo colorem videt se videre'.

[208] *CDA* disp. 6, q. 4, n. 4, t. 2, 504: 'Nulla potentia sensitiva cognoscit proprie et actu distincto suam propriam operationem'.

[209] In Proposition 14 of his *Elements of Theology* Proclus says: 'All that is capable of reverting upon itself is incorporeal'. He (*The Elements of Theology*, 18–9) explicates this as follows: 'For it is not in the nature of any body to revert upon itself. That which reverts upon anything is conjoined with that upon which it reverts: hence it is evident that every part of a body reverted upon itself must be conjoined with every other part–since self-reversion is precisely the case in which the reverted subject and that upon which it has reverted become identical. But this is impossible for a body, and universally for any divisible substance [. . .]'

There can always be only one part of an extended whole applied to a distinct part of this whole, as in the case of a paper that is turned back on itself. But Suárez justifies this argument by saying that no material power can perfectly reflect upon itself because its act is not included in the overall ambit of its sensible object and no power can be intentionally carried outside its own formal object. Neither the external nor the internal senses can immediately revert to their own acts. Neither of these can be perfectly reflexive. Moreover, their operations are, ontologically speaking, nobler than their proper sensible objects. Contrary to the intellect, whose *radius operandi* is not limited since its object is being as such (*ens ut sic*) and thus can be related to all beings as to its proper objects,[210] the senses' objects are generically limited. Only a spiritual power, such as the intellect, can perfectly and reflexively cognize its own acts and through their apprehension inferentially know other entities such as habits, powers, species and the soul. Only the intellect can cognize the quiddities of these acts and inspect them more thoroughly. Since the external senses cannot perceive the sensible species and the sense organs, they *a fortiori* will not be able to sense their own acts which are ontologically higher than the organs.[211]

In opposition to Aristotle's *On Sleep and Waking*, Suárez formulates his second conclusion: 'No sense through its own act can perceive an act of a distinct sense through its own sensible species and as part of its proper object'.[212] As a higher sensory power the common sense cannot perceive the external senses' acts through their own species. If the common sense could *per impossibile* do this, the external sense would have to imprint two sensible species in the *sensus communis*, namely the species of white colour and the species of seeing this white colour. Consequently, these two species would have to trigger two distinct acts in the internal sense power, since one and the same power can elicit two distinct acts related to two distinct objects. However, this is at variance with our experience. Contrary to Scotus,[213] Suárez argues that if the common sense apprehended an act of the external senses through its own species, the cognition would have to be perfect in the abovementioned 'quidditative' sense. The common sense would have to be able to form a quidditative cognition of the first-order act. But if this were the case, the common sense would have to conceive the quiddity of perceptual acts. Accordingly, we would have to be able to explicate what this or that vision is like based on mere inward seeing. Again, this does not accord with the significant imperfection of the sensory powers and their objectual limitation to their proper sensible objects.[214]

While the attribution of proper reflection is denied both to the external and the internal senses, concomitant self-awareness is necessary at all cognitive levels, that is, at the levels of the external senses, of the internal senses and of the intellect. In his

[210]For the adequate object of the intellect, see *CDA* disp. 9, q. 1, t. 3, 67–78.

[211]*CDA* disp. 6, q. 4, n. 4, t. 2, 504–6.

[212]*Ibid*., n. 6, 508: 'Nullus sensus potest proprio actu percipere actum alterius sensus per propriam illius speciem, et tamquam partem proprii obiecti'.

[213]For this difference from Scotus, see Sect. 5.1.

[214]*Ibid*., n. 6, 508–10.

third conclusion, Suárez makes it clear as follows: 'Every sense power somehow perceives its own act but not through a reflection but imperfectly and in a quasi-exercised act'.[215] This self-perception is built into all cognitive acts, including acts of the external senses. Albeit a veridical perception of extramentals constitutes a basic point of departure for Aristotelian-scholastic philosophers including Suárez, this does not mean that beside this apprehension of external objects we do not simultaneously *experience* that we sense.[216] By the mere exercise of first-order attentional acts we *ipso facto* are non-intentionally and without a mediating species aware also of the existence of the concomitant acts. This concomitant self-awareness accounts for what can be called Suárez's subjective facet in his philosophy of perception.

Suárez seems to be also implicitly claiming that we cannot be wrong in this internal experience. Although, as we shall see more in Sect. 3.9, God can insert a sensible species in the organs and represent a state of affairs to our visual power that is not the case here and now, He cannot insert vital acts in us and cannot deceive us in our cognition that the vital sensory acts of a distinct person are ours. As said in Sect. 2.2, this would be at odds with their essentially vital character. By concomitantly perceiving these acts I simultaneously cognize that these acts are extant and that they are mine. As has been repeatedly said, the vital acts of which I am necessarily conscious can be (co-)caused only by me, that is, from within, or from my soul which is their intrinsic principle. Even in beatific vision, in which the blessed in heaven seem to be most cognitively passive because assisted by the glorified light (*lumen gloriae*) than in any other cognitive situation they cannot be inserted to my powers from outside by God. This conclusion about concomitant self-awareness thus can be regarded as a systematic consequence of Suárez's teaching on the soul's vital attentional presence in the acts of the various powers. This presence is what causes the fundamental internal experience of one's own acts.[217]

Having shown that the perceiver indirectly or concomitantly experiences her own perceptual acts of both senses through both powers, what is the situation like with the internal sense and its cognition of the operations of the lower external senses? How does the common sense cognize the lower-order acts of the external senses? Although this topic pertains largely to Chap. 5, in which the interior sense, its object and the origin of its species are *ex professo* treated, due to the tight interconnection of these kinds of self-cognition it is necessary to complement the exposition of the

[215] *Ibid.*, n. 7, 510: 'Omnis sensus percipit aliquo modo actum suum, non per reflexionem, sed imperfecto modo et quasi in actu exercito'.

[216] Christian Rode regards Suárez's same-order theory of self-consciousness in the context of the unity of the soul (see also Sect. 2.4) as 'tatsächlich Bahnbrechendes' (Rode, 2015, 376). Rode points out the pivotal factor of the soul's attention which he characterizes an 'innere Erfahrung' of cognitive acts. Like Marleen Rozemond (2012), Rode stresses that Suárez's theory transcends Aristotelian psychology based on cognitive division of labour and that his theory of the soul's substantial attention is grafted onto traditional Aristotelian philosophy.

[217] The simultaneous existence of this double apprehension, one related extrospectively, another introspectively, is not contradictory because sunlight also represents both colours and itself. For this illustrative comparison, see Suárez, *De divinia substantia*, book 2, ch. 18, n. 9, 115.

external senses' perceptual awareness by a brief presentation of the common sense's cognition of these acts already at this place.

As it has been said in the second conclusion, the acts of the external senses do not imprint their own species on the internal sense. The common sense does not have a higher-order cognition of these acts. Such self-awareness would exceed the capacity of the material powers including the common sense. However, are the acts of the external senses apprehended by the common sense after all in a special way? In the last conclusion of *CDA* disp. 6, q. 4, Suárez shows that they are. The interior sense cognizes the operations of the external senses in a special way that is different from concomitant apprehension. There must be such an apprehension because both human and nonrational animals have a memory not only of the external objects but also of the perceptual acts through which they apprehended them. The apprehension of these acts must somehow arrive at the interior senses because both rational and perfect nonrational animals have a memory of these acts. Only with a memory of past acts can they be cognitively complete in their sentient nature.[218] A perfect animal must know whether an act was beneficiary or harmful for it. It must know this in order to pursue or escape the object producing such acts in the future.[219] Without this cognition it could hardly be able to preserve its life. But how can this self-awareness be spelled out? If there are no special species representing these acts in the internal sense, how can humans and higher animals remember them? Suárez agrees that sensory memory is related to the acts of the external senses. However, he is also clear that it cannot be regarded as an effect of the common sense's special capacity somehow exceeding the power of the external senses. The common sense cognizes these acts only through a *modified* species of the external sensibles themselves.[220] Where does the species of the interior sense representing a red colour of a certain size with a shape get this representational modification related to acts of the external senses? Employing again the ontology of modes for him typical Suárez proposes an analogical explanation with the perception of distance according to which the sensible species received in an organ is modified according to the distance from which it has been multiplied.[221] The sensible species multiplied from the sensible objects that are near to the beholder affect her sight more strongly than those that are emitted from more remote objects. The analogy can be formulated as follows: as sight perceives the distance of an object because the sensible species emitted by this object are modified in this or that way, so the interior sense's species represents the acts of the external senses since

[218]For the sensory memory in perfect and imperfect animals as related to the query about the subject (s) of the internal senses, see Sect. 5.2.2.

[219]*Ibid.*, n. 8, 510–2.

[220]*CDA* disp. 6, q. 4, n. 8, t. 2, 510: 'Sensus interior speciali modo cognoscit operationem sensuum externorum, non per proprias species, sed per species sensibilium externorum, quasi modificatas in ipsis sensibus externis'.

[221]For this explanation of the perception of distance through a *modus*, see Sect. 4.2.4.

it is modified after the model of the external senses's operation. As we shall see in Sect. 5.1, the acts of the external senses constitute a quasi-exemplary cause for the interior sense's formation of its species. Accordingly, the common sense is not to be regarded as a power that imports an additional cognitive aspect to the external senses' cognition. It merely 'copies' the apprehension of the lower sense.[222]

3.8 Sensible Objects, Perceptual Judgment and Errors of Sense

The subjective element in philosophy of perception comprises not only concomitant self-awareness, but also the existence of illusory perceptions. The sensible species received in a power often encodes an information that does not perfectly match the sensible from which it was emitted. Following a rich portfolio of examples of perceptual illusions adduced already by ancient *academici*, Peter Auriol laid an emphasis on the methodological significance of nonveridical perceptions in the theory of perception and determined the later development in the scholastic philosophy of perception.[223] In his *Scriptum on Book I of the Sentences*, Auriol presents a series of visual experiences to argue for the conclusion that acts of the sight place the perceived objects in intentional being, or, as he says, in 'apparent being' (*esse apparens*). A stick looks bent in water; iridescent colours appear on a dove's neck in particular light; the trees on the coast seem to be moving for an observer in a sailing boat; things reflected by a mirror appear to be behind the mirror; afterimages of sunlight arise from looking at the sun; a fire circle in the air made by revolving a

[222]*CDA* disp. 6, q. 4, n. 8, 512. For a limitation of the interior senses's cognitive functions (as compared to Aquinas) in Suárez, see Chap. 5.

[223]No doubt, the renaissance of Auriol ('the Eloquent Doctor') in Second Scholasticism was stimulated by Cardinal Constantino Sarnano's edition of his *Commentariorum in primum librum sententiarum pars prima* (1596). Mastrius calls Auriol the 'Promptuarium Neotericum' ('a Repository of Ideas for Recent, to wit Jesuit, Authors'). See Bartholomeus Mastrius, *Philosophiae ad mentem Scoti cursus integer*, vol. 4, *Metaphysica*, pars prior, disp. 6, q. 1, n. 33, 234. Sven Knebel (2009, 330) says that Auriol was very much alive in mid-seventeenth century but that 'Aureol's influence over the centuries is a matter we are certainly not yet prepared to ascertain adequately'. Since this edition was published long after Suárez had finished the first and complete version of his *CDA* and since it is unlikely that he employed Auriol's manuscripts, he must have known him only indirectly. It is highly probable that he became acquainted with Auriol from the *Defensiones theologiae* of his doctrinal antipode, Johannes Capreolus (ca. 1380–1444). In the majority of remarks related to Auriol in *CDA* and *DM* Suárez quotes him from Capreolus's text. For Capreolus's extraordinary familiarity with Auriol, see also Pierre Bayle, *Dictionnaire historique et critique*, vol. I, 401–2. This and the next section give evidence that Suárez was also familiar with Auriolian views, even though he did not explicitly mention him in *CDA*.

burning stick; double vision; and others.[224] In all these examples either a real thing appears to be different from the way it really is or a nonexistent thing appears to be existing at the moment of sensation.[225] Without engaging in an extensive discussion of Auriol's *esse apparens*, that is, whether its postulation leads to perceptual representationalism and phenomenalism, or to non-naïve direct realism, and whether the ontological status of this *esse* amounts to intramental or extramental being,[226] the abovementioned illusory perceptions constitute a theoretical challenge for Suárez especially in relations to these questions: how do these illusory perceptions arise? What is their cause? What are the general conditions of veridical sensation, being the normative case of perception for Suárez? Does the basic reliability of our external senses differ in respect to the different kinds of sensibles? In order to present Suárez's replies to these questions two theories constituting the immediate background of his theory must be laid out, though. One is represented by the Aristotelian theory of three kinds of sensibles, the other is related to his doctrine of perceptual judgment deriving from his theory of the substantial attention of the soul. The first question poses two further sub-questions bearing upon the manner of the perception of common and incidental sensibles. While Suárez's replies to the question regarding the perception of the common sensibles will be presented in this section, the second query will be left for Sect. 5.3.1 as it is directly connected to the intellect.

(1) As an Aristotelian, Suárez distinguishes three kinds of sensibles. While two are per se sensible objects that directly affect the power, the third are per accidens (incidental) sensibles that, as being accidentally conjoined to per se sensibles, do not affect the power directly. Yet, as sensibles they are also *sensed* together with the per se sensibles.[227] *Sensibilia* per accidens can be twofold. First, they can be sensibles per accidens for *all* sensory modalities including the external and internal senses. An example of this is the material substance. A material substance does not immediately affect the powers. It affects them only through its accidents.[228] Second, some incidental sensibles can be such only for *some* senses. The sweetness of milk is not a sensible per accidens for taste but only for sight. The odour of an apple is not an

---

[224]Petrus Aureolus, *Scriptum super primum Sententiarum*, vol. II, d. 3, s. 14, art. 1, n. 31, 696–8. In a parallel passage Auriol presents five experiences, some of which are not indicated in the passage considered in the third distinction. For them, see Aureolus, *ibid.*, *Prooemium*, vol. I, s. 2, art. 3, nn. 81–6, 198–9. For a translation of the second section of the Prologue, see Pasnau (2002b, 178–218).

[225]While the first kind of illusory perception is treated in *CDA* disp. 6, q. 3, the second one, largely at the supernatural level of a hypothetical divine intervention, is analyzed in *CDA* disp. 6, q. 5. The latter will be treated in the next section.

[226]For these problems in Auriol, see Tachau (1988, 85–112), Pasnau (1997, 69–76), Lička (2016), and many other authors. For other secondary literature on Auriol's theory, see also the next section.

[227]If they were not sensed, they could not be counted among the sensibles. This is also why God cannot be considered to be an incidental sensible. He can be naturally known only by an intellectual inference from sense experience. See *CDA* disp. 6, q. 1, n. 3, t. 3, 456. For Suárez these three sensible objects exhaust all the possibilities of how sensibles can affect the external senses.

[228]As Aristotle states in *On the Soul*, the son of Diares is an incidental sensible because he is accidentally united to this white colour, which only is sensed per se. See Aristotle, *On the Soul* II.6, 418a20–3,103.

incidental sensible for smell but for taste. Following Aristotle, sensibles per se are twofold. They are the proper sensibles and the common sensibles. Embracing the teleological 'coaptatio' of the external senses to the proper sensible objects irreducible to 'quantitative' common sensibles,[229] Suárez distinguishes between sensibles per se *primo*, or the proper sensibles, and sensibles per se *secundo*, or the common sensibles. While the former affect a power immediately through their proper species, the latter do this only indirectly through the species primarily representing the first sensible object. It is a proper sensible to which a particular power is directed. It is also a proper sensible that primarily specifies a particular sensory modality. Sight is teleologically directed to colours (and light); hearing is correlated to sounds; smell is proportioned to odours etc. All these modalities are (transcendentally) directed to these sensibles. They would be ordained to them even if they did not exist.[230] In contrast to the common sensibles, the proper sensibles can be perceived only through a single sense. Sight is not primarily directed to the detection of the common sensibles, of which (*pace* Aristotle) there are six (or five), that is, magnitude, shape, motion, rest, number and unity.[231] Although size and shape constitute the proximate subject of the proper sensibles (white colour cannot exist without a shaped quantity as its subject), they are not the proper objects of any particular modality.[232]

While the first sensible per se is cognized directly through the proper species impressed on the power and substance is distinctly cognized only inferentially by the intellect, how is it with the 'medial' common sensibles, such as size and shape? Suárez mentions three theories related to the question of the manner of their perception. (a) According to the first, these sensibles are common since they are the sensibles per se only for the common sense. As advocated by Alhacen (Ibn al-Haytham),[233] they generate no imprint in the external senses but only in the

[229]As regards this *coaptatio*, see *CDA* disp. 3, q. 2. For the irreducibility of the proper sensibles to what early modern philosophy called primary qualities in Suárez, see Chap. 4.

[230]For a parallel *coaptatio* of the soul to its body as that which does not have to exist, see *CDA* disp. 1, q. 3, n. 11, t. 2, 104.

[231]For six common sensibles, see Aristotle, *DA* III.1, 425a15ff, 143; for five, see *DA* II.6, 418a17–8, 103. In the second passage the Stagirite does not name unity, which he thus takes to be reducible to number of which it is the principle. For this explanation, see *CDA* disp. 6, q. 1, n. 9, t. 2, 462.

[232]For this division, see *CDA* disp. 6, q. 1, nn. 1–2, t. 2, 454–6; n. 11, t. 2, 466–8.

[233]Instead of five/six common sensibles, Alhacen lists even 22 of them. Apart from the ones mentioned by Aristotle he names also distance, spatial disposition, corporeity, continuity, discontinuity or separation, roughness, smoothness, transparency, opacity, shadow, darkness, beauty, ugliness, similarity and difference. In the third chapter of his second book of *De aspectibus*, 3.16–7 (in Smith (2001b), vol. 2, 431), he says that these 'intentions' are apprehended by the faculty of discrimination (*virtus distinctiva*): '[. . .] not everything that is perceived by sight is perceived through brute sensation; instead, many visible characteristics will be perceived through judgment and differentiation in conjunction with the sensation of the form that is seen. However sight does not possess the power to differentiate; the faculty of discrimination differentiates these properties'. All these (in Alhacen's parlance) 'intentions' are discriminated by what he calls the final sensor (*ultimus sentiens*), or the faculty of discrimination, which he locates at the front of the brain. Despite only implicit traces of Aristotelianism in Alhacen–'[. . .] Alhacen felt no obligation to cite sources by name', Smith (2001a, vol. 1, xxv)–(and also Suárez in his exposition of this view)

common sense.[234] In line with the general tendency to reduce the number of the cognitive functions of the interior sense's apprehension,[235] Suárez is quick to reject this view by saying that the interior senses can perceive only what has already been sensed by the external senses. With the exception of composed images, such as chimeras and golden mountains, and action-oriented intentions, such as those of enmity, the internal sense can perceive nothing beyond the 'data' acquired from the external senses. The common sense does not possess a capacity to detect 'hidden' intentions that have not been detected earlier by the external senses.[236] (b) The second view derives from a metaphysical distinction between two grades (*rationes*) in the external senses.[237] While the first is a *ratio* that is proper only to a particular sense, the second is a *ratio* that is common to all the external senses. On this view, the common sensibles imprint their own species on the external senses but these do not receive them insofar as they are such, namely *qua* sight and hearing, but only insofar as they are external senses in general. This is why the species of the common sensibles received in them under the common metaphysical aspect of 'being an external sense' is called common. Suárez is not hesitant to reject this view as well. Common sensibles are not common according to universal abstraction and predication but according to a *real* community according to which things, such as size, shape, etc. are cognized by (at least) two external senses. It is sight *qua* sight, not *qua* an external sense, that perceives the common sensibles.[238] (c) The third view holds that the common sensibles also produce their own likenesses that are received in the external senses. The only difference between these two is that the species of the proper sensibles are received in a single sense, while those of the common sensibles are received in multiple modalities.[239] Suárez's negative approach to this view derives from his emphasis

---

compares the *virtus distinctiva* (conceived as a judgmental, comparative, inferential and deductive power) to Aristotle's common sense (Smith 2001b, vol. 2, 538, note 42). In general, Suárez does not often refer to Alhacen or other authors of the perspectivist tradition by name. With several exception, such as Alhacen, the Polish mathematician Witelo (born ca. 1230) and the Franciscan John Peckham (1230–1293), he cites them generically by using the name 'perspectivi'. Considering additionally the absence of Roger Bacon (1214–1294) in his references to *perspectivi*, it may be said that Suárez's knowledge of them was not in-depth.

[234] *Ibid.*, n. 4, 456–8.

[235] For this claim of mine, see especially Chap. 5.

[236] *CDA* disp. 6, q. 1, n. 4, 456–8. In a similar vein, Suárez also rejects the theory according to which the *vis aestimativa* or the human *vis cogitativa* detects the *species nonsensatae*, such as those of enmity, that are conceived as really distinct from the sensed species. For this, see *CDA* disp. 6, q. 2, n. 15, t. 2, 490–2; see also Sects. 5.1 and 5.2.2.

[237] For this theory, see Jandun, *Super libros Aristotelis De Anima quaestiones subtilissimae*, II. q. 18, 39–40^{v}.

[238] *CDA* disp. 6, q. 1, n. 5, t. 2, 458.

[239] This theory was advocated by Scotus. See Joannis Duns Scotus, *Ord.* IV, d. 12, p. 2, q. unica, nn. 217–9 (Vat. 12: 386): '[…] sibi not potest [quantitas] fieri primo preaesens nisi per speciem, quae non potest causari in intellectu nisi prius species objecti fuerit in sensu'. However, this view was controversial among Baroque Scotists. Mastrius and Belluto considered the opposite view,

on the abovementioned ‘coaptatio’. If the common sensibles produced their own species in the external senses, they would have to affect them in the same way as the species representing the proper sensibles do. Accordingly, they would have to be of the same sensible character (*ratio*) as proper sensibles. Useless to say, this would subvert a crucial principle according to which a single sensory modality can be immediately modified by a single proper sensible or by a single cluster of specifying sensibles. Moreover, if the common sensibles emitted their own species received in the power, there would be only two ways to conceive the species. It would have to be (i) the same in all the powers, or (ii) different in them. Both are implausible. The first option is implausible since the *temperamentum* of each external sense modality is proportional to its own proper sensible and to the character of the species emitted from a sensible. Therefore, the species of a common sensible could not be one and the same for all the powers. The second option is not acceptable either since an identical natural agent, such as size, cannot emit and cause species of diverse characters. On each occasion it would have to cause a species of the same kind.[240]

In line with Aquinas's approach,[241] the only remaining option of how the common sensibles can be sensed is that they are conceived as *modes* of the species that represent the proper sensibles.[242] However, this modification is not the same for all the common sensibles and their derivatives, such as distance.[243] As for magnitude and shape, these two common sensibles are more ‘internal’ than others. Sight (more clearly) and touch (less clearly)[244] apprehend them by means of an intrinsic modification of the species that represent the proper sensible. This intrinsic modification is due to the intimate connection between the proper sensible and size (shape). Since the external senses are entities located and extended in their organs and the sensible objects are extended entities too, there can be contact between these two only if the mediating entity (species) is extended as well. This modification is clear from the fact that a difference in size makes a difference in our incoming impressions. A big object makes a bigger impression in the power than a small one, because the former emits more species than the latter. Since shape immediately follows magnitude, the species of a proper sensible will also represent an object's shape through its intrinsic modification. Contrary to these ‘intrinsic common sensibles’, the remaining common sensibles are sensed in a more ‘external’ and mediated way, namely in relation

---

namely that the common sensibles do *not* emit their own species, to be more probable. For this, see Mastrius and Belluto, *Philosophiae ad mentem Scoti cursus integer*, vol. 3, *In libros De anima*, disp. 4, q. 5, art. 2, n. 60, 73.

[240] *CDA* disp. 6, q. 1, n. 7, t. 2, 460–2.

[241] For Aquinas's position, see *STh*. I, q. 78, art. 3, ad 2, 254: ‘Et ideo sensibilia communia non movent sensum primo et per se, sed ratione sensibilis qualitatis; ut superficies ratione qualitatis’.

[242] *CDA* disp. 6, q. 1, n. 8, t. 2, 462.

[243] For a perception of distance, see *ibid*., n. 9, 462.

[244] For a comparison of the external senses in respect to the apprehension of the common sensibles, see also Sect. 4.1.

to other sensibles. While number is perceived since the manifold objects impress more species in the organs, unity is perceived without this multilateral imprint. The perception of distance is dependent on the perception of size that is actually or potentially interposed between two things or between an observer and an object. Finally, while the perception of motion is correlated with changes of these interposed magnitudes, the common sensible of rest is sensed when a thing is perceived without them.[245]

(2) Suárez's theory of perceptual errors also requires a presentation of the second abovementioned premise, namely the theory of perceptual judgment. Only if a form of judgment is available to us can we speak of truth and falsity. This is why Suárez assumes the existence of judgment already at the level of the external senses.[246] Although Suárez relegates several quasi-intellectual operations, which in Aquinas's conception of the cogitative power are associated with sensorial judgment, to the intellect, he acknowledges what he calls non-complex judgment in a broad sense (*iudicium latissime sumptum*). This kind of judgment differs from judgment in the proper and strict sense (*iudicium proprie et stricte sumptum*), which is based on the composition or division of two extremes, such as subject and predicate, premises and conclusion at a more complex level of syllogism. Judgment in the proper sense consists in the explicit attribution of our representations created by comparing two extremes to a represented extramentally existing object. No sense can generate this kind of judgment since this operation, like that of perfect reflection, exceeds the capacity of a material power. This does not mean, however, that a less perfect variant of judgment cannot occur in the external senses and in simple apprehension.[247] If simple apprehension is nothing else than an active and vital conception drawing the objects to a power (*tractio rei cognitae ad potentiam*),[248] it must also somehow include judgment. This judgment must be nothing else than simple apprehension's judgement *in actu exercito*. In fact, as such it is a necessary part of all our cognitive acts including those of the external senses. If object *X* were not judged, *X* would not be known. If sight apprehends a white colour, it *ipso facto* (imperfectly) judges 'hoc esse album'. This judgment is 'assimilated' (*imbibitur*) in *all* our acts of perception.[249] In line with what has been said in Sect. 3.6, the attribution of this imperfect judgment to sight and other external sensory modalities and its connection with

---

[245] *Ibid.*, n. 10, 462–6. Suárez's theory of visual spatial perception will be considered in Sect. 4.2.4.

[246] See *CDA* disp. 5, q. 6, nn. 12–18, t. 2, 424–36; *DM* disp. 8, s. 3, n. 7, t. 1, 285; disp. 9, s. 1, nn. 14–6, t. 1, 317–8. For this notion of perceptual judgment in other authors in late medieval philosophy (especially in Ockham and Wodeham), see Tachau (1993) and Perler (2008a). For this judgment as being related to the *virtus discriminativa* in the perspectivist tradition (Alhacen, Roger Bacon, John of Peckham, Blasius of Parma), see Silva (2017).

[247] *Ibid.*, n. 11, 422.

[248] *Ibid.*, n. 5, 416.

[249] For this important term 'imbibere' in Zabarella's tenet of sensation, which in many respects resembles that of Suárez, see South (2017, 737). Suárez himself bases the existence of such perceptual judgment on Aristotle's claim from *Posterior analytics* II.19, 99b35–6, 257: 'They [all animals] have an innate faculty of discrimination, which we call sense-perception'.

perceptual discrimination can be spelled out as follows: when it apprehends white and black colour through two acts at the same time, sight at the same time judges that white colour is not black colour. The external senses can generate two simultaneous distinct acts caused by two distinct species of the same genus. Sight can bring about a vision of white colour and a vision of black colour with the implicit judgment that white colour is not black colour. Of course, sight cannot judge that a white colour is not a sweet taste, since it is not capable of comparing or dividing sensibles belonging to different senses. This can be done only by the common sense which is capable of unifying the data from the external senses and has a wider *radius operandi*. Sight also cannot produce a special *third* act of judgment by means of which it would divide both colours and through which it would judge that one is not the other. Unlike the intellect, sight cannot produce this additional complex act. It can judge only by apprehending.[250] It is not surprising that in *DM* Suárez also speaks about this incomplex truth and this incomplex falsity in respect to simple apprehension, both at the level of the intellect and at the level of the senses. The *notitia simplex* of the external senses participates in the notion of truth.[251] However, he conceives both truth and falsity in simple apprehension in occasionalist and causal terms (*causaliter et occasionaliter*). As we have seen, sight often perceives external sensible things differently from the way they really are and gives occasion to the interior sense and ultimately to the intellect to formulate a false judgment based on this illusory perception that does not correspond to reality. If external sense operations were not controlled and corrected by a higher cognitive instance, sc. by the intellect, which is the only power that can evaluate incoming information against the background of the available information, they would give occasion to the intellect to formulate false judgments in the proper and strict sense.[252]

Having presented two background theories of the sceptical issue in Suárez's philosophy of perception I may now address the topic itself. As usual, the starting point in his debate is represented by a doctrinal polarity, this time between the views of academic sceptics[253] on the one hand and the Epicureans on the other. While the former espouse a position according to which no sense can have certain judgment of any sensible object, the latter endorse what can be called a dogmatic statement according to which the external senses receiving *eidola*, that is, atomic figure-like configurations of sensibles flying through the air and banging into one's eyes, are always right.[254] As elsewhere, Suárez opts for the middle position (*via media tenenda est*). This middle way is here represented by Aristotle according to whom

---

[250] *CDA* disp. 5, q. 6, n. 12, t. 2, 424.

[251] *DM* disp. 8, s. 4, n. 6, t. 1, 291.

[252] *DM* disp. 9, s. 1, n. 16, t. 1, 318; see also *CDA* disp. 6, q. 3, n. 1, t. 2, 494.

[253] For academic scepticism as one of the important starting points in medieval debates about scepticism, see also Perler (2006, 17–9).

[254] Obviously, Suárez assumes that Epicurus's view substantially differs from that of Democritus, for whom sensible qualities were not real qualities of bodies and arose only in a percipient. For Epicurus, on the contrary, they are properties of macroscopic bodies, even though their possession is explicable in terms of properties of and relations amongst the individual atoms of which the

the external senses cannot go wrong (or can err only exceptionally) in the perception of the proper sensibles but only in the common sensibles.[255] Regarding sensible objects per accidens, the issue is clear. Since these are distinctly cognized only inferentially by the intellect, it is no wonder that we often err in them. A scarecrow in a field taken to be a man is a good example of such illusory perception. However, as Suárez notes, Aristotle's claim about the crucial epistemological distinction between the proper and the common sensibles in respect to epistemological (in)fallibility is not without difficulty. In fact, we can go wrong not only in our judgment about size, shape and movement but also about the proper sensibles, such as colours.[256] The perception of changing colours reflected from the neck of a dove is an example.

Suárez formulates his position in four conclusions, all of which are based on the (omnipresent) crucial distinction between general consideration (*in communi sumptum*) and particular consideration (*in particulari sumptum*), this time applied to the proper and the common sensibles. In line with the teleological directedness of the particular senses to their proper sensible objects, his first conclusion is not difficult to anticipate. Applying the general framework, there can be no perceptual error about a proper and adequate sensible object, since a particular external sense cannot be carried outside of its object. Sight cannot go wrong when judging that a thing is lucid; hearing cannot err while judging that there is a sound; smell cannot be amiss when judging that there is an odour.[257] This essential cohesion of the powers with their objects is not only the reason that sight cannot confuse colour for sound. As we have seen, it is also an argument for Suárez's claim that the external senses cannot reflexively cognize their own acts and cannot be related to them as to their proper objects.[258] In his second conclusion, still applying the general framework of consideration, he formulates the statement of the fundamental parity in regard to the proper sensibles and the 'intrinsic' common sensibles, namely magnitude and shape. What holds for the proper sensibles according to the general consideration holds also for these common sensibles. Sight cannot err in judging that a thing is extended or shaped.[259] In the cases of the remaining 'external' common sensibles, such as

macroscopic bodies are composed. For this difference between Democritus and Epicurus, see O'Keefe (2000).

[255] For this distinction, see Aristotle, *DA* III.3, 428b18–26, 163: 'The perception of proper objects is true, or is only capable of error to the least possible degree [. . .] perception of the common attributes which accompany the concomitants to which the proper sensibles belong (I mean, e.g., motion and magnitude); it is about these that error is most likely to occur'.

[256] *CDA* disp. 6, q. 3, n. 3, t. 2, 496.

[257] *Ibid.*, n. 4, t. 2, 496: 'Circa sensibile proprium et adaequatum non potest falli sensus, ut visus non potest falli [iudicans] [esse] rem lucidam, [nec auditus iudicans nunc esse sonum]'.

[258] *CDA* disp. 6, q. 4, n. 24, t. 2, 506.

[259] *CDA* disp. 6, q. 3, n. 5, t. 2, 498: 'Circa aliquod sensibile commune adaequatum, vel in communi sumptum, etiam non potest falli sensus'.

motion and rest, the situation is different. In them sight can get mistaken more easily. This can be illustrated with the aforesaid example of the apparent movement of trees seen from a sailing boat.[260]

However, the epistemic situation significantly changes when the proper and the common sensibles are considered 'in particulari', that is, in respect to their specific differences. When considered in this way (so Suárez's third conclusion goes) the external senses can be mistaken with respect to both kinds of sensibles.[261] Sight can be false while judging that a thing is of this or that size. It can err in judging that object *X* is of this or that shape. A round coin that is seen obliquely from a large angle is sensed as elliptical. In the same way, sight can go wrong in its judgment that the neck of the white dove is red, blue or green. However, following Aristotle's basic claim about the disparity between the proper and the common sensibles, Suárez insists that there is an important criteriological difference in sight's proneness to error. In his last conclusion he asserts that the external senses can after all be more easily deceived in respect to the common sensibles than in regard to the proper sensibles. In line with Aquinas[262] he says that while in apprehension of a common sensible sight can err per se, in relation to the proper sensible it can err only per accidens.[263] What does the Jesuit mean by this distinction? In principle he intends to say that sight can be mistaken about the common sensibles even though *all* actual and potential obstacles of veridical perception have been cleared away. Suárez's statement that we can go wrong about the proper sensibles only per accidens assumes the existence of criteria determining when our perception is veridical and when it is not. This list gives us a litmus paper to differentiate between these two cases. Unlike the eight conditions of veridical perception introduced by Alhacen,[264] Suárez is much terser and lists only three clusters of conditions. First, it is the good disposition of the organ. A sick person whose tongue is covered with a bitter juice will take honey to be bitter. This obstacle can be removed if the organic indisposition is removed. When the person has been cured, she will be able to do epistemic justice to the gustatory perception of the honey's sweetness. Second, it is the good disposition of the medium. If the air is foggy, if there is twilight and the light is bad, or if it rains, sight often goes wrong not only about the common sensibles but also about the proper sensibles. Third, it is the good position of the object in respect to the percipient. An object which is small, distant, or seen under a large angle, is often seen falsely. From a certain angle and under an inadequate light the species emitted from a dove's neck through

---

[260] *Ibid.*

[261] *Ibid.*, n. 6, 498.

[262] For this distinction in Aquinas, see *STh.* I, q. 17, art. 2, c., 220; *STh.* I, q. 85, art. 6, c., 342.

[263] *CDA* disp. 6, q. 3, n. 7, t. 2, 500.

[264] Smith (2001b), vol. 2 (Alhacen, *De aspectibus*, book 3, ch. 3, 588–9: 'The conditions, moreover, according to which a visible object is perceived as it actually exists number eight: first, distance; [then] facing orientation; light; some bulk; opacity; transparency in the air; time; and a healthy eye. When all of these conditions are met a visible object will be correctly perceived; if the object lacks any of them yet is still perceived by sight, then it will not be correctly perceived'.

the medium with its refraction of light will bring an information to sight that will not match the object. In general, if these conditions and obstacles are appropriately considered, we cannot go wrong in regard to the apprehension of the proper sensibles and their kinds. With this claim Suárez underlines the fundamental reliability of our external senses. But even when all these potential obstacles have been cleared, we still can be wrong in our sensations and form false incomplex judgments about the common sensibles that will give rise to false judgments in the proper sense. The reason for this epistemological distinction lies in the manner of the alternation caused by the proper and by the common sensibles. If all potential impediments are removed, a proper sensible will act as naturally as it can toward its correlated sense. It will produce its own likenesses in a power, to which it is naturally ordered.[265] Falsity in the simple apprehension of the external senses is primarily caused by a defect on the side of the impressed sensible species. The defect can be caused by means of an imperfection of the object, of the medium, or of the organ. Our sensation can only be in error due to these defective species which are somehow deviated in their way from the sensibles they represent. If *per impossibile* sensible objects came to be apprehended *without* the mediation of the impressed species, error of the senses would be substantially reduced.[266]

Considering this defect on the side of the species, does explaining illusory perceptions also require that we assume a special ontological realm of intentional or apparent objects in which our perceptual acts are terminated, which some scholars attribute to Auriolian 'apparent being'?[267] In both *CDA* and *DM*, Suárez is silent in the context of illusory perceptions about this kind of being. We read this silence to imply that, as above in the case of our author's theory of the quasi-objective being in perceptual acts (Sect. 3.5), Suárez does not think that these intentional objects should be analyzed by reference to a special being or *tertium quid* inserted between the acts and extramental reality. Accordingly, he aims to avoid representationalist consequences. All that seems to be needed to explain perceptual error are real entities. They are the sensible species, the medium, the organ, the position and distance of an object and perceptual act *qua* quality with its perceptual *content* that aspectually represents an object seen from a certain (subjective) point of view. In fact, it is Suárez's emphasis on the real identity of entitative and representative being of both intentional species and cognitive acts conceived as qualities having a perceptual content that is capitalized in his theory of perceptual error.[268]

---

265 Suárez is not interested in a foundationalist theory of justification. Given the external senses' teleological directedness to the proper sensibles he clearly assumes what can be called a reliabilist approach to the external senses' cognition. In principle, when all the conditions are met, the external senses at the natural level lead mostly to true judgments about extramental objects.

266 *DM* disp. 9, s. 2, n. 1, t. 1. 321; n. 10, 324.

267 For a representationalist reading of Auriol, see Tachau (1988, 85–112).

268 For some interpreters this perceptual content, or the object seen under a certain aspect, just *is* the apparent being of the object. Due to this perceptual content a sensible object 'gains' a new relation to the cognitive power, and this is precisely what Auriol means by 'putting an object in apparent

3.9 The External Senses' Intuitive Cognition of Absent Objects?

Auriol's examples of perceptual illusions testify not only to the possible fallibility of sight and indirectly perhaps also other sensory modalities, but also to the fact that visual acts as operations of intuitive cognition (henceforth: IC) need not necessarily terminate in external objects. Contrary to Suárez's characterization of IC conceived as necessarily terminating in external objects,[269] Auriol indicates that the external senses' cognition can be *of* nonexistent and absent things. Importantly, it was Scotus's 'existential' theory of IC, followed by Suárez, which became the background against which Auriol elaborated his own theory.[270] For Scotus the difference between intuitive and abstractive cognition (henceforth: AC) can be expressed as follows: while AC abstracts from extramental existence and nonexistence and

---

being by the (active) cognitive power'. For them this third-person (metaphysical) perspective is not the whole explanatory story, though. The second part is the first-person (phenomenological) viewpoint. This perspective represents the conscious and phenomenal component of the cognitive experience. It is from this point of view that the *esse apparens* can be also said to be an object of our perception. For this exposition of Auriol's view conceiving him as a non-naïve direct realist, see Lička (2016). I concede that this methodological distinction can, in principle, be applied to Suárez too. On the other hand, however, it is impossible not to see that, when compared to the metaphysical context, in *CDA* the phenomenological perspective is only secondary. No matter how fundamental this approach may be for Suárez (and I think it is), he should be seen rather as someone who only formally 'stakes out' a theoretical subjective space, of which the filling was the work of later, more explicitly phenomenologically-oriented authors. At the same time I cannot fail to think that Suárez deliberately avoids this term, since especially from the time of Capreolus's critique he is well aware of the potential negative representationalist and sceptical consequences of Auriol's view.

[269] Apart from the abovementioned definitions of intuitive and abstract cognition in *CDA* (see *CDA* disp. 5, q. 1, n. 14, 288; disp. 5, q. 5, n. 2, 370–2), see also *DM* disp. 8, s. 1, n. 5, t. 1, 276: '[. . .] intuitiva quae terminatur ad rem prout in se existit, ut est, verbi gratia, visio beatifica ipsius Dei, quae potest dici vera cognitio Dei, quandoquidem per illam agnoscitur Deus prout est in se' and *DM* disp. 30, s. 15, n. 40, t. 2, 182: 'Rursus reperitur in nobis scientia intuitiva et abstractiva, quae quatenus dicunt cognitionem rei existentis, vel non existentis [. . .]'

[270] Scotus's understanding of intuitive and abstractive cognition became a standard point of departure in epistemological debates in late medieval philosophy. For his theory, see Scotus, *Quodl.* q. 7, n. 8 (Vivès 25: 290): '[. . .] cognitio abstractiva possit esse non existentis, aeque sicut et existentis, tamen intuitiva non est nisi existentis, ut existens est [. . .]'; see also *Quodl.* q. 6, n. 8 (Vivès 25: 243–4) and q. 13, n. 18 (Vivès 25: 525). See also Scotus, *Reportata Parisiensia*, III, d. 14, q. 3 (Vivès 23: 359): 'Contradictio est igitur quod sit cognitio intuitiva in genere proprio, et quod res non sit, quia species non potest sufficere ad cognitionem intuitivam sine praesentia rei [. . .]'; *Ord.* II, d. 3, p. 2, q. 2, nn. 318 et 321 (Vat. 7: 552–3): 'Primo distinguo de duplici intellectione: potest enim aliqua esse aliqua cognitio obiecti secundum quod abstrahit ab omni exsistentia actuali, et potest esse aliqua eius secundum quod exsistens et secundum quod praesens in aliqua exsistentia actuali [. . .] Primam voco 'abstractivam' [. . .] Secundam [. . .] voco 'intellectionem intuitivam' [. . .]' As regards Scotus's theory of IC as a type of apprehension necessarily related to actual existence, see Langston (1993, 14). On IC in Scotus as a type of cognition with an import of existential certainty, see Tachau (1988, 75). Among recent surveys of Scotus's teaching, see Cross (2014, 43–5) and Chabada (2005, 64–100).

immediately terminates in a mental representation (AC is called also scientific cognition since it applies to the necessary essences and not to the contingent existence), IC terminates only in an extramentally existing thing.[271] An astronomer's knowledge of a future eclipse is an example of AC. IC thus necessarily attains its object in so far as it exists and is present *hic et nunc*. Only if IC is conceived 'existentially' in this way, can it become the basis for the formation of propositions about the contingent existence of objects. Both corporeal and beatific vision are commonplace instances of IC. Neither is possible without existent extramental endpoints, that is, without an external sensible or without God. Moreover, as Scotus's pupil Francis of Meyronnes (ca. 1285–ca. 1328) stated, an act of IC cannot be an entirely absolute entity: it must be a real relation, that is, the relation in which the subject and its endpoint exist.[272] Not even a miracle can dissolve this real relation with its external *relatum* and leave the act of IC intact. The cases of perceptual error prompted Auriol to criticize this 'existential' characterization of the acts of IC and AC as provided by Scotus and his followers.[273] Real existence cannot be the specific difference of IC. Speaking about external sense perception, an act of IC can occur and often occurs without a present sensible object. Two sorts of arguments are given by Auriol for his conclusion. One amounts to examples of perceptual errors and illusions that have already been mentioned in the previous section. In the context of *CDA* disp. 6, q. 5 'Whether a sense power can cognize in the absence of an object' Suárez takes up one of Auriol's examples, sc. the phenomenon of afterimage.[274] When we have looked at the sun or a luminous object and then close our eyes, we continue to have, at least for a while, the vision of light. As Auriol stresses, this afterimage is not an act of the internal sense but of the visual power, that is, of the external sense.[275] In addition to arguing from perceptual acts directed to nonexistent objects under normal or common conditions of perception, Auriol, like the majority

---

[271]For the IC and AC of the external senses in late medieval philosophy as constituting a historical background of Suárez's view, see Heider (2016c).

[272]Franciscus de Mayronis, *In primum Sententiarum (Conflatus)*, Prologus, q. 18, f. 10vbQ): '[. . .] non potest fieri notitia intuitiva sine reali praesentia obiecti [. . .]'; *ibid.* (f. 11raB): '[. . .] Deus potest supplere vicem in obiecti in ratione moventis [. . .] non in ratione terminantis'. For an exposition of Meyronnes's theory of epistemological realism, see Maurer (1990).

[273]As regards Auriol's critique of Scotus, see Tachau (1988, 89–100).

[274]*CDA* disp. 6, q. 5, n. 3, 518–520. This experience was well-known already to Aristotle and Augustine. For Aristotle, see *On Dreams*, ch. 2, 459b8–12, 355: '[. . .] when we turn from sunlight to darkness; the result is that we see nothing, because the movement produced in our eyes by the light still persists'. For Augustine, see *The Trinity* 9.1.4, 306 (transl. by Hill): 'It often happens that when we look at some lights for a little while and then close our eyes, certain luminous colors continue to revolve in our vision, changing their hues and gradually becoming less bright until they cease together'.

[275]As Pasnau (1997, 188–9) notes, this sort of argument is even more urgent today when we know that light is not transmitted instantaneously. The vision of the afterimages seems to be an entirely common thing. We perceive distant parts of the universe which existed already millions of light years ago. For this time lag argument in favour of the sense-datum theories, see also Fish (2010, 15–6).

of fourteenth-century authors, employs an a priori argument or a thought experiment based on the theological factum of divine omnipotence and the substitutability of secondary causes by the primary cause.[276] According to this argument, if *X* is an absolute (non-relational) entity that is only efficiently dependent on its secondary cause, then *X* can be caused by God's concurrent causal action. In contrast to Scotus and Meyronnes, Auriol makes clear that an act of IC, like any other cognitive act, is just a non-relational quality depending on its parent sensible object as on its efficient cause alone. Consequently, a sense perception of an absent and nonexistent object cannot be regarded as an impossible state of affairs. The possibility of sensing an absent and nonexistent object brings Auriol to a 'correction' of Scotus's view. The specific difference of IC and AC can no longer be grounded in the extramental existence of the sensible object or in indifference to existence and nonexistence of the latter, but only in the different psychological or phenomenological modality on the part of the cognitive act itself. Consequently, an act of IC must be defined, according to Auriol, as 'a direct presentive cognition of what it extends to, objectively actuating and (so to speak) positing existentially'.[277] Accordingly, the intentional modality of IC comprises (at least) three conditions bearing upon the manner in which the sensible is represented to the cognitive power: (1) the sensible object is represented directly; (2) the object is represented *qua* present; and (3) the operation of IC posits the existence of its object. An act of AC, on the contrary, fails to meet these conditions.

This Auriolian setting in several respects methodologically and doctrinally determines Suárez's systematic and historical context of *CDA* disp. 6, q. 5. First, the issue of whether an act of IC of the external senses can be of nonexistent and absent objects is treated in two main contexts. One involves the question of whether this state of affairs can happen *naturally* or under the normal perceptual circumstances. In this context the query can be reformulated in the following way: is it nomologically possible for the external senses to cognize or be intentionally related to an object in its absence? The other involves a miraculous scenario: is this state of affairs metaphysically possible? Can it occur through a divine intervention? Second, as said in Sects. 3.5 and 3.8, for Suárez a cognitive act as such is an absolute entity that is apt to refer to an extramentally existing object. This means that an act of cognition *qua* quality can exist without a real thing. This act can be terminated only spiritually or objectively. Such termination of cognitive acts does not require the existence of external objects. Third, beside the object's 'ratio terminative,' a cognitive object also has a 'motive force'. As such it is also a principle that moves the power by means of the impressed species generating, as we know, the efficient causality toward the sense powers. However, if that is the case, the whole question of

[276]For the significance of thought experiments in medieval philosophy, see Perler (2008b). For the application of this argument in late medieval philosophy, see Woods (1987); Perler (2009); and Perler (2014b, 387–91).

[277]For this translation, see Pasnau (2002a, 2002b, 205).

the possibility of the external senses' IC of nonexistent and absent objects boils down to the following query: can the external senses conserve the sensible species of an object in its absence as the interior senses and the intellect do? Clearly, if they can, the external senses are capable of IC of absent objects; if they cannot, they are not able to intuit them.[278]

Before presenting his view on the issue of the logical possibility of the external senses' IC of absent objects Suárez approaches the question of the natural possibility of this perception. He is clear that 'Naturally speaking, no external sense can perceive unless a sensible object causing a [sensible] species is present'.[279] Unlike the sensible species of the internal senses and the intelligible species, the sensible species of the external senses depend in their becoming, being, and conservation naturally on the "parent" object. Responding to the experience of the afterimage that was supposed to affirm the external senses' (at least brief) IC of nonexistents, Suárez notes that the visual species *cannot* exist without its sensible object in vision, not even briefly and in a weakened form. The example of the visual afterimage must be rejected since it can properly be applied only to the internal sense, not to vision. Auriol's example can also be attacked with a counterexample. If we look at the sun only for a moment and then we close our eyes, no sensible species of light will remain in our eyes since no afterimage has actually been produced. With the continuation of vision the sensible species is not changed and it does not become more intensified or 'rooted' in the visual power. The sensible species come to exist in an instant as perfectly as they can by means of the natural agency of the sensible objects if no obstacles, such as the indisposition of the organ or the medium, prevent their emission. If the afterimage in vision is not produced momentarily, however, it will not be produced by a lingering vision either. In fact, the experience of afterimages results from our erroneous phantasy. It comes from the internal sense *qua* phantasy and is caused by lingering attention. This visual power's attention causes the *species phantastica* and is followed by the production of an imaginative act. Therefore, it cannot be said that the species of light is identical with the act and that the species of light has been imprinted on the eyes so strongly that it remains in them even after the light is removed. It is the interior sense *qua* phantasy that judges that it sees the original object and not an image caused by the sensible species of the interior sense that was left in this power after the original vision had ceased. Since the phantasy's attention does not last for a long time, the appearance of light is not long and thus it is not corrigible by the will either.[280]

Nevertheless, the original clear-cut negative stance to the natural possibility of the external senses' IC of nonexistents seems to be substantially relativized by the

---

[278] For these *preliminaria*, see *CDA* disp. 6, q. 5, n. 2, t. 2, 518.

[279] *Ibid*., n. 4, 520: 'Naturaliter loquendo nullus sensus exterior potest sentire, nisi praesente obiecto sensibili causante speciem'.

[280] *Ibid*., n. 4, 520–2. Clearly, the interior sense can cognize an absent object (see *CDA* disp. 6, q. 5, nn. 7–9, 524–8). For a similar reply to this example, see also Adam Wodeham's reply to Auriol in Woods (1982).

following two qualifications.[281] First, the presence of the object cannot be construed as "facial" presence to the perceiver, because, as we have seen, the species can alter the eye by means of reflection from a mirror and the perceiver can thus see an object from the back. Second, as said in Sect. 3.8, the necessary presence of the sensible object in the perception does not mean that the sensible quality exists always in exactly the same way as it is perceived and that our cognition always matches the sensible object as it exists independently of our knowledge. Having said that, should we claim that Suárez, after all, agrees with the claim that there can be, naturally speaking, IC of an absent object in the external senses, as there can be a perception apprehending objects differently from the way they exist? On the one hand he seemingly agrees. Suárez himself admits that 'quodammodo sensus percipit id quod est absens'.[282] On the other hand, however, he does not hesitate to state that this claim is not 'contra dicta', that is, against the abovementioned negative conclusion. In fact, he seems to want to have it both ways. But why does Suárez think that it is not 'contra dicta'? I believe that the reason is to be found in the aforesaid 'quodammodo'. Suárez distinguishes between two senses of perception of absents. One can be termed perception of absent objects in a strong sense, the other can be called perception of absent objects in a weak sense. While the former excludes the existence of *any* external object or truth-maker, the latter rules out only the existence of an object that would be a true match to an apparent object. It is beyond doubt that the adverb 'quodammodo' entirely rules out the possibility that there can be a perception where no extramental truth-maker exists at all. Indeed, this is the sense of the *status quaestionis* in *CDA* disp. 6, q. 5, which differs from that of *CDA* disp. 6, q. 3 (the subject-matter of the previous section). The variable colours on the dove's neck and the broken stick immersed in water have their extramental truth-makers, sc. the light and the species refracted from the dove's neck and from the stick. Clearly, if the absence of an object is understood in this strong sense, there is no such absence in those two cases. There is one, only if it is understood in the weak sense. Only if we take the external objects themselves (leaving aside the intermediate light) on one side and the apparent properties on the other, which can be considered to be without extramental counterparts, can we speak (*naturaliter loquendo*) about the external senses' IC of nonexistent objects. Only in this restrictive and narrow sense will external sense perception be of an absent object. However, this is not at odds with the abovementioned conclusion.

In the context of divine miraculous intervention no such qualifications are necessary since Suárez is clear: 'By means of absolute power it can happen that external senses perceive an absent thing'.[283] This voluntarist conclusion, inspired by the general tenet of late medieval nominalism, is typical of Suárez. If the sensible species and the powers are absolute entities and the latter can be terminating only spiritually, and if the sensible species generate only efficient causality, and, finally,

---

[281] *CDA* disp. 6, q. 5, n. 5, t. 2, 524.

[282] *Ibid.*

[283] *Ibid.*, n. 6, 524: 'De potentia absoluta fieri potest, ut per sensus externos cognoscatur res absens'.

given the principal substitutability of secondary causes by the primary cause, there is no conceivable contradiction in the divine conservation of the sensible species in the absence of sensible objects. The presence of species in the cognitive faculty attended by the (attentive) soul is actually all we need for the production of a perceptual act. If God conserves the species of an absent object in the eyes of the perceiving Peter, Peter can see an absent object. Suárez exemplifies this conclusion with the aforesaid instance of the reflection of the visual species in the mirror. God can cause the species representing "Paul-in-Rome" to be reflected in a mirror situated in Salamanca. Consequently, in the mirror in Salamanca Peter will see "Paul-in-Rome." Given his presence in Salamanca, Peter will see an absent object. Clearly, what can be caused by God in the mirror can be effected directly without the mirror in the eyes and without the existence of a sensible object.

How does Suárez evaluate this very possibility of objectless sensation caused by divine intervention? Considering his Existential View of IC, he says, not unlike the late Ockham,[284] that in the course of divine intervention the original act of IC turns into an act of AC. However, such transformation is not without difficulty. An objection to a switch from IC to AC may go as follows: if "Paul-in-Rome" were miraculously displaced and seen as "Paul-in-Salamanca," turning the original IC into AC, Peter would, no doubt, be deceived by God, which, useless to say, is something absurd for a theologian like Suárez. Moreover, it seems that this metamorphosis from IC to AC, as Chatton argued, would occur with the preservation of the "qualia" of

[284] Ockham endorses the possibility that an act of IC (an absolute entity for him) of a tree standing in front of me without its existence be divinely conserved, since for the Venerable Inceptor all really distinct absolute entities are, in principle, separable at least by God. Contrary to Suárez, the early Ockham makes his famous claim about the IC of nonexistents. Given the possibility of the aforesaid divine manipulation (as we have seen, not with the species which Ockham denies but with cognitive acts), the Franciscan philosopher says that I cannot go astray in thinking that I see a tree even though no tree exists, since immediately after the annihilation of the tree and the conservation of my perceptual act I come to have an evident act of IC of a nonexistent tree *qua* nonexistent. Accordingly, I will not be misled to falsely judge 'There is a tree'. In his critique of Ockham's teaching, Walter Chatton argued that if one holds that God can cause whatever is free of contradiction, He can certainly actualize a state of affairs in which there will be an act of IC of a tree even though there is no tree present. Accordingly, after such divine intervention the original act of IC will, however, still be an act of IC, this time without its extramental object. Even though, under the pressure of Chatton's objections, he abandoned the theory of an IC of nonexistents, he still advocated a position of cognitive realism, which would have made him immune to sceptical questions. As Ockham explains, and this is where the similarity to Suárez lies, a 'new' cognitive operation, produced after the divine intervention, will not be an act of IC, but rather a creditive act of AC. If the coherence of this abstractive act with other (naturally received) judgments, based on IC, were not sufficiently considered, it might ultimately lead to the false judgment 'There is a tree'. However, if this (in)coherence is considered, and this is where the possibility of epistemic defence lies, no false judgment will follow. For Ockham's IC of nonexistents, see Ockham, *Scriptum in librum primum Sententiarum (Ordinatio)*, book 1, prol. q. 1, art. 1, 31.10–5. For Chatton's denial of Ockham's IC of nonexistent objects, cf. Gualterus Chatton, *Reportatio et Lectura super Sententias*, Prologus, q. 2, art. 3, 98.19–20 and 102.133–5. For Ockham's view regarding the change of the original act of IC to an act of AC, see Ockham, *Quodlibet* V, q. 5, 498.72–6. For a lucid exposition of Ockham's and Chatton's positions and the latter's critique, see Perler (2006, 253–80).

IC. Clearly, this indiscernible change will result in the external senses' fallibility that can ultimately jeopardize the certainty of the Aristotelian *scientia*. Should we thus say that God as an omnipresent divinity can be, like the Cartesian *genius malignus*, a deceiver causing in us the cognition of nonexistent objects without any possibility for us to epistemically defend ourselves? Unfortunately, Suárez's reply to this objection is very brief:

> Not in the least. He would only permit our senses to err, for God produces true species in the eye, but it is the human who deceives herself in judging that a thing is present.[285]

Despite the brevity of this reply a plausible answer can be reconstructed. As I have argued in the previous section, Suárez distinguishes between two kinds of judgment. One, judgment in the strict sense, is the collation (or division) of one thing with another, namely a predicate with a subject. The other, judgment in a broad sense, is nothing but simple apprehension. Since simple apprehension is a vital act, it can judge, so to say, concomitantly or in *actu exercito*. Vision can spontaneously judge that it sees, for example, the colour white. As we have seen, this judgment is involved in the cognition of the external senses. In an implicit link to this distinction it may be said that when Suárez says that not God but man would be the cause of error, he seems to mean that God can allow an error to happen only with respect to judgment conceived *in a broad sense*. Nevertheless, this spontaneous verdict of the sight can be corrected and revised by the reflective judgment of higher cognitive powers, namely the intellect. How can the intellect perform the correction? I believe that there are at least two possibilities. One of these is to proceed by means of a closer (reflective) introspection of a new (transformed) act, that is, by examining the act of AC. In *CDA* disp. 3, q. 2, as we have seen, when treating the issue of the specification of the powers of a soul, Suárez establishes the validity of two principles: (1) each power is intrinsically specified and distinguished by means of its adjustment (*coaptatio*) to its acts related to its object[286]; (2) every operation *qua* quality is specified by its object, to which it is directed and in which it terminates.[287] However, given that an operation is necessarily directed to its (extramental) object, as Suárez's Existential View suggests, then if IC is necessarily directed to the extramental counterpart, and an act of AC is directed merely to the "imaginabile" or "scibile" by abstracting from contingent existence, then the miraculous removal of the extramental sensible will have to be manifested introspectively in a noticeable change in the new act. The character of this new act, seen from the position of the higher-order act, will not be the same as the character of the former act of IC. Accordingly, on the reflective (intellectual) level, Peter will be able to stand up to the spontaneous judgment exercised by his sight concerning the existence of 'Paul-in-Salamanca' by detecting the distinction between the acts of IC and AC. The

[285] *CDA* disp. 6, q. 5, n. 6, t. 2, 524: 'Minime, sed permitteret tantum sensum falli, nam Deus veras species producit in oculo, ipse autem homo se fallit iudicans rem illam esse [sibi] praesentem'.

[286] *CDA* disp. 3, q. 2, n. 8, t. 2, 88.

[287] *Ibid.*, n. 20, t. 2, 102.

second option that seems to me to be available to Suárez is that Peter will employ the criterion based on the coherence of the propositions. Petr will be able to recur to the comparison of the current judgment about Paul's existence in Salamanca by comparing the proposition 'Paul is in Salamanca' with other judgments that he considers reliable. He could reject the judgment by realizing that it does not cohere with other judgments that he takes for certain, such as 'Paul is in Rome', 'Paul is seriously sick there', 'The journey from Rome to Salamanca takes several weeks and the sick Paul could not endure it' etc. Both strategies show that Suárez has an epistemic resource that the deceived Peter can use to inspect his deception.

Concluding, Suárez's position on the IC and AC of the external senses can be summarized in five points. First, absolutely speaking, natural sensation of absent and nonexistent things is impossible. Such sensation is only possible in a certain respect and in specific cases. Second, miraculous perception of nonexistent sensibles is, logically speaking, a possible state of affairs, caused by divine manipulation of the sensible species. Although the sensible species naturally depends on its sensible object, God can conserve the sensible species without its "parent" sensible. Third, given the validity of the Existential View of IC, IC, and AC differ *ex parte obiecti* and *not ex parte subiecti*. There is no faculty that cognizes only intuitively or that knows exclusively abstractively since the intellect and the internal senses can also cognize intuitively. The first entity apprehended by the intellect is a material singular existent here and now. Fourth, given Suárez's Existential View of IC, the sensation of an absent object cannot be a case of IC, but only of AC. Fifth, phenomenologically speaking, it may well be that for Suárez acts of IC and AC differ in the presence (absence) of the quality of the phenomenal consciousness and that the intellect or a higher cognitive power will have enough resources to acknowledge that what is *prima facie* perceived as an instance of IC is actually a case of AC. Such discernment can also be done through the comparison of a proposition based on the new (transformed) act with other judgments that the percipient considers to be reliable enough. Although he does not explicitly say which of these options is preferable, it may be said that in this context Suárez adopts a reliabilist position.

Chapter 4
External Senses II

Having treated the cognition of the sentient part of the soul and of the external senses considered 'in communi' (*CDA* disp. 5–6), in *CDA* disp. 7 Suárez embarks on the large subject matter of the particular sensory modalities. Unlike the two preceding disputations, *CDA* disp. 7 also extends its thematic focus to natural sciences, such as optics and anatomy. Despite the scientific catch of this disputation Suárez's bent remains philosophical. A large part of disp. 7 engages in philosophical topics concerning the issues of nature, subject, cause and kinds of the particular proper sensibles and the media in which these sensibles are spread. This is not surprising considering that Suárez individuates and differentiates the senses especially through reference to their objects. Considerable space in *CDA*, disp. 7 is also devoted to the topic of the sensible species of the particular sensory modalities and their multiplication *in medio*. One of the crucial questions raised in this disputation involves the relation of the sensibles' intentional and natural actions and affections of the media and organs. A significant part of the disputation is dedicated also to Suárez's denial of Aristotle's general claim about the necessary existence of a medium interposed between the organ and the objects. This denial brings Suárez close to a theory of tactile introception, namely to a view according to which touch can sense not only the first qualities of external objects but also the internal states of one's own body called dolorogenic qualities. In general, this disputation zooms in on the particulars and fleshes out what had been presented earlier in the more general disputations.

Following Suárez's model, I will advance in what can be called a bottom-up procedure. First, for every modality an analysis of the adequate object, viz. its nature and kinds, cause and subject will be given. Second, for each external sense I aim to present Suárez's analyses of the medium, its kinds and its transducibility of the sensible qualities and the sensible species. Third, against the background of the Aristotelian and Galenic views I will lay out Suárez's physiological views related to the particular sense organs. However, before entering on these topics I will explicate Suárez's views of the individuation of the external senses and their various ranking.

D. Heider, *Aristotelian Subjectivism: Francisco Suárez's Philosophy of Perception*, Studies in the History of Philosophy of Mind 28,
https://doi.org/10.1007/978-3-030-67341-3_4

4.1 Individuation and Ordering of the External Senses

Unlike contemporary philosophers, Suárez does not address the issue of the individuation of the senses straight away. In *CDA* disp. 7, q. 15 he approaches this question in the context of his investigation of the necessity and sufficiency of the classical five senses. Since counting requires the individuation of units, the query about the number of the senses cannot be separated from the topic of their individuation and differentiation. In *CDA* disp. 3, q. 2 Suárez introduces his criterion of the specification of the powers, which is largely based on what is today called the world-based criterion of the individuation of the external senses.[1] In his two conclusions he says that each faculty is intrinsically specified through its coaptation (a) to its operation conceived *qua* quality which (b) is then related and individuated through reference to its formal object.[2] Having received the species and produced an act, the cognitive power receives this act. As receiving the act, it must be specified through it. Accordingly, powers are specified by acts since nature or God provides the soul with all the instruments necessary for the exercise of its proper acts, including powers. In order for the human being to see colour God gives sight to the soul; in order for the human being to hear sound God gives hearing to the soul etc. An act conceived *qua* quality specified by its formal object is the reason *why* the soul has the power it has. A craftsman imitates nature by producing a tool according to a set goal, that is, according to the tool's operation toward a certain object.[3] It does not matter that acts and objects are external and posterior to the powers. It does not matter if an object does not exist at all. Final causes do not have to actually exist to be causally efficacious. The powers' intrinsic ordering to inanimate objects does not pose a problem either. The ontologically higher sight is related to the ontologically lower colour not because of the colour but on account of its own perfecting through the production of visual acts.[4] These formal objects are at best manifested in the so-called being of objects (*esse obiectorum*) or in the being of objects *qua* cognized.[5] These formal objects are manifested in two ways. First, they show up in the manner in which a power attains its object. The intellect attains its object differently from the way the will does. Second, they are manifested in the way in which an object affects

[1]For the world-based criterion of the individuation of the external senses and its various versions based on the nature of things and their properties as contrasted with the introspective criterion grounded in the "qualia" of the acts immediately accessible to the perceiver, see Matthen (2015). For the view that the latter criterion is to be preferred in the theory of the individuation of sensory modalities, see Grice (1962). We will see below that Suárez qualifies this world-based criterion, though.

[2]*CDA* disp. 3, q. 2, n. 8, t. 2, 88, and n. 11, 94.

[3]Ibid., n. 8, 90.

[4]Ibid., n. 26, 108.

[5]Ibid., n. 9–10, 90–4. As is well-known, these *formalities* or *formal objects* can differ in the degree of universality. A more perfect and universal power has a more universal object, a less perfect power has a less universal object. In this way, the external senses' apprehension differs from that of the internal sense because they have more limited objects. Ibid., nn. 12–3, 94–6.

(*immutat*) a power.[6] This second criterion is crucial in Suárez's theory of the individuation of the external senses.

Having sketched the general background of the issue of the individuation of powers, we can now turn to *CDA* disp. 7, q. 15. Following what can be called 'folk psychology', Suárez aims to argue that there are precisely five senses in humans and perfect animals.[7] There are no fewer and no more than five external senses since there are no fewer and no more than five sensibles or clusters of sensibles that per se *primo* affect the powers. Each of these can be perceived only by a single modality. Suárez's substantiation based on the objects' affection of the powers comes from Aquinas's *STh*. I, q. 78, art. 3, c.[8] At the outset of the body of this article called "Are the five exterior sensory powers appropriately distinguished?" Aquinas rejects three views according to which the explanation of the distinctions between the exterior sensory powers is to be taken (a) 'from the organs', (b) 'from the medium', and (c) 'from the diverse natures of the sensible qualities'. Aquinas shows that none of these can satisfactorily explain *why* we have just the five senses. The first view cannot be right because the organs exist because of the powers and not vice versa. The difference of the powers cannot be explained by the difference of the organs. The second tenet is denied in a similar way. Diverse media are attributed to the diverse powers in so far as these media are appropriate for the powers' acts. Finally, Aquinas also rejects the last view by claiming that the investigation of the natures of the sensible qualities is irrelevant since natures can be investigated only by the intellect. The criterion must be taken only from what belongs to the senses per se and not from what is considered by the intellect. Since the exterior senses are passive powers (as we have seen, this holds for Suárez's first and also the third stage of perception) and as such they are susceptible to being affected in this or that way by this or that sensible, the differences between the external senses must be first of all taken from the manner through which the senses are affected by the sensible qualities. In this context Aquinas introduces a basic distinction between two kinds of change, namely between natural and spiritual (intentional) change which have already been treated in Chap. 3. While in natural change the form of a thing that effects a change is received according to its natural being (a wall receiving the form of red turns red and instantiates the form of redness), a change is spiritual insofar as the form of a thing that effects this change is received according to its spiritual being (a pupil receiving the species of red does not turn red and does not instantiate this form).

After distinguishing these two affections Aquinas formulates the following 'deduction' of the five external senses:

[6]Ibid., n. 13, 96.

[7]*CDA* disp. 7, q. 15, n. 1, t. 2, 750. Obviously, Suárez has in mind rational and *perfect* nonrational animals. He shows that there are numerous imperfect animals that have less than five senses. For an unequal "distribution" of the external senses among perfect and imperfect animals, see *CDA* disp. 7, q. 15, nn. 8–10, t. 2, 760–4.

[8]For an analysis of this article, see also Pasnau (2002a, 172–7 and 180–4).

> Now in certain sensory powers, for example, the sense of sight, there is only a spiritual change, whereas in the other sensory powers the spiritual change is accompanied by a natural change either on the part of the object alone or on the part of the organ as well. On the part of the object, the natural change is with respect to place in the case of sound, which is the object of the sense of hearing. For sound is caused by vibration and movement in the air. And as for alteration, in the case of odour, which is the object of the sense of smell, a body has to be altered in some way through heat in order to give off an odour. On the part of the organ, a natural change is involved in the sense of touch and the sense of taste; for instance, when a hand touches something hot, it itself becomes hot; and the tongue is moistened by the moistness of various tastes. By contrast, the organ of the sense of smell and the organ of the sense of hearing do not undergo any natural change in sensing, except incidentally.[9] [...] The organ of the sense of taste, however, is not necessarily affected with a natural change by the quality that is its proper object, in such a way, namely, that the tongue itself becomes sweet or bitter; instead, it is affected by the preparatory quality on which taste is based, viz., moistness, which is an object of the sense of touch.[10]

Although Suárez significantly modifies some doctrinal points from this quote, regarding the differentiation of the exterior senses he basically falls in line with Aquinas. For him, the justification of the existence of the five modalities based on the various ways of the affection of the organs and the media is a *probable* view. He makes it clear that this criterion cannot be searched for only in the objects themselves but also in the objects' relation to the powers. An information about a colour is transmitted through a different kind of transducers than an information about a sound. There are five possible ways media and organs can be affected. First, in order for the senses to be able to produce operations their organs must be affected *intentionally*. As highlighted, the reception of the species is a sine qua non condition of the production of a sensory experience in general. Suárez begins with the most 'spiritual' sense, sc. sight. An object, namely colour (not light), can affect sight only intentionally, that is, without the admixture of natural action. Colours do not cause natural changes in the immediate medium and in the pupil. Second, following Aquinas, Suárez states that spiritual affection can be accompanied by natural change either only on the side of objects, or also on the side of organs. The first can happen in two ways. (i) A quality can be produced by a local motion, as it happens in sound

[9]Thomas de Aquino, *STh*. I, q. 78, art. 3, c., 254: 'Sed in quibusdam sensibus invenitur immutatio spiritualis tantum, sicut in *visu*. In quibusdam autem, cum immutatione spirituali, etiam naturalis; vel ex parte obiecti tantum, vel etiam ex parte organi. Ex parte autem obiecti, invenitur transmutatio naturalis, secundum locum quidem, in sono, qui est obiectum *auditus*: nam sonus ex percussione causatur et aeris commotione. Secundum alterationem vero, in odore, qui est obiectum *olfactus*: oportet enim per calidum alterari aliquo modo corpus, ad hoc quod spiret odorem. Ex parte autem organi, est immutatio naturalis in *tactu* et *gustu*, nam et manus tangens calida calefit, et lingua humectatur per humiditatem saporum. Organum vero olfactus aut auditus nulla naturali immutatione immutatur in sentiendo, nisi per accidens.' English translation in Aquinas (2010, 72).

[10]Ibid., ad 4, 254–5: 'Gustus autem organum non immutatur de necessitate naturali immutatione secundum qualitatem quae ei proprie obiicitur, ut scilicet lingua fiat dulcis vel amara; sed secundum praeambulam qualitatem, in qua fundatur sapor, scilicet secundum humorem, qui est obiectum tactus.' English translation in Aquinas (2010, 74).

originating through a percussion of two bodies, or (ii) a quality can be caused by an alteration, namely by heating, as it occurs in odours. Third, in line with Aquinas, Suárez adds that natural change can affect both objects and organs. This double alteration can be again twofold. (i) Natural affection of the organ can proceed through the same quality represented by the species. This *ratio immutandi* is typical of most tangible qualities. (ii) Natural affection can proceed by means of a quality that differs from the quality represented by the species. In this way sapours affect the power of taste. Since gustatory objects must be moistened by saliva to be causally concurrent in the production of gustatory experience, the quality that attains the tongue need not be the same as the tasteable quality represented by the species.[11]

For Suárez, these spiritual affections with the various kinds of 'admixtures' of natural changes in intentional actions show that there are *no fewer* than the five senses. At the same time Suárez says that they show us that there can be *no more* than the five senses. We do not have an experience of other (latent) modes of alterations than the abovementioned ones. Contrary to what Aquinas said about the impossibility of the argument based on organs, Suárez also employs the argument 'from the distinct organs'. There are no more than the five senses because we do not know about any other sense organs. Moreover, these capacities are fully sufficient for perfect sentient life. We do not need more than these five senses. Since the sentient nature occurs in humans in the most perfect grade, a perfect sentient nature cannot comprise more than the five exterior senses which are instantiated in humans.[12]

There are two major objections against this commonsensical view. Both come from Aristotle.[13] The first of these caveats states that there must be fewer than the five senses, namely four, because taste and tasteables are reducible to touch and tangibles: 'The tasteable is a kind of tangible'[14]; 'Taste, in fact, is itself, as it were, a sort of touch'.[15] Second, it seems that, on the contrary, there must be more than the five senses because there are at least two kinds of tactile power.[16] In the beginning of *On the Soul* II.11, when referring to objects in his definition of the senses, Aristotle says:

> For if touch is not one sense but several, there must be several kinds of tangibles. It is difficult to say whether touch is one sense or more than one [. . .] For every sensation appears to be concerned with one pair of contraries, e.g., vision is of white and black [. . .] but in the tangible there are many pairs of contraries, hot and cold, dry and wet, hard and soft [. . .][17]

[11] *CDA* disp. 7, q. 15, n. 1, t. 2, 750.

[12] Ibid., n. 2, t. 2, 750–2.

[13] For the problems with touch and taste in Aristotle, see Sorabji (1971, 68–78). For an analysis of these two objections in Suárez, see Heider (2016d, 99–100).

[14] Aristotle, *DA* II.10, 422a8, 125.

[15] Aristotle, *Parts of Animals* II.10, 656b37–657a1, 179–81.

[16] *CDA* disp. 7, q. 15, n. 3, t. 2, 752.

[17] Aristotle, *DA* II.11, 422b18–28, 129.

There must be at least two pairs of basic tactile qualities, namely hot/cold and dry/wet, from which all others can be derived.[18] If we start from the premise that powers are specified by formal objects, we seem to be led to the conclusion that there are two kinds of touch. One kind of tactile power will be based on what we today call 'thermoreceptors', the other will be grounded in 'mechanoreceptors'. The former will be related to thermal properties, the latter will perceive mechanical properties connected with the state of matter, to say the least.

In his reply to the first objection Suárez shows that almost all who have commented on Aristotle's *On the Soul* claimed that touch and taste are two distinct senses. Taste cannot be reduced to touch since their proper sensibles differ. Moreover, with an implicit reference to the abovementioned distinction in the manner of natural change between taste and touch, he says that their ways of affection differ too. Moreover, considering their different organs they cannot be the same. While the organ of touch is spread all over the body, the organ of taste is located in a single part of the body, sc. the tongue. A correct reading of Aristotle's 'Gustus est quidam tactus' thus requires a distinction. The copula 'est' in the proposition can be taken either *formaliter* or *praesuppositive*,[19] that is, in the sense that touch is necessary but not sufficient for the operation of taste. If we take the copula in the first sense of formal identity, the sentence is false. Albeit the tongue perceives tangibles, its operation is not that of tactile perception. However, if the meaning of the proposition is taken in the second way, namely as a necessary condition, the proposition will be true. Taste can perceive only if the organ is touched by a tasteable. We cannot taste without touching.[20]

In his reply to the second objection Suárez says that there is only a unique sense of touch. Following the argument from the way of organ immutation he says that there is only a single way in which the tangible sensibles of hot, cold, wet and dry affect the organ. Moreover, referring to the biological criterion of the organ's unity he says that their organ is unique too. Contrary to absolutizing the world-based criterion, however, he notes that to get a single sense it is not necessary to assume a single pair of contrary qualities. If the existence of a single pair of contrary qualities were the unique principle of the senses' individuation, there would have to be two powers of sight and two powers of smell since, as we shall see below, the proper sensible of sight are both colour and light, and the proper sensible of smell are both nutritive and nonnutritive ('aesthetic') odours, such as those of flowers.[21] Simply speaking, if the premise 'For a specifically unique power there must be a specifically unique pair of contrary properties' is taken as the only criterion, it is not true.[22]

[18]For the proper sensible of touch, see also 4.5.2.

[19]This general distinction will be important below for Suárez's analysis of the proper sensibles of various sensory modalities.

[20]*CDA* disp. 7, q. 15, n. 4, t. 2, 752–4. Moreover, it may be added that these modalities must differ as there are imperfect animals that have touch but no taste. See *CDA* disp. 7, q. 15, n. 8, t. 2, 760–2.

[21]*CDA* disp. 7, q. 10, n. 7, t. 2, 696–8.

[22]*CDA* disp. 7, q. 15, n. 5, t. 2, 756–8.

Aquinas's survey of the five modes of affection by the proper sensibles is far from a disorderly account. It constitutes the 'absolute' ordering of the external senses according to the degree of their 'immateriality' or, which is the same for Suárez, according to the degree of their corporeal subtlety. The more 'immaterial' an external sense is, the nobler it will be and the higher position in the ontological *scala* of the senses it will occupy. The earthier a given sense is, the lower position it will take. This criterion will not be restricted to the elemental character of a sense organ (for example, the more fine-grained, or airy, organ of the sense of hearing is ranked higher than the terrene organ of touch), but it will also be applicable to the nature of the proper sensibles and to the manner in which the sensibles affect the medium and the organ. Suárez calls this comparison 'absolute' because it abstracts from the subject in which the powers inhere and does not confront the human and the beastly senses. Regarding the absolute comparison, Suárez thus endorses the same ranking as Aquinas.[23] The most perfect sense is sight, the second is hearing, the third is smell, followed by taste and touch.[24] Sight is the noblest sense since its object is the least corporeal.[25] Its object involves not only colour but also light. As medieval imagery of light based on metaphysics of light shows, light stands closest to the realm of immateriality. Sight is affected by colour and light in the most "spiritual" way. As both Aquinas and Suárez say, in the reception of colours the organ of sight and its medium are affected *purely* intentionally. The inherence of a coloured object in the eye disallows sight's operation. As compared to the other senses, sight's *radius operandi* is largest. In an instant it can attain planets and stars. Moreover, its organ is of the most sophisticated composition.

Although *sonus* is ontologically an imperfect entity (as a transient entity it seems to be less perfect than the permanent quality of the other four senses), in its *esse sensibile* it makes a quality superior to the qualities of odours, sapours and tangibles. Sound propagation *in medio* is more 'immaterial' than the smoky evaporations of odours. While a sound often affects the sense of hearing through a local movement of air (as we will see, it *can* also affect the power purely intentionally), an odour also often affects the sense of smell through a material change together with heating up the organ. The *sensorium* of hearing (the inner ear located behind the eardrum) can be evaluated as more 'immaterial' than the pupil. While the organ of hearing is composed of air, the pupil consists of water, viz. the crystalline humour. However, the elevation of the airy element above the watery one should not be regarded as a reason to reverse the current ordering. The main criterion to assess the perfection of the sense organs is not the predominance of this or that element but rather the quality of the elements' *temperamentum*.[26]

---

[23]See the text of the *corpus* of *STh.* I, q. 78, art. 3 following the above-quoted text.

[24]*CDA* disp. 7, q. 16, n. 2, t. 2, 764.

[25]An exception to this 'visiocentric' encomium was the French philosopher and mathematician Charles de Bovelles (1479–1566), one of a few who prioritized hearing over seeing. See Fragenberg (1991).

[26]*CDA* disp. 7, q. 16, n. 3, t. 2, 766–8.

In this absolute comparison smell is ranked higher than touch and taste. As we will see in 4.4.2 and 4.5.2, odour can affect its power purely intentionally too. Smell's object is less 'earthy' than tasteables and tangibles. Finally, without an explicit argument for the priority of taste to touch Suárez somewhat enigmatically concludes: 'Et eisdem rationibus gustus superat tactum'.[27] Why does he think so? It seems that the priority of taste is due to its greater 'rareness'. As we have seen, some imperfect animals can be nourished only tactually. Living tactually constitutes the minimal and necessary level of sentient life.[28] This argument 'from rarity' holds also for the organs. The organ of taste is 'less universal' than the *sensorium* of touch. While the organ of touch is spread all over the body, the organ of taste is located in a single part. From the absolute point of view, *tactus* must be evaluated as the lowest and basal sense because it is the universal sense (*sensus universalis*) which also exists in the organs of the other senses. This is also why all these organs can be injured by a strong natural affection.[29] At the same time the same ordering must be advocated in respect to the powers subjected in a human composite. This relative line-up must be identical with the absolute ordering since the absolute arrangement follows the specific perfection of these senses. However, the senses with the highest specific perfection exist in rational animals.[30]

An entirely distinct conclusion will be reached from comparing the particular human senses with those of other animals. If they are compared in this relative way going across different kinds of animals, the least perfect human sense will be smell. Albeit the scope of human smell is wider than that of dogs and vultures (humans discern both nutritive and nonnutritive odours) and thus in an unqualified sense (*simpliciter*) human smell is assessed as superior to nonhuman smell, in a certain respect (*secundum quid*), that is, in respect to the sensation of distant odours and the capacity to localize causal sources, human smell is much less perfect than that of nonhuman animals.[31] On the other hand, the most perfect senses according to this comparison will be touch and taste. The human organs of these two senses are most perfect and most delicate. They can distinguish numerous sensible nuances that nonhumans cannot. Only among humans can we find degustators of wine and gourmets of food. By a process of inference, the middle position in this proportional ranking with other animals is taken by sight and hearing. Human sight and hearing

---

[27]Ibid., n. 2, 766.

[28]For this see *CDA* disp. 2, q. 2, n. 5, t. 1, 154.

[29]*CDA* disp. 7, q. 15, n. 8, t. 2, 760–2.

[30]*CDA* disp. 7, q. 16, n. 6, t. 2, 772.

[31]Ibid., n. 4, 768–70. Regarding the nonhuman smell of nonnutritive odours, Suárez says the following: '[. . .] tamen odores alios, qui delectant sine coniunctione ad sapores, ut sunt odores rosarum, etc., alia animalia vel nullo modo percipiunt, vel imperfecte, neque in eis delectantur, homo vero perfectissime percipit hos odores, qui perfectissimi sunt et delectatur plurimum'. (ibid., 770). For perception of nutritive and nonnutritives odours by human and nonhuman animals, see also 4.4.1.

are by far outdone by dogs, eagles etc. as far as the acuteness of perceiving objects is concerned.[32]

Beside these comparisons related to the external senses' dignity and practical utility (action), Suárez *ex professo* introduces a comparison indicated in the famous passage from the beginning of Aristotle's *Metaphysics*:

> All men by nature desire knowledge. An indication of this is our esteem for the senses; for apart from their use we esteem them for their own sake, and most of all the sense of sight. Not only with a view to action, but even when no action is contemplated we prefer sight, generally speaking, to all the other senses. The reason of this is that of all the senses sight best *helps to know things, and reveals many distinctions* [my italics].[33]

In these famous introductory sentences of the *Metaphysics* Aristotle distinguishes two reasons for our love of the external senses in general and of sight in particular. We love our senses (a) because of their practical utility, that is, for life-preservation and sensual pleasure, and (b) because they are at the service of our knowledge. While the first reason is common to human and nonhuman animals, the second seems to be proper only to humans who have a rational soul and an intellect. In this quote it is this second criterion that makes vision the most preferable and the most loved sense of all the sensory modalities. We take the highest delight in vision since it is the most useful 'tool' for acquiring knowledge. Accordingly, Aristotle's passage gives Suárez an occasion to compare the senses according to the criterion of utility for knowledge. In his explication Suárez proceeds by distinguishing two modes by means of which we can acquire knowledge. One is the acquisition of knowledge by learning (*per disciplinam*), the other is the acquisition of knowledge through discovery (*per inventionem*). It is with respect to the second kind that sight is to be preferred to the other senses. As we have seen, sight detects the most information and most clearly from our environment, especially as far as the common sensibles, such as size, shape, movement, and number are concerned. Nevertheless, if acquisition by learning is considered, it is hearing which is to be preferred to sight. Following Aristotle's *De* sensu *et sensato*,[34] Suárez says that it is our hearing of spoken words through which we best grasp meanings and concepts signified by *voces*. True, we can capture concepts also through script. However, through hearing words we get vivacity and clarity (*vis ac claritas*) that we do not get in reading. This is why debates and workshops can sometimes be more useful for knowledge than reading books. The priority of hearing is also corroborated by the fact that we find erudite people among the blind and not among the deaf. However, when compared with the acquisition of knowledge *per inventionem*, the extraction of knowledge *per disciplinam* is only incidental. The latter depends on the former. Without discovery no learning would be possible. There would be nothing to learn and nothing to teach.

---

[32]Ibid., n. 6, 772.

[33]Aristotle, *Metaphysics* I.1, 980a22, 3.

[34]Aristotle, *On Sense and Sensible Objects* ch. 1, 437a4–17, 219.

So, if the external senses are compared according to acquisition through inventing, sight and touch exceed hearing.[35]

In order to compare sight and touch, Suárez introduces two additional criteria. These senses can be confronted in two ways. They can be compared (i) according to their being *signs* of the powers that are better/worse for the acquisition of knowledge, and (ii) as *instruments* that serve the discovery of knowledge.[36] As it was for Aristotle, touch is for Suárez by far the best indicator of *bonum ingenium*. He refers to Aristotle's claim according to which '[. . .] men of soft flesh [are] well endowed with intelligence'.[37] However, this seems to be in collision with Aristotle's other formulation from the beginning of *Problem* 30[38]:

> [. . .] all those men who become extraordinary in philosophy, politics, poetry, or the arts are obviously melancholic, and some to such an extent that they are seized by the illnesses that come from black bile [. . .].[39]

In this passage it is the melancholic complexion caused by black bile that determines the corporeal conditions of human ingenuity. In line with the Galenic humoral doctrine this complexion is incompatible with tactile sensibility. According to the Galenic doctrine, however, black bile is composed of the qualities of cold and dry. But these two qualities are not compatible with the optimal physiological disposition required by touch. On the contrary, the cold and the dry lead to tactile 'desensibilization'. The melancholic complexion and tactile acuteness seem to be contradictory. Therefore, touch cannot be the best sign of intelligence. In his reply, Suárez, in principle, agrees with the claim that an excessive amount of black bile leads to desensitizing the tactile power and to dullness and stupidity. However, he adds that a moderate amount of this black bile heated by blood constitutes the best disposition for the *bonum ingenium*. A moderate portion of black bile will dispose the body to better diffusion of the animal spirits necessary for all sensorial operations parallel to intellectual activities. This admixture will reduce an undue sanguineous vivacity of the blood, which (as we know) is unfitting and often disturbing for peaceful study and contemplation.[40] Accordingly, the melancholic complexion will be not only a sign and *effect* of intellectual work, but also the disposition for intellectual work.[41]

---

[35] *DM* disp. 1, s. 6. n. 11, t. 1, 55–6.

[36] For this distinction, see *DM* disp. 1, s. 6, n. 12, t. 1, 56 and *CDA* disp. 7, q. 16, n. 8, t. 2, 776.

[37] Aristotle, *DA* II.9, 421a26–7, 121.

[38] Apparently, Suárez takes the *Problemata* as Aristotle's genuine text. Today we know that, indeed, a work of this kind was written by Aristotle because he refers to it in his other (authentic) works. Nevertheless, at present we also know that what is today known under this title is not his text but a later compilation. For this, see *The Works of Aristotle*, ed. W. D. Ross, vol. 7: Problemata, ed. E. S. Foster, 1927, vii.

[39] Aristotle, *Problems* XXX.1, 953a10–4, 155. See also ibid., 954a30–3, 163.

[40] *CDA* disp. 7, q. 16, n. 9, t. 2, 776–8.

[41] For this, see *CDA* disp. 11, q. 2, n. 15, t. 2, 356. As regards the relation between genius and melancholy in Suárez, see also García Cuadrado (2017b).

Concerning the instrumental character of vision and touch it is not surprising to see Suárez saying that sight exceeds touch in many ways. (1) It is superior from the point of view of the cognition of the number of objects. Unlike touch it can perceive distant objects, such as planets and stars. It discerns figures, motions and other common sensibles better and in a wider scope than all the other external senses. (2) Since light is not moved by local motion, the visual power perceives its objects in an instant. (3) Moreover, sight imprints its objects on the phantasy more firmly and durably than the other senses do. Memory retains visual images better than images of the other senses. (4) Visual experience (*experimentum*) is more certain than experience of the other senses.[42] Although sight goes wrong about many things, especially if perceiving distant objects, if the objects of both senses are properly disposed (see the abovementioned criteria of veridical perception), sight will be less prone to errors than touch. As regards the proper and the common sensibles, if properly applied to the powers, visual perception is more certain than tactile perception.[43]

4.2 Sight

Since sight enjoys the highest degree of 'spirituality', Suárez starts his exposition with it. In *CDA* disp. 7, qq. 1–5 he deals with five main topics in relation to the visual power. They are the nature of light and its diffusion (4.2.1), the ontological status and nature of colours (4.2.2); the total (adequate) object of vision (4.2.3); the query regarding the direction of vision related to the traditional issue of extramission and intromission (4.2.4); and the topic of the visual organ (4.2.5).[44]

4.2.1 Suárez's Naturalistic and Qualitative Concept of Light

The visual process cannot be understood apart from the issue of the nature of light and its propagation. The topic of light constitutes the basic point of departure for every visual theory. In the title of *CDA* disp. 5, q. 1 'Quidnam sit lumen?', Suárez implicitly refers to the well-known scholastic distinction between light *qua lumen* and light *qua lux*.[45] In his *De opere sex dierum* he characterizes these two 'lights' in the following way:

> '[...] philosophers properly distinguish between light as *lux* and light as *lumen*: they call the light as *lux* the light which is per se and quasi-intrinsically conjoined to a substance and is a quasi-source and principle of light as *lumen*; they call light as *lumen* a splendour or the

[42] *CDA* disp. 7, q. 16, n.7, t. 2, 772–6. For Suárez's analysis of *experimentum*, see 5.3.2.

[43] *DM* disp. 1, s. 6, n. 12, t. 1, 56–7; *CDA* disp. 7, q. 16, n. 10, t. 2, 778.

[44] As regards these topics, see also Heider (2017b).

[45] For this distinction, see also Lindberg (1976, 113, 133–5, and elsewhere).

quality (similar to light as *lux*) that intrinsically approaches the body in which it inheres [. . .]'[46]

Light as *lux* and light as *lumen* are distinguished both through their location and causal relation. While light as *lux* inheres in a luminous source, such as the sun, fire, and is the luminous source or cause of light as *lumen*, light as *lumen* (its effect) inheres in a medium having a transparent character, such as air, water or glass. In *CDA* Suárez adds another distinction between lumen *qua* ray and lumen *qua* luster. If *lumen* is a quality in the air which is radiated in a rectilinear line to the eye, it is called a ray (*radius*); if it is reflected from a shining body it is called luster (*splendor*).[47] Since in *CDA* disp. 7, q. 1 Suárez devotes major attention to light as *lumen*, without this caveat we would use the term 'light' largely in the sense of *lumen*.

Without referring to any advocate, Suárez starts by introducing an eliminativist account of light in a single sentence. According to this view light is nothing more than a manifestation of colour, or in a broad sense it is a vehicle that manifests knowledge. As is well-known, the term 'light' in this broad sense is applied also to cognitive powers, such as the phantasy and the intellect. In line with the medieval tradition of 'metaphysics of light', Suárez agrees that light-imagery can be employed in this sense. The etymology of the word 'phantasia' supports this application and meaning. 'Phantasia' is taken from the Greek *φάνος* (apparition), which is derived from *φάος* (light). The phantasy thus makes visible or manifests what is not present here and now.[48] The agent intellect compared to light is also a manifestative force that makes the intelligibles in these images visible to the passive intellect by illuminating phantasms.[49] Knowing this manifestative force of light Suárez makes clear that light cannot be limited to this aspect. We often see light not seeing a colour, as when we see, for example, celestial bodies. Moreover, if light were not a real entity, it could not manifest material things at all.[50]

Having established that light is a *real* entity that cannot be explained away as a manifestative force, and having thus replied to the question 'An lumen sit?', Suárez approaches the second question, that is, 'Quid sit lumen?'. He rejects three theories. First, he turns down the view according to which light is a body (*corpus*).[51] Following Aquinas,[52] Suárez introduces three arguments against this position.

[46]Suárez, *De opere sex dierum*, book 2, ch. 2, n. 3, 94: '[. . .] adverto philosophos accurate distinguere inter lucem et lumen: nam lucem vocant, quae per se, et quasi ab intrinseco conjuncta est alicui substantiae, et quasi fons est, et principium luminis lumen autem vocant splendorem, seu qualitatem similem luci, ab illa defluentem, et corpori, in quo inest [. . .]'

[47]*CDA* disp. 7, q. 1, n. 4, t. 2, 554–6.

[48]*CDA* disp. 8, q. 1, n. 5, t. 3, 18.

[49]*CDA* disp. 9, q. 8, n. 14, t. 3, 226–8.

[50]*CDA* disp. 7, q. 1, n. 1, t. 2, 552.

[51]For this view, see, e.g., Simplicius (1544), *Commentaria in quatuor libros de coelo Aristotelis, In II De coelo*, com. 29, 18–9.

[52]Suárez copies the arguments from Aquinas's *STh*. I, q. 67, art. 2., c., 164.

(1) If light were a body, then two bodies, that is, light and the illuminated object, would have to co-exist in one place. However, in the natural order such physical compenetration is not possible. (2) If the effluence of light were similar to the egression of substantial corpuscles from a luminous object, light would have to traverse space in time by local motion. Albeit it might seem that in a short stretch light spreads in an instant since it is radiated in an imperceptible amount of time, in light propagation over a big space a noticeable temporal interval would have to be observed. Experience teaches us that it is not the case. The sun's illumination in a whole hemisphere, that is, between east and west, is not perceived in time. The spreading of light must be (in contrast to what we learn from physics today) conceived as an instantaneous change of the medium. (3) If light were a body, it would have to be generated and corrupted. It would have to be substantially changed. At dusk, the form of light would have to get corrupted and from the potentiality of prime matter a new substantial form, namely that of darkness, would have to be educed. In such case darkness would have to be something positive. This is at odds with the scholastic theory according to which it is a privation. It would also be absurd to say that light is not generated and not corrupted but created and annihilated. If this were the case, God would have to intervene in all-natural changes as the immediate cause. No doubt, this would affect the relative causal autonomy of the (secondary) natural causes.[53]

After disapproving the corpuscularian account of light Suárez disclaims another view according to which light is the *substantial form* of the sun. In explaining this denial, he shows that this view cannot be correct since no substantial form can be sensed per se. However, light *is* sensed per se.[54] Having rejected these two tenets he arrives at his first conclusion: 'Light is an accidental act that pertains to the category of quality'.[55] If it is neither a body nor a substantial form, and provided it is a real entity, it can be nothing else than an accident. However, there is no other kind of accident available than quality. Only a quality can be intensified and remitted, only a quality can be sensed per se *primo*. Despite the obvious plausibility of this view, it was Scotus who had formulated a different view, though. He claimed that light *qua lumen* is nothing else than the intentional likeness of light *qua lux*.[56] Accordingly, light cannot be a quality having a natural being since its properties are in fact identical with those of the intentional species. Several arguments were brought up for this statement. (1) Like the sensible likeness by virtue of which we perceive sensibles while not seeing the species themselves, we perceive the sun through light without seeing light itself. Light as *lumen* does not constitute the per se visible but

[53]*CDA* disp. 7, q. 1, n. 1, t. 2, 552.

[54]Ibid., n. 2, 554.

[55]Ibid., n. 3, 554: 'Lumen est actus accidentalis ad praedicamentum qualitatis pertinens.'

[56]For this view of Scotus's, see *Ord.* II, d. 13, q. unica, n. 16 (Vat. 8: 232): 'Hoc modo dico (c) quod lumen proprie est intentio sive species sensibilis ipsius lucis. Quod probatur: quia si non esset intentio, tunc sensibile positum supra sensum prohiberet sensum, quia quod est tantum sensibile et non ratio sentiendi, si ponatur supra sensum, impedit sensationem [. . .] et ita lumen suprapositum oculo, impediret ipsum videre; hoc autem est falsum [. . .]' See also Cross (2014, 37–8).

only that by which (*quo*) we apprehend the sensible object. Therefore, our sight does not terminate in *lumen*. (2) Moreover, as Aristotle says, a sensible superimposed on a sense does not cause sensation.[57] However, if light is placed on an organ, it does co-cause sensation. Therefore, it can be only the sensible species, namely of the sun. (3) Like the intentional species, light exists only if the illuminating source, namely the sun, exists. Therefore, light *qua lumen* can only be the sensible species of light *qua lux*. (4) Like the sensible species, light as *lumen* does not have a contrary opposite. If it were a natural and physical quality, it would necessarily have an opposite. (5) In analogy to the sensible species, which can co-exist in one place of the medium despite their crossing, rays of light can co-exist in this manner too. Two sources of light cause two shadows in the same place. It may be concluded that light *qua lumen*, rather than being a natural and physical entity, is a more subtle entity which is nobler than a physical sensible quality and as such has spiritual being.[58]

Suárez is not impressed by these arguments based on the analogy between light and the sensible species. He is sure that an excessive emphasis on this analogy leads to an undue 'denaturalization' of light.[59] Suárez shows that these 'analogies' between the intentional species and light are only fictitious: 'Light is a quality that has a proper real and natural being'.[60] In his reply to the first argument he leaves no doubt that light *is* the object of seeing. Unlike the species it denominates its subject. We say that Peter is illuminated whereas we do not say that Peter who receives the sensible species of green colour is green. Through the reception of the species Peter does not turn green. Besides, light cannot be the species since it generates material effects, such as heating and drying.[61] In his refutation of the third and fourth arguments he notes that it is not true that all natural qualities necessarily have contrary opposites. He adds that some can also exist only if their agent is operative. As an example of the latter Suárez mentions sound. In his counterargument to the fifth argument Suárez notes that it is not correct to state that two light rays can exist in the same place like two overlapping sensible species. There is no experience to substantiate this view. What we really experience is that the efficiency of two light sources is bigger than the efficiency of a single luminous source. But this greater

---

[57]For Suárez's exposition of this Aristotle's claim, see 4.5.3.

[58]*CDA* disp. 7, q. 1, n. 5, t. 2, 556–8.

[59]While a kind of reasoning that regards *lumen* as the intentional species of *lux* can according to Vasco Ronchi (1970, 61) be considered a thought that is proper to 'the metaphysics of light with incursions even into the theological field', those, on the contrary, who attribute a material nature to *lumen* and study it as an object *sui generis* are to be taken as physicists. Accordingly, Suárez should be ranked among the latter. Ronchi (ibid., 65) also shows that with late medieval philosophers, such as Buridan, Ockham and Oresme, *lumen* as physical light and the analysis of the physico-physiological process, which was typical of Alhacen and other Arabic philosophers and scientists, gained in value in the problem of vision.

[60]*CDA* disp. 7, q. 1, n. 8, t. 2, 558: 'Lumen est qualitas habens proprium esse naturale et reale.'

[61]Ibid., n. 6, 558.

efficiency is not due to the multiplication of two distinct rays but to the constitution of a single light with a higher intensity.[62]

Suárez's naturalistic and qualitative account of light is further manifested by his last complex conclusion partly summarizing what has been said above. It is composed of three partial conclusions[63]: (1) light of the air is a quality that can be seen as an object (*terminus*); (2) light *qua lumen* is not visible per se but only by means of its species[64]; (3) light *qua lumen* is not the sensible species through which we see the sun.[65] (1) We all experience that we perceive light of the medium in different 'qualities' when we see it at noon, at dawn or at twilight. At noon we see light as brighter than at twilight and at dawn. Moreover, if we slant our eyes we can see sunrays in the air. (2) Given Suárez's universal employment of the sensible species in all sense modalities, it is not surprising to see him saying that light can be seen only through a representational likeness. We can discern a figure of light in the air and its more and less illuminated parts. However, we could not see these parts through a particular light *realiter* attaining the eyes since this particular light represents all these distinct parts. Such a particular light would not represent all the different parts and grades of the illumination. Besides, we have an abstractive cognition of light. We can imagine and remember a light that we have seen before and its intensity. However, in order to know light in this abstract way we must have its visual species in the phantasy. (3) The sensible species of light *qua lumen* does not represent the sun since the medium can be illuminated in the same way regardless of whether a luminous object is of this or that shape. It must be represented by its own proper species only. Rather than a representation of light *qua lux*, light *qua lumen* is to be conceived as a communication of a luminary source, in the same way as a heat is the communication of a thermal source. Moreover, if light were the species through which we perceive the sun, once illuminated our sight would have to see the sun as well. This is not always the case.[66]

What is the formal effect of light conceived as a real quality and what is its subject? In line with Aristotle Suárez says that the formal effect of light *qua lumen* is to make medium which is transparent (*perspicuum* or *diaphanum*) in potency transparent in act. A luminous object reduces a subject, such as the air, from being diaphanous in potency to being diaphanous in act.[67] Regarding the subject of light

---

[62]For an explicit rejection of the overlap of the distinct rays, see *DM* disp. 5, q. 8, n. 12, t. 1, 192–3. For Suárez's rejection of Scotus's arguments, see *CDA* disp. 7, q. 1, n. 11, t. 2, 564–6.

[63]Suárez's 'naturalization' of light can be seen also in his claim that although light as *lux* of the sun is a celestial quality and light as *lumen* is an elemental quality, both are of the same sort. For this, see *CDA* disp. 7, q. 14, n. 14, t. 2, 568–60.

[64]This is also Suárez's reply to Scotus's second argument.

[65]*CDA* disp. 7, q. 1, n. 9, t. 2, 560: 'Lumen aeris est qualitas visibilis, quae videtur tamquam obiectum visus; et non videtur per se, sed per species sui; neque etiam est species per quam sol videtur'.

[66]*CDA* disp. 7, q. 1, n. 9, t. 2, 560–4.

[67]Ibid., n. 12, 566. For the passage about the relation between light and the transparent, see Aristotle's *DA* II.7, 418b3ff, 105.

Suárez asserts that there are two main genera. One is bodies that are completely transparent in their depth; the other are opaque bodies. While the former, such as air, water and glass, receive light thoroughly, the latter receive it only on the surface.[68]

4.2.2 Elemental Character of Colours

The formulation of the *status quaestionis* from *CDA* disp. 7, q. 2 concerning the proper sensible of sight, that is, colour, shows that the question of the ontological status of colour was not debated only in early modern and contemporary philosophy but that it was an issue discussed also in ancient philosophy and by scholastics including Suárez.[69] This question is marked by a challenge based on a reductionist account of colours. On this approach, colours are nothing but structural properties having quantitative characteristics, such as shape, size, texture, at the surfaces of opaque bodies, which account for the various ways in which light is refracted and reflected to the perceiver. On this view, colours are nothing else than the light received and reflected by the surface of these bodies. Accordingly, there is no qualitative difference between light and colour. The manner of reception and reflection at the surface of the body correlates with the perception of particular colours. Colours are construed in terms of the object's quantitative properties that through the reflection of the light cause sensations in a percipient. Although this position comes close to one advocated by Descartes,[70] this account is not without its medieval predecessors. It was, for example, Nicole Oresme (c.1320–1382)[71] who can be regarded as advocating a similar view.[72] Aristotle's definition of colour from *On Sense and Sensible Objects*, namely 'the limit of the transparent in a defined body', seems to confirm this reductionist view. A defined body is an opaque body (an undefined or unterminated body is a transparent body) in which a vision is terminated. The perspicuity that exists on the extremity of the body is nothing else

[68] Ibid., n. 13, 566–8.

[69] For three distinct theories of colour in Descartes, see Wolf-Devine (1993, 44–49). For a contemporary debate involving physicalism (a qualitative aspect of colour of which we are conscious is a physical quality of sensibles), dispositionalism (a qualitative aspect of colour is a disposition of sensibles to produce certain colour experience) and subjectivism (a qualitative aspect of colour is exemplified by perceivers), cf. Ross (2015, 411–9). For Aristotle's view, see below in this section.

[70] See his *Dioptrics*, AT VI, 118: '[. . .] l'obiet V, que ie suppose, par exemple, estre rouge, c'est a dire estre disposé a faire que les petites parties de cete matière fubtile, qui ont esté seulement poussées en lignes droites par les cors lumineux, se meuuent aussi en rond autour de leurs centres, après les auoir rencontrés, & que leurs deux mouuemens ayent entre eux la proportion qui est requise pour faire sentir la couleur rouge [. . .]' See also Wolf-Devine (1993, 45–6).

[71] For Oresme, see Maier (1967, 30) and Pasnau (2017, 57).

[72] Suárez does not quote Oresme as an advocate of the view according to which colour is not essentially distinct from light. He cites many other authors, such as Avicenna, Avempace, Al-Farabi and also the Greek and Latin commentators on Aristotle, such as Alexander of Aphrosidias, Themistius, Giles of Rome and Cajetan.

than a received light. This reading can be confirmed by the fact that in spring we see that colours come from the mere termination of light in a dense cloud (body).[73]

Although Suárez does not abandon the basic position according to which colours are grounded in light actualizing the transparent, he is far from reducing colours to quantitative properties existent on the surfaces of visibles. He is sure that colours are *real* and *permanent* qualities in coloured bodies, even though these bodies are not illuminated at all. As such they result from the mixture of elements composing the bodies.[74] (1) It is inconceivable that Peter ceases to be a white man at sunset and starts to be a coloured man at sunrise. Colours are not visible in the dark, indeed. In order to become visible, they need to be illuminated. However, their visibility, which is a property (*proprium*) unique to colours, is something different from their essence. (2) Sight can clearly distinguish between light and white colour in an object. It can apprehend two bodies one of which is white and the other black in colour, both equally illuminated. (3) Colour differs essentially from light since white colour can remain identical while the degree of light has changed. As experience shows, the gradual decrease in light at sunset can be compensated for by the perceiver's approximation to a coloured body. With this approximation the perceiver can have the same perception of the same colour in the same 'phenomenal' quality even with a significant diminution in light. The variation of the latter with the invariant of the former indicates that they are mutually distinct. (4) If colours were identical to light, they could not differ specifically. There could not be more than one kind of colour since things identical to a third thing, in this case light, are identical too. Such a claim would no doubt relativize the *sortal* status of colours and as such would be at odds with our commonsensical view. Colours could not be specifically different if their only difference were based on the accidental distinctions of structural quantitative properties on the side of a subject. (5) Moreover, as the example of the moon shows, light can be received in a terminated body and it need not always cause colours.[75]

If colour is a permanent quality of objects and provided that it originates from the mixture of elements in a coloured body, what is this elemental mixture precisely like? In his exposition, in which he follows Aristotle's *De* sensu and Aquinas's *Commentary* on this text, Suárez starts from two main principles accounting for the resulting of colours.[76] These principles apply to light and the transparent, which is the subject of light. While light can be taken as the formal principle, the transparent can be regarded as the material principle in the constitution of colours. The causal proximity of light and colour is evident from Aristotle's statement according to which light is the colour of the transparent.[77] Clearly, the difference between this first 'colour' of the transparent and the colour conceived as a real and permanent

[73] *CDA* disp. 7, q. 2, n. 1, t. 2, 570–2.

[74] Ibid., n. 2, 572.

[75] Ibid., n. 3, 574.

[76] Aristotle, *On Sense and Sensible Objects* ch. 3, 439a13–439b18, 229–31; for Aquinas, see his *Sentencia libri De sensu et sensato*, ch. 1, ch. 6, 34–5.

[77] Aristotle, *DA* II.7, 418b11–2, 105.

quality inherent in a coloured body lies in the fact that the former is caused by an external luminous source entirely permeating the body while the latter is brought about by and within a coloured opaque object. In line with ancient physics Suárez distinguishes four basic elements, namely water, earth, fire and air. These are composed of various combinations of first qualities. Importantly, these elements can also be classified according to their degree of diaphaneity. Diaphaneity caused by airy and watery elements can be found in different degrees in all bodies. Besides diaphaneity, which is due to the elements of air and water, there is also the element of fire, which accounts for lucidity and the element of earth, which is responsible for opacity. While fire is the source of an object's lucidity, water and air are best suited for the reception of light and the element of earth impedes and 'chokes' the fire's illumination. A mixture of these four elements with these contrary features of opacity and clarity can be found in all bodies. The more air and water a mixture contains, the more it participates in diaphaneity and the more it disgregates (scatters) sight. Conversely, the more earth it possesses, the more opaque it is and the more it congregates (unites) sight. Since every resulting mixture is always composed not only of a single element, such as earth, but also of other elements in greater or lesser degree, it always retains traces of diaphaneity. Importantly, the element of fire inherent in the body which is more or less 'choked' by the opacity of earth does not persist in the mixture formally *qua* fire. In fact, this element gets transformed to a new quality, sc. to colour. Accordingly, colour is to be regarded as an *emergent* or *supervenient* property of the abovementioned mixture of elements in a body.[78]

These three aspects, that is, light, perspicuity and opacity in their various fusions and proportions, which give rise to permanent colours in objects such as trees, grass, and snow, can also produce colours that are more ephemeral than the colours in (dense) material objects. Here Suárez has in mind, first of all, the colours of clouds and rainbows, namely colours arising in the air. No matter how difficult it is to find a criterion for distinguishing between true and apparent colours, Suárez asserts that the colours of a rainbow are to be regarded as true colours, even though they are significantly less permanent than the colours of trees, grass, snow etc. The main reason for this realist evaluation lies in the affirmation of a structural causal similarity between permanent and ephemeral colours. Both the rainbow and true colours seem to originate by a certain blending of light, transparency, and perhaps also some opacity. On the other hand, if there is no such similarity and if colours vary quickly depending on the variation of a percipient's position and perspective, as it happens with the colours of a dove's feathered neck, they must considered to be only apparent colours.[79]

---

[78] In the context of this transformation, interestingly, Suárez does not employ the term 'emergence' but the word that seems to imply a change for the worse, namely 'degenerate' (*degenerat*), to degenerate toward a new quality (colour). For Suárez's account of this mixture and the emergence of colour from this mixture, see *CDA* disp. 7, q. 2, n. 3, t. 2, 574–8.

[79] *CDA* disp. 7, q. 2, n. 8, t. 2, 582–6.

No matter how Aristotelian his view of colours is, Suárez rejects the abovementioned characterization of colour defined as 'the limit of the transparent in a defined body'.[80] The main reason seems to be that it concedes too much to the abovementioned reductionist understanding of colour. He agrees with Aristotle that a defined body (*corpus terminatus*) is a body in which diaphaneity is mixed with opacity. However, unlike Aquinas who follows Aristotle,[81] Suárez makes it clear that colour pertaining to the accident of quality (and this might be called its 'generic' feature) cannot as its subject (the 'specific' difference of colour as such) have only the extrinsic surface of a coloured body. Colour cannot be a quality that exists only on the surface of a coloured body. As the examples of milk, snow, blood etc., show, colour exists also *within* these objects. Although these 'internal' colours are not visible, they are still colours *in actu*. It is not necessary to divide a coloured object to get a colour *in actu*. Concluding, Aristotle's definition from *De* sensu should not be regarded as the proper definition of colours but only as an explanation of colour in relation to us, the beholders. Instead, a proper definition should be formulated as follows: Colour is a secondary (in the scholastic sense!) quality, which is the act of a mixed body inasmuch as it has perspicuity limited (dampened) by opacity.[82]

A mixture with the prevalence of light caused by fire and a low degree of opacity caused by the presence of the elements of air and water gives rise to the first (extreme) colour, namely white. A mixture with the predominance of opacity caused by earth and a minimal amount of light, which is brought about by earth, gives rise to the contrary colour, namely black. White colour disgregates sight since it contains an abundance of light penetrating throughout the transparent. It is light that is disgregative in the highest degree, even to the degree that it can damage sight. Black colour, on the contrary, congregates sight because it contains opacity, which is a congregative force to an extreme degree. Metaphorically speaking it may be said that black colour 'concentrates' us or pulls us inside. This is also why the congregative black is the colour of funerals. However, since fire does not produce only light but also heat, it can sometimes happen that by exsiccating earth it can

[80]For this definition, see Aristotle, *On Sense and Sensible Objects* ch. 3, 439 b11–2, 231.

[81]Thomas de Aquino, *Sentencia libri De sensu et sensato*, tr. 1, ch. 6, 38: 'Sicut autem corpora intrinsecus quidem habent superficiem in potentia et non autem in actu, ita etiam intrinsecus non colorantur in actu set in potentia, quae reducitur ad actum facta corporis divisione'.

[82]*CDA* disp. 7, q. 2, n. 3, t. 2, 578–80. While the Aristotelian primary qualities are the most basic explanatory principles of the four elements, the secondary qualitites, such as colour, odour, flavour, texture, etc., are caused by a mixture of the first ones. Unlike the secondary qualities, the first qualities are also the primary causal agents in the physical realm. While a red colour cannot make a wall red, fire can make water hot through its heat. This holds also for Suárez: '[. . .] color non sit activus sui similis [. . .]' (*CDA* disp. 5, q. 2, n. 16, t. 2, 316); the same applies to sapour and odour: '[. . .] sapor non est qualitas activa, sicut neque odor [. . .] per humorem sapori coniunctum agat in rem insensibilem'. (*CDA* disp. 7, q. 13, n. 3, t. 2, 722). Clearly, this 'sui similis' does not indicate an intentional likeness but a real quality. Importantly, the explanatory priority of the first qualities does not imply that the secondary qualities are reducible to them. For Suárez and many other scholastics the secondary qualities are irreducible *real* items of the world. For an excellent exposition of the primary and the secondary qualities in scholasticism, see Pasnau (2011b).

cause black and not white colour. This happens, for instance, when the first quality of heat dries up the human yellow bile and causes black bile.[83]

4.2.3 Adequate Object of Vision

In Aristotelian-Thomistic philosophy colour is often deemed to be the only formal object of vision. It seems to be regarded as the *total* or *adequate* object of sight.[84] As indicated above, strictly speaking, this is not Suárez's position. Sight can see per se *primo* not only colours but also light. Clearly, this is not only light *qua lumen* but also light *qua lux* constituting the visible. Following Aristotle, Suárez refers to a group of colourless luminous objects, such as fire, stars (the moon) and animals with 'fiery' parts, such as cats' eyes, the scales of fish, which the Stagirite called luminous things.[85] The object of sight is not only colour and light as *lumen*, that is, light that is not seen as inherent in an object but as a transient entity coming from a luminary source, but also light qua *lux*, which is 'fixed' to a lucid thing.[86] However, since there is a prevalent opinion according to which cognitive (and appetitive) powers must be specified by a single adequate total formal object and must be per se *primo* related to it, and since they cannot be directed only to a partial object or a plurality of objects, it seems that sight must be (i) per se related either to something *common* to light and colour (this common element cannot be defined as a visible since this would result in a circular definition[87]), or (ii) only to one of these objects. A different issue conditioning the reply to this question is treated in *CDA* disp. 7, q. 3, namely which illumination is necessary to elicit an act of vision. In other terms, the question to be answered is that of the necessary conditions of the visibility of colours and luminous objects in relation to the illumination of an object, medium and power/organ. To put it tersely: Is it sufficient for a visible object to be illuminated *ex parte obiecti*, or do we have to assume an illumination *ex parte medii* too?[88]

---

[83] *CDA* disp. 7, q. 2, nn. 5–6, t. 2, 580–2. It is through additional mixtures that the medial colours, namely light-blue, orange, yellow, red, green and purple, are produced. These colours are called 'medial' because they are situated between two extreme colours, sc. black and white colour. They are constituted by different mixtures of the features of the two extreme colours in various numeral proportions. For these medial colours, see ibid., n. 7, 582.

[84] No doubt, some of Aquinas's formulations go in this direction. See Thomas de Aquino, *STh.* I-IIae, q. 10, art. 2, c., 86: 'Sicut autem coloratum in actu est obiectum visus, ita bonum est obiectum voluntatis'; *STh.* I, q. 19, art. 3, c., 235: 'Sicut et quaelibet alia potentia necessariam habitudinem habet ad proprium et principale obiectum, ut visus ad colorem; quia de sui ratione est, ut in illud tendat'.

[85] Aristotle, *DA* II.7, 419a3–6, 107.

[86] For the distinction between the vision of something as a terminus (a coloured wall) or as a way (via) or medium (illuminated air) to a terminus, see *CDA* disp. 7, q. 3, n. 1, t. 2, 586–8.

[87] See *DM* disp. 44, s. 11, n. 64, t. 2, 714; cf. also *CDA* disp. 7, q. 3, n. 7, t. 2, 596.

[88] Ibid., n. 1, 586.

Suárez puts forward two conclusions in which he aims to clarify the nature of the interrelationship between these two visibles terminating vision, namely between colours and light conceived as *lux*, in their regard to this topic of the adequate object of sight. In his first conclusion he formulates a tenet in respect to colourless lucid things as follows: 'Lucid things can be seen through their own light; in order to be visible they do not require an illumination of the medium'.[89] Following the abovementioned distinction between natural and intentional affection, lucid things generate twofold activity. They rectilinearly diffuse *lumen* and multiply the species through the medium up to the organ of the perceiver. It might seem logical to say that lucid things need to illuminate the medium and that only through this lighting can they multiply the visual species up to the organ. But Suárez is of a different opinion. He shows that illumination of the medium is not necessary for their 'being seen' and for multiplication of the species. The light of a cat's eye is so weak that it cannot illuminate the medium incident between it and the perceiver and it is too weak to be visible in broad daylight too. It quickly gets 'stifled' by sunlight. The truth of the claim that an illuminated medium is not necessary for an imprint of its species is even clearer if we consider what Suárez says in *TDA*. In *TDA* he explicitly takes the view according to which the visual species of light can be multiplied and diffused *farther* than the natural entity of light as a more probable theory.[90] Being a less perfect entity, the visual species of light can attain more remote places than light itself.[91] Accordingly, unlike touch and taste, there is no necessary subordination of the multiplication of the species of light to the diffusion of the real quality of light in the way that the former must be necessarily accompanied by the latter.[92]

Although the actual illumination of objects is not necessary for colours to be colours *in actu* (they are such even in darkness), in order to be visible they must be per se illuminated. In the first part of his second conclusion, Suárez declares that, contrary to luminous things, in order to be visible colours must at first be per se illuminated.[93] Without being illuminated they cannot emit visual species and cannot imprint intentional likenesses on the organs. Even though colours are visible without

---

[89]Ibid., n. 2, 588: 'Res lucida videri potest ratione suae lucis tantum, neque requiritur lumen in medio ut videatur'.

[90]*TDA* book 3, ch. 16, n. 3, 668.

[91]This further emission of sensible species holds for audible and olfactory species too. For hearing, see *CDA* disp. 7, q. 8, n. 7, t. 2, vol. 2, 670–2 (see 4.3.2); for smell, see *CDA* disp. 7, q. 11, nn. 3–4, t. 2, 702–6 (see 4.4.2). It is also one of the main reasons for Suárez to posit sensible species in these external senses.

[92]*CDA* disp. 7, q. 3, n. 3, t. 2, 590. The separability of these two actions is important for Suárez also for a theological reason. This "separability" in the propagation of real sensible qualities and sensible species is by him fully utilized in his treatment of the sense operations of the blessed. Since impassibility is one of the main features of glorified bodies, it is the multiplication and the reception of species that will guarantee the necessary perfection of the sense perception of resurrected human beings in heaven. For this, see Francisco Suárez, *Expositio et Quaestiones in quinque priores primae secundae divi Thomae Aquinatis quaestiones materiam de beatitudine continentes* (forthcoming 2021), 'Disputatio de quidditate humanae beatitudinis', q. 2, 132–4.

[93]*CDA* disp. 7, q. 3, n. 4, t. 2, 590.

illumination, they are visible only 'remotely'. In order to become visible 'proximately' (and it is this proximate visibility that is expressed in Aristotle's definition of colour as 'the limit of the transparent') they need to be lit beforehand. The more the sensibles are lit, the more they are visible and the better they emit the visual species. Since for Suárez colour is light 'choked' by the opacity of terrene objects, unlike luminaries it cannot illuminate itself sufficiently in the way to become a proximately visible object. In order to be visible colour requires being lit as a necessary complement. Only by this empowerment coming from an external source of light can a colour constitute a perfectly visible object. Without this complement it is not a proximately visible object.[94] Consequently, replying to the basic question of the adequate object of sight, it may be said that if a colour is incipiently visible due to its diminished or 'choked' light and made perfectly (proximately) visible by virtue of an external light, it is clear that light, which exists both in coloured objects and in the medium, becomes a common warrant and connecting point in the unity of the total object of sight.[95] From this point of view, Suárez's reply to the initial question of *CDA* disp. 7, q. 3, being fundamentally related to the character of visibility, about the adequate object of sight seems to make perfect sense: The adequate object of sight is light and everything (*res illuminatae*) that is manifested by it.[96] The fact that lucid and illuminated things are seen only in broad daylight (colours), only in darkness (cat's eyes) or in both (as in the case of fire) comes as accidental to this unifying aspect of the object's visibility based on actual illumination.[97]

In the second part of his second conclusion, in line with the principle of parsimony largely employed throughout *CDA*, Suárez states that albeit this per se illumination of colours is a necessary condition of their visibility, this does not hold for the medium. It is not necessary for the medium to be illuminated per se since it is sufficient only for an object to be illuminated. For the medium it is sufficient to be illuminated *ex consequenti*.[98] What does it mean? I take Suárez to be saying that the illumination of the medium is only a consequence of the illumination of the visible. If a coloured object is illuminated, its immediate surrounding becomes consequently more or less illuminated too. In a structural analogy to lucid things, it holds that while the first illumination is a necessary condition in respect to the first stage of the production of visual operation, the second illumination, that is, that of the medium, constitutes a dispensable circumstance that can contribute to better

---

[94] Ibid., 590–4.

[95] A similar reasoning can be found in Aquinas's *Sentencia libri secundi De anima*, c. 14, 130.

[96] *CDA* disp. 7, q. 3, n. 7, t. 2, 596

[97] Ibid., n. 8, 598.

[98] Unlike Suárez, Manuel de Góis, the author of the Coimbran Commentary on the *De anima*, advocated a different view. Góis says that the existence of the actual transparent or the illuminated medium is necessary for vision. Only if the medium is illuminated can the species begin to exist and be multiplied in it. Only under this condition can the species attain the organ. If only a part of the whole medium between a coloured thing and the organ were illuminated, the sensible object would be visible only within this part of the medium due to its species's limited multiplication. For this, see Collegium Conimbricensis, *In II lib. De anima*, ch. 7, q. 4, art. 2, 229–30.

vision but is not necessary. Even though there might in fact never happen to be illumination of objects without a consequent illumination of the medium, this makes no difference for Suárez's conclusion which considers this illumination of the medium to be unnecessary for the proximate per se visibility of colours. According to the third part of this second conclusion, admitting this dispensability of the prior light in the medium, *a fortiori* the redundancy of the power's illumination must be granted.[99] As spectators in a theatre we know that we can perceive actors even though we are in total darkness ourselves.

4.2.4 Extramission, Intromission and Spatial Perception

Although Suárez regards the reception of visual species as the first necessary stage in the process of visual act production, an ancient question to be answered is whether an act of vision arises at this first stage solely through the reception of species, or whether another complement in the form of extramisssion coming from the percipient's eyes needs to be added. There is the ancient query as to whether this intromission (reception) of species is not to be supplemented or even replaced by the extramission or emission of visual spirits (as Galenists claimed), visual rays (as Euclidians stated) or inner light (as Platonists advocated). Although this theory prima facie does not appear very convincing, in ancient philosophy it was a fairly dominant theory of vision.[100] According to Plato,[101] Galen,[102] Augustine[103] and other ancient authors of the tradition of geometrical optics, such as Euclid and Ptolemy, vision does not run through the mere reception of something from outside but (also) by the emanation of visual rays, which are composed of the fiery element, from the eyes toward a visible object.[104] From the fact that the ancients who

[99] *CDA* disp. 7, q. 3, n. 4, t. 2, 590–2.

[100] For the fact that the extramission postulate, according to which visual perception consists in the perceiver extending outwards to the visible object, was not a rare and eccentric feature also in the thirteenth century, see Lička (2020). With respect to Francisco Vallés and his defence of the extramission theory, the same can be cum *grano salis* said about the sixteenth century too.

[101] For an exposition of Plato's view and that of his forerunners, see Lindberg (1976, 3–6).

[102] As regards Galen's view, see Siegel (1970, 46–7).

[103] In this connection Suárez does not mention Augustine. The reason might be that he knows that his theory is not clear. As noted by many scholars, Augustine, on one hand, seems to have endorsed a version of extramission theory according to which the soul can perceive external objects only if the pupil sends forth visual rays that reach external objects (see *The Greatness of the Soul*, 23.43–4, 66–7). On the other hand, however, he says that an external object causes bodily changes in the organs and that the soul perceives the objects by reacting to these bodily changes (viz. ibid., 25.48, 73). For this doctrinal discrepancy in Augustine, see also Toivanen (2013b, 136) and Pasnau (1997, 131).

[104] The reason for this physiological claim was often experiential: when pressure is exerted on eyeballs, fire seems to flash out in the eyes. For a critique of this naïve explanation of the phenomenon, see Aristotle, *On Sense and Sensible Objects* ch. 2, 437a23ff, 219–21.

advocated the theory of extramission considered perception to be essentially an active process, one could argue that the extramission theory need not be too far from Suárez's active view of perception. One could say that this theory of extramission might constitute the physiological counterpart to his philosophical and psychological theory of perception based on the crucial role of the attentive soul stretching out to sensibles, especially if we take into account that Suárez speaks about a physiological complement grounded in the theory of the effluence of animal spirits from the brain to the sense organs disposing the sensorial organs for perception. One might say that these animal spirits just do not stop in the eyes but simply proceed out of the eyes toward the visible objects.

Suárez opens *CDA* disp. 7, q. 4 'Whether vision comes to proceed through intromission or extramission' with a presentation of four possible versions of the extramission doctrine, largely explaining the contact between an observer and an object in haptic terms.[105] These versions can be divided into two main genera, each divided into two kinds. While (1) one genus can be called the non-reflexive variant, (2) the other can be termed the reflexive version. (1a) According to the first kind of the first genus of extramission theory, the eyes receive nothing from the sensibles. Intraocular fire is said to emanate from the pupil and attains the object by altering the medium. In the object this fire intermingles with the visual species which are emitted from it. Without coming back to the eyes it constitutes a vision of the object in its very presence. (1b) According to the second kind of the first genus, the visible object at first emits species up to the eyes where they get mixed with visual spirits. As soon as these spirits connect with daylight, the air acquires a visual capacity. Much like an extrapolated optical nerve, a point in the air will become a sort of prosthesis of the eyes through which the visible object comes to be visually 'touched'. Accordingly, the air turns to be 'sentient' and the percipient acquires an extended visual organ in the medium by which she can measure a sensible by 'touching' it.[106] The reflexive genus of the extramission doctrine also comes in two varieties. (2a) According to one, when the eyes emit light or visual spirits to a visible object, they blend with the object's visual species and *return* to the eyes to produce an act of vision there. This version is similar to the theory of Plato's *Timaeus* who says the following:

> Now whenever daylight surrounds the visual stream, like makes contact with like and coalesces with it to make up a single homogeneous body alligned with the direction of the eyes. This happens wherever the internal fire strikes and presses against an external object it

---

[105] For an illustrative scheme of these versions, see Castellote (1962, 147); for a similar classification, see Lička (2020).

[106] Vallés, *Controversiarum medicarum et philosophicarum libri decem*, book 2, ch. 28, 107: 'Multiplicari a coloribus species eiusdem naturae cum illis, usque ad pupillam oculi, & rursus exire animales spiritus usque ad externam superficiem pupillae, & illinc lumen quoddam coeleste, eiusdem generis cum lumine solis, quod solum per se non potest illustrare aerem, cum externo lumine coire [. . .] Iuxta quam sententiam recipit aer influxum virtutis videndi illustratione illa, non aliter quam nervus, & ita est nobis aer pro medio coniuncto, in quo videmus. Visio itaque unica agitur in aere & oculo. Haec enim duo media unum fiunt. Unde constat quodammodo, aerem esse sentientem, quia videmus per aerem.'

> has connected with. And because this body of fire has become uniform throughout and thus uniformly affected, it transmits the motions of whatever it comes in contact with as well as of whatever comes in contact with it, to and through the whole body until they reach the soul. This brings about the sensation we call 'seeing'. At night, however, the kindred fire has departed and so the visual stream is cut off. For now it exists only to encounter something unlike itself.[107]

Through 'a single homogeneous body' the likenesses of visible objects are emitted and reach the soul. Only in the soul does sensation take place. (2b) According to the second kind of this reflexive approach, the visual spirits do not reach the visible object but mingle with the visual species at a point in the medium. From this point in the medium they come back to the eyes, in which vision is realized.[108]

Having presented these versions, in a direct link to Aristotle and most scholastics Suárez clearly rejects the extramission theory in favour of the intromission theory.[109] In the first stage of the three-stage sensorial process, vision proceeds *only* by means of the reception of visual species. If there is an activity on the side of the power, it cannot be spelled out by means of a materialistic account of extramission.[110] Suárez introduces four arguments. (1) The visual rays are *redundant*. The impressed species emitted from the sensible objects through the air and received in the attentive cognitive power is actually all that is needed for the production of a visual act. No emission of visual flux proceeding from our eyes is necessary. Air, water and other suitable media can be 'moved' by coloured objects without being previously disposed by the percipient's emission of visual rays. Likewise, the organ and the power

---

[107] Plato, *Timaeus*, 45b–d. For the emission of sensible species up to sight in Plato, see also *Theaetetus*, 153e–154a.

[108] For the presentation of these versions, see *CDA* disp. 7, q. 4, n. 1, t. 2, 600–4.

[109] For Aristotle's rejection of the theory of extramission, see *On Sense and Sensible Objects*, ch. 2, 438a27–438b16, 225–7. In *Generation of Animals* (V.1, 780b35–781a1, 505), however, Aristotle seems to be more open to the theory of extramission. He says that 'Animals with prominent eyes do not see well from a distance, but those with sunken eyes placed in a hollowed recess are able to see things at a distance, because the movement does not get scattered into space but follows a straight course. It makes no difference to this which of the two theories of sight [whether the extramission or the intramission] we adopt'. Suárez is clear that the last sentence does not mean that Aristotle grants the plausibility of extramission theory. What the Stagirite only aims to say is that the theory of extramission is not necessary to explain the abovementioned phenomenon of better vision in animals with sunken eyes and worse vision in animals with prominent eyes. An explanation based on the intramission account is immediately available. Sunken eyes are better in the reception of sensible species because they are more focused. The species are not dissipated in them as they are dissipated in prominent eyes. This is also the reason why we screw up our eyes when we want to see something in the distance (see *CDA* disp. 7, q. 4, n. 5, t. 2, 613–4). The critics of the traditional materialistic notion of extramission, however, were not recruited only from among authors advocating an intromission theory, let alone a more passivist account of perception. Olivi, a radical proponent of an activist view, also endorsed what he called the *virtual* or attentional extension of the soul (*aspectus*) to visible objects and could not make sense of a theory employing visual rays or spirits coming from the eyes. For Olivi's critique of this theory, see Toivanen (2013b, 136–7), Pasnau (1997, 169–70) and Lička (2020).

[110] *CDA* disp. 7, q. 4, n. 2, t. 2, 604.

can be moved too.[111] (2) The ontological status of the visual rays is *obscure*. There are two possible ways to understand their nature. They can either be substances or accidents. However, neither of these is possible. If they were substances, they would have to be (a) either generated in the air, or (b) they would have to come to be through a local movement caused by the eyes. The first alternative is impossible. Every substantial generation would have to entail the corruption of a preceding substance. A substantial generation of *X* is posterior to a substantial corruption of *Y*. In this case the corrupted substance *Y* would have to be the air. Clearly, no such experientially verified corruption is the case. Further, if (b) were the case, vision would have to be of a temporal duration. This contradicts the instantaneous character of seeing, as advocated by Suárez. Moreover, if the substantial visual spirits came out of the eyes, on sunny days these tenuous visual rays would be eliminated by the more intense daylight. On windy days, these particles could not spread in a rectilinear line to an object. They would be easily dispersed. Accordingly, one would see an object which she did not intend to see since the visual species of an object standing entirely out of the percipient's visual field could be brought to the beholder by the wind. If the visual rays were substantial particles, vision would be impossible in impermeable media, such as glass. The visual spirits could not penetrate glass. If the emitted spirits were accidents, they would be identical with light. However, if light rays did emanate from the eyes, vision of colours would be possible even at darkest night. However, this is not only denied by Plato, it is also at odds with Suárez's account.[112] (3) Importantly, the aforesaid non-reflexive genus of extramission theory, according to which vision occurs somewhere in the air, is at a flagrant variance with what Suárez stresses throughout *CDA*, sc. with his theory of the essentially *vital* and *immanent* nature of sensation. It is utterly absurd to claim that the air, being a prosthetic eye and soul-less entity, can *see*. Moreover, considering the version of extramission theory according to which light reaches only a medial point in the medium somewhere between the sensorial power and the object, no emission would be necessary. If this emanation is not necessary for this last stretch of the medium, namely from the object to the medial point, it will be dispensable for the first part of the medium (from the power to that medial point) too.[113] (4) Lastly, (2b) is *unintelligible*. It makes no sense to state that the visual rays come only to a certain point in the medium from which they return to the eye. If there is no impediment in the medium, no such reflection from this point is possible. The emission of visual spirits from the eye is not the action of a free agent who can 'decide' to turn back whenever and wherever it wills.[114]

Despite these cogent arguments Suárez is quick to note that the extramission theory is backed up by a relatively strong argument. He cites Vallés who argues for the theory of extramission by referring to what he regards as a weak spot in the

[111]Ibid.

[112]Ibid., n. 2, 602–6.

[113]Ibid., n. 3, 608.

[114]Ibid., n. 4, 610.

Aristotelian view of perception. It is the question of the perception of the common sensibles, such as size and distance. In contrast to the theories of early modern philosophers, Aristotelian-scholastic philosophy holds that these sensibles are perceived by the *external senses* and not acquired by a rational inference.[115] Beside the five classical common sensibles Suárez mentions also distance, of which he says that it is reducible to size. As it has been said in 3.8, he characterizes distance as a size actually or potentially inserted between two things.[116] But how can there be an intention of *relational* thing, such as distance, in visual species primarily representing the absolute qualities of colour and light?[117] Suárez's contemporary Francisco Vallés was clear that Aristotelians do not have a satisfying reply.[118] He rejects the view that there is a special visual species of distance. However, he also denies the claim that distance is perceived by means of a modification of the species representing colour (Suárez's position). Neither perfect nor imperfect species can cause a vision of proximity or distance. The only thing these species can cause is a vision of colour with varying perfection. Accordingly, size and distance can be seen only if sight somehow steps out of itself and 'touches' a visible thing through the medium. Only when it has successively touched the object can it reliably 'measure' the distance and size of an object seen.[119]

In his reply Suárez agrees that the size of a sensible object is seen neither through the size of the visible species (a small species can represent a mountain or the sun), nor by virtue of a special species. Nevertheless, as said in 3.8, he insists that size and distance are represented through modifications of the visual species representing colours. Despite this predictable and already formulated conclusion, however, Suárez agrees that the issue of spatial perception is 'hot'. However, he adds that it is no less problematic for extramission theoreticians. Considering Vallés's version of this theory, according to which vision occurs outside sight, Suárez notes that there

[115]Despite the prima facie clear implausibility of extramission theories, the issue of spatial perception was the main reason for its perseverance in the history of philosophy and in the history of science. For extramission theories and spatial perception, see Hilbert (2015, 260). On different models of spatial perception including a medieval one, represented by Alhacen, and modern ones, exemplified by Descartes and Berkeley, see Yrjönsuuri (2014).

[116]*CDA* disp. 6, q. 1, n. 9, t. 2, 462.

[117]Olivi criticized the theory of the perspectivists's multiplication of species *in medio* for this reason. This theory does not account for perception because 'the size and distance from the percipient of an exterior object could not be sensed by means of any species received from it'. (Olivi, II *Sent.*, q. 73 [III, 82]). For Olivi's critique, see also Tachau (1988, 42–3).

[118]Galen also complained that Aristotle failed to explain how we distinguish the position, size and distance of objects seen. For this, see Lindberg (1978, 364).

[119]Vallés, *Controversiarum medicarum et philosophicarum libri decem*, book 2, ch. 28, 106: '[...] (si nihil de oculis prodiit ad rem videndam) non aliter potest videri [proximity], quam aut specie alia, aut illius eiusdem speciei modo alio. Sed horum alterum est ridiculum: alterum falsum [...] intensior species intensum videri faciet colorem, non propinquum: remissior remissiorem, non distantiorem'.; ibid., 107: 'Visio itaque fit intrantibus speciebus, quo colores sentiantur, & afficiantur proprium visus instrumentum: & facultate exeunte ad dimetiendum distantiam & magnitudinem'.

are actually two main options regarding the place where vision can occur. Vision can arise (a) either in direct contact with the object perceived, or (b) at some point in the medium. Nevertheless, if (a) were the case, spatial perception would have to fairly reliable. It would be almost impossible to be wrong about the apprehension of size and distance. If (a) were valid, we would be able to sensorily measure size and distance of visible objects very reliably without a mistake. But, clearly, as we have seen in 3.8, this is not the way how our vision proceeds. If (b) were the case, a question would be how to explain the perception of a distance that remains between a given point which was reached by a visual ray and the visible object. The problem would thus remain also for the theory of extramission. All in all, if Vallés denies the theory according to which the distance and size of visibles are perceived through a modification of the species, it is not Suárez but rather Vallés who ought to search for a plausible *explanans* of spatial perception.[120]

In his explanation of the perception of size and distance Suárez does not fully embrace the traditional *explanans* based on a visual cone or pyramid either. According to this theory, advocated not only by representatives of geometrical optics called *perspectivi*, but also by Albert and Aquinas,[121] visible objects appear bigger or smaller because of the size of the angle formed by the visual species sent forth from the visible objects to the perceiver. According to this theory, the visible species are emitted from objects in the form of a pyramid, with the visible object as its base and the eye as its vertex. The longer the lateral sides of this cone are, the lesser the angle is. Provided that this reduction of the visual angle obtains in the case of the same object, it is clear that the object appears smaller and more distant to us than it did before.[122] Despite the prima facie common plausibility of this view, Suárez does not share it without reservations. He formulates three objections to it. (1) If we judge size by the visual angle alone, then small objects perceived close to our eyes should be perceived as larger than big objects seen from afar. In this way my hand that is placed just in front of my eyes would be perceived as being of identical size or even bigger than a high mountain or the sun on the horizon. Since both objects subtend the same angle, the resulting visions will be of the same size. This seems to make

---

[120] *CDA* disp. 7, q. 4, n. 7, t. 2, 614–6.

[121] Thomas de Aquino, *Sentencia De anima*, book 2, l. 15, nn. 8–9, 133–4: '[...] omne corpus videtur sub quodam angulo cuiusdam trianguli vel magis pyramidis, cuius basis est in re visa et angulus est in oculo videntis [...] cum res visa sit maior quantitate, quam pupilla, oportet quod proportionaliter diminuendo pervenit immutatio visibilis, usque ad visum; manifestum est autem, quod quanto latera trianguli vel pyramidis sunt longiora, dummodo sit eadem basis, tanto angulus est minor, et ideo quod a remotiori videtur, minus videtur, et tanta potest esse distantia quod omnino non videtur'. Clearly, this geometrical approach based on the notion of a visual cone has its roots in Euclid's and Ptolemy's geometry. For Euclid and Ptolemy, see Lindberg (1976, 11–7) and Smith (2001a, xxviii–xxxvii). Adopting this approach does not imply an avowal of this or that theory: It can be employed by advocates of both the intromission and the extramission theory. In fact, this reference to the theory of a visual cone is one of Suárez's very few references to geometrical optics in *CDA*. This gives us evidence that Suárez did not have much interest in the tradition of geometrical optics.

[122] *CDA* disp. 7, q. 4, n. 8, t. 2, 616.

sensory perception an overly fallible instrument. (2) If we strictly adhered to this view, we would be forced to accept the consequence that the doubling of the distance of an object necessarily entails the reduction of the object's size to half. However, this proportionality does not accord with our experience. Sometimes doubling the distance can bring about only a small change in the perception of the object's size. (3) Within certain distance the variation of the visual angle does not influence the perception of size. Within certain distance in which a sensible is sufficiently close a visible object can be seen as being of its size regardless of a shift in the angle of the visual cone.[123]

In his 'calibration' of the fallibility and the infallibility of spatial perception Suárez refers to a theory which he regards as least problematic, based again on his metaphysics of modes. Although his theory might be regarded as less plausible, if compared to other theories,[124] Suárez's aim is clear: He wants to satisfy the Aristotelian claim according to which the common sensibles of size and distance are sensibles per se. As such their perception cannot result as late as from inferential processes attributable to the interior senses or the intellect. This is why he makes clear that within a certain distance the visible object is capable of producing a perfect likeness of itself or, in other words, a likeness whose modification corresponds relatively accurately to the object's size. This species will be apt to represent its own size relatively perfectly. However, beyond this distance the sensible likeness emitted from an object fades out and sight becomes more often liable to the production of imperfect perceptual acts as far as assessing size is concerned. It will assess things as smaller than they really are. The size of the sun will be perceived as the same as the size of a palm. Simple as it may sound, an object appears to be of this or smaller size only owing to the reception of a perfect or imperfect visual species.[125] Mutatis mutandis, the same holds for the perception of distance which is, according to Suárez, reducible to magnitude.[126] All too unproblematically he says that it is not difficult to recognize a distance of an object that imprints its species in the eye perfectly. If the sensible likeness and its impression are less perfect, the object is cognized as more remote and an error in the perception of distance is more likely. Clearly, the vision of distance largely depends on the quality of the medium.

---

[123]Ibid., n. 9, 618.

[124]For a different theory advocated by the perspectivists headed by Roger Bacon, see Mantovani (2020). For Bacon et al., the human cognition of common sensibles, such as size and distance, proceeds by 'vision by syllogism' operated by the interior senses. Accordingly, size perception requires one 'to consider the angle and the length of the visual pyramid and to compare these with the base of the pyramid, which is the visible object'. (ibid., 136). Size and distance perception cannot be attained through 'vision by naked sense', but only through a syllogistic process. Mantovani concludes that while for Bacon there is a significant difference related to the manner and the powers of the apprehension of the proper and the common sensibles, such as size and distance, for Aristotelians (it can be added, including Suárez) the cognition of both sensibles is attained by the same power, namely by sight (ibid., 131).

[125]Strikingly and laconically enough, Suárez says that the perspectivists explain this current perfection of the sensible species through the angles of the visual pyramid.

[126]*CDA* disp. 6, q. 1, n. 9, 462.

At dark night we can go wrong in the perception of the distance of luminary objects more easily than in daytime.[127]

4.2.5 Organ of Vision and 'Double Vision'

While in the context of his critique of the theory of extramission Suárez refers to the medical tradition represented by Galen and Vallés as to a negative point of reference, in *CDA* disp. 7, q. 5 'What and of which quality is the organ of the visual power and the power itself' this tradition becomes authoritative for him. Not Aristotle but Galen and Andreas Vesalius, a Dutch anatomist and physician, are to be followed in anatomical and physiological matters. It is not Aristotle's fault that his views of the eye's anatomy have become outdated. It is the demerit of the time in which he lived and in which no dissections of the human body were performed.[128] As mentioned above, one of the main 'corrections' of Aristotle's physiological statements made by a number of physicians and philosophers long before Suárez is his refusal to locate the principle of sensation in the heart. The heart is the organ of sensory appetite and not of sensation.[129] The heart is neither the corporeal principle or root of sensation, nor the organ of touch. The 'root' of *sentiendi* is the brain and the organ of touch is (above all) skin of the whole body.[130] Although the query in *CDA* disp. 7, q. 5 is formulated in physiological terms, it is philosophically framed: Given that we have two organs of sight, do we have one vision, or two visions?[131]

The answer to the question regarding the organ of vision and its character calls for an outline of the anatomy of the human eye. For Suárez up-to-date anatomy is to be found especially in the second book of Vesalius's *De fabrica corporis humani* (1543).[132] In this book Vesalius shows that the substance of the eye having

---

[127]Ibid., n. 10–1, 618–20.

[128]*CDA* disp. 7, q. 5, n. 2, t. 2, 624.

[129]*CDA* disp. 11, q. 1, n. 7, t. 3, 332.

[130]*CDA* disp. 6, q. 6, n. 6, t. 2, 534. For the tactile organ, see 4.5.4.

[131]In connection with so-called stereoscopic vision this issue is also raised by contemporary philosophers. Since our eyes have slightly different points of view due to their position, the two retinal images are slightly different. In line with the topic of the previous section, this binocular disparity is also an important source of information about the distance of objects. Although the visual differences are small, the end result significantly improves depth perception of one's surroundings. The human visual system, as well as the systems of some animals that are sensitive to the successive perception of these differences, can employ them in the performance of a number of visual tasks including spatial perception. For this stereoscopic vision, see Hilbert (2015, 263–4).

[132]For the muscles of the eye, see Vesalius, *De humani corporis fabrica*, book 2, ch. 11: *De oculorum musculis*, 239–42; as regards the optical tunics and humours, see ibid., book 7, ch. 14 "De oculo visus instrumento," 646–650 (for a useful anatomical picture of the eye, see ibid., 643). As C.U.M. Smith, Eugenio Frixione, Stanley Finger and William Clower (2012, 85) say the following: 'This [Vesalius's] magnificent volume is regarded, with reason, as the foundation work of modern anatomy. It is based on his personal experience with probably hundreds of human dissections and

an oval shape contains five tunics, three humours and seven muscles. Passing over the names of the muscles that do not serve vision immediately because the blind men also have them, Suárez provides the reader with the names of the humours and the tunics. At the exact centre of the oval substance there is the pupil (*pupilla*) consisting of colourless crystalline (glacial) humour. This pupil is flattened both front and back into a lenticular form. The transparent liquid in the pupil is served by two surrounding humours, namely by vitreous and albuminoid humour. The former, located in the interior part with respect to the pupil, is called vitreous humour. It is called vitreous because it is soft and glistening. It is like a mass of glass that has been melted in a furnace. Since the pupil is not able to nourish itself by blood, it needs the vitreous humour to provide it with nutrition.[133] The other humour, which is white, is called albuminoid (sometimes also aqueous) humour. It resembles egg white. This humour protects the crystalline humour. Besides these two assisting humours, the central crystalline humour is supported by five tunics. The innermost tunic, immediately protecting the pupil, is called the arachnoid membrane. It is called arachnoid for its similarity to a spider web. There follows the recticular membrane interwoven with many fibrils. Then comes the uvea tunic that makes our eyes of this or that colour. This tunic leaves a foramen through which the species can get to the pupil. The penultimate is the cornea which is hard and transparent. It encircles the pupil from all sides and constitutes, as it were, a rampart to protect it. Lastly, there is the outer conjunctive layer which is robust enough to shelter all the interior humours and tunics of the ocular cavity.[134] Finally, there are also two optic nerves, one right, the other left. While the left one originates in the left part (hemisphere) of the brain, the right one descends from the right hemisphere. Both nerves are channels through which animal spirits flow from the brain to the eyes. As said above, they physiologically condition vision and emanate especially when attention is strained. This is also

---

similar numbers of animal vivisections [. . .] With the Fabrica we have finally broken through from the medieval world of copyists and flights of anatomical fancy to the 'light of common day.' David Lindberg (1976, 173) says that Vesalius's *De humani corporis fabrica* is '[. . .] a detailed description of the eye which clearly reveals a more intimate firsthand knowledge of ocular anatomy than any of his predecessors could boast'. Michael Edwards (2012) says that in physiological matters related to the anatomy of the eye the late Suárez (namely, the Suárez of *TDA*) was influenced not only by Vesalius but also by Girolamo Dandinus (1554–1634), an Italian Jesuit, and his gigantic *De corpore animato* (1610). Suárez quotes Dandinus directly in *TDA* book 1, ch. 11, n. 8, 545, where he deals with the issue of the intellect and its immortality, and in *TDA* book 3, ch. 18, n. 1, 673 where he *ex professo* treats the issue of the organ of sight. However, it is editor Alvares who introduces this paragraph with the following heading: 'Consule Dandinum, lib. 4, De partibus heterog. a c. 38'. Moreover, since the part of *TDA* discussing the organ of sight does not differ significantly from *CDA* where of course no reference to Dandinus is made and since this part was not revised by Suárez but by Balthasar Alvares, it is clear that Suárez's views of the anatomy of the eye in *CDA* were influenced rather by Vesalius than by Dandinus.

[133]Suárez notes that Aristotle nowhere distinguishes the vitreous humour from the crystalline humour. See *CDA* disp. 7, q. 5, n. 1, t. 2, 622. However, again, in this matter not Aristotle but authors from the medical tradition, this time Galen, are to be followed. For the distinction between these two humours, see Galen, *De usu partium corporis humani*, book 10, 285ff.

[134]*CDA* disp. 7, q. 5, n. 1, t. 2, 620–4.

why a brain or nerve injury affects external sensation. When the brain is injured, it does not release these animal spirits to the eyes and to the other sensory organs. Both visual nerves, which also carry the sensible species to the brain, intersect and together make up what anatomists call the common nerve or optic chiasma.[135]

As Suárez notes, some medieval authors, such as Avicenna, assigned the proper organ of vision just to this optic chiasma.[136] We can have a unique vision only because of this intersection. This intersection is the place where sight is to be located. The visual images coming from both eyes somehow fuse at this intersection. If the pupil and not this intersection were the proper organ of vision, we would get two visions of the same thing.[137] Suárez rejects this Avicennian view. Following Aristotle and Galen he declares that the pupil and not the optic crossing is the proper organ of the visual power.[138] He introduces two arguments for this conclusion.[139] (1) The first one comes from the character of the crystalline humour. The adequate organ of the visual power must be the crystalline humour since only this matter and not a coloured stuff is the most suitable receptacle for the visual species. The glacial humour is soft, diaphaneous and colourless. If it were, for example, black, it would not be suitable for the reception of the sensible species of white colour. Even though the reception of white colour, in its essential aspect, is 'spiritual', in the material powers the assimilation concerns not only the powers but also the organs which are the medium through which the species are received in the power. This is why this reception also requires a proper disposition on the side of the organ according to which a receiving organ does not and cannot possess the properties of the received qualities. The organ of hearing, that is, the air in the inner ear, must be motionless to be perceptible of sounds subjected in the airy waves. Also, the tongue must not have a bitter coating to be able to sense the sapour of honey properly. Since touch perceives all the first qualities without which its organ cannot exist, it cannot be deprived of them. However, it must be somehow free of them by being well-balanced, that is, by being not too cold and not too hot to be able to detect tactile qualities as objectively as possible. The same holds for the pupil. In order to perceive colours the pupil must be colourless and diaphaneous.[140] However, this dispositional character is not exacted from the other ocular parts surrounding the pupil. Some of

---

[135] Ibid. n. 7, 630; see also *CDA* disp. 6, q. 6, n. 6, t. 2, 534–6.

[136] Avicenna, *Liber de anima seu Sextus de naturalibus*, p. 3, ch. 8, 268–9. This question of whether the visual power is to be located in the glacial or crystalline humour, or at the junction of the optic nerves, was a commonplace issue in late medieval philosophy. For its explicit treatment in Blasius of Parma (d. 1416), see Lindberg (1976, 131).

[137] *CDA* disp. 7, q. 5, n. 8, t. 2, 630.

[138] Aristotle, *History of Animals* I.9, 491b20–1, 39: 'In the inner part of the eye we have (a) a fluid part, the instrument of sight, called the pupil'. For Galen, see *De usu partium*, book 10, 285: 'Dictum a nobis praeterea est, quod ipse crystallinus humor est primum videndi instrumentum'; Aristotle, *DA* II.7, 418b26–7, 165: 'It is the colourless which is receptive of colour, as the soundless is of sound'.

[139] *CDA* disp. 7, q. 5, n. 3, t. 2, 626.

[140] For this see *CDA* disp. 5, q. 2, nn. 27–30, t. 2, 330–6.

these parts can be and are coloured. Properly speaking, they are not the adequate visual organs. They only serve and conserve the pupil which is the adequate organ of vision. Being soft and aequous the pupil also best receives (not retains!) its forms. Following Aristotle's *De* sensu, the organ of sight cannot be fiery. When the eye is damaged or decays, the aqueous humor flows out.[141] The organ of sight must be of a watery disposition, rather than of a fiery or airy one, since only as such it is more apt for the reception of the visual species.[142]

An objection referring to the abovementioned experience of the cat's eyes glittering at night can be adduced to this claim. Does not this after all lead to the conclusion that the composition of eyes is fiery?[143] Suárez gives a twofold answer to this question. In his first exposition, he refers to Aristotle's explanation from *De* sensu.[144] Not only feline but also human eyes and the heads of certain fish glitter because of their smoothed surface. This phenomenon of gleaming is explained by the eyes' surface and quick movement. This is also explained by an experiment according to which when we press our eye rapidly, fire seems to flash out, which happens also at night and darkness. By pressing our eye quickly, a single eye becomes double in a way. One part will be seen, the other part will be seeing. This does not happen if we press the eye slowly. It is not due to an intraocular fire existing in the eyes, but only due to the special character of their surface and rapid motion that leads to the eye's 'duplication', which explains this vision of light in the eyes.[145] (2) The other argument is structurally similar to an experiential evidence often utilized by Galen. Galen employed this kind of argument (see 5.2.2) in his theory of the three interior senses. The ancient physician argues for the existence of three distinct internal senses based on the existence of three distinct brain ventricles. Each is the seat of a distinct interior sense. An injury of this or that ventricle leads to a damage corresponding to this or that cognitive function. Analogically, concerning the visual organ, experience teaches us that when the pupil is irrecoverably injured, vision is irrecoverably impaired too. This does not obtain if some other external parts of the substance of the eye are impaired.[146]

However, if the pupil and not the optic chiasma is the adequate organ of sight, what is the natural purpose of this optic crossing? Coming back to the main question, how can we have a single vision, if there are two pupils in which two powers of sight are located? Why does a thing not appear double? Suárez is clear that this intersection of the optic tubular nerves is ordained by nature not for getting a unique vision but for a certain compensation. Once a single eye is impaired and one becomes single-eyed, the optic chiasma will secure that the animal spirits which normally

---

[141]For this example, see Aristotle, *On Sense and Sensible Objects* ch. 2, 438a18, 225.

[142]Aristotle, *Generation of Animals* II.10, 656b2–3, 177: '[...] the sense-organ of sight is identical in its nature with water, which of all transparent substances is the easiest to keep confined'.

[143]*CDA* disp. 7, q. 5, n. 6, t. 2, 628.

[144]See Aristotle, *On Sense and Sensible Objects* ch. 2, 437a30–b10, 221.

[145]*CDA* disp. 7, q. 5, n. 6, t. 2, 628–30.

[146]Ibid., n. 3, 626.

flow from the brain to both eyes will come in a higher degree to the second functioning eye. By that channeling the power of the healthy eye will be strengthened and reinforced. Accordingly, it will compensate for the lost force of the broken eye. This can be proved by a simple experiment: If one eye is closed, or if one eye is covered by hand, the other one will see more sharply.[147] Suárez thus agrees that the visual power comprises two powers since it contains two distinct organs.[148] He agrees that all binocular vision even under optimal conditions produces two visions. However, he does not agree that this necessarily results in the double appearance of a thing. It does not do so due to the equal positions (*situs*) and due to the fact that both eyes are always intentionally affected in the same way.[149] It is not often the case that one eye has a concave shape while the other has a convex one, or that one is situated significantly higher than the second. Normally, the eyes are equally affected by the visual species since they exist in the same line. Only in rare or abnormal cases, in which the eyes are inequally posited, can a single thing appear double.[150]

4.3 Hearing

In the order of perfection, the second sense treated by Suárez is the sense of hearing. From the point of view of the history of ideas, hearing was always regarded as 'the second sense'.[151] Since ancient times up to early modern period, the majority of philosophers and scientists[152] regarded auditive perception as 'the small brother of visual perception'. As shown in 4.1, this evaluation can be substantiated in manifold ways. Apart from the abovementioned reasons, it is also possible to mention the origin of the corresponding sciences, namely optics and acoustics. Unlike optics, which got its impetus from the work of Alhazen and other Arab authors, acoustics as a systematic physiological and psychological research into hearing did not become a separate science until the seventeenth century when it began to get its shape with the

---

[147]Setting aside whether this actually is the case, Lindberg (1976, 36) mentions Hunain Ibn Ishaq (d. 877) who defends the validity of this experiment. However, this view can also be found in the pseudo-Aristotelian *Problems* XXXI.4, 957b21–2, 185.

[148]This is the view held by Aristotle in *Parts of Animals* II.11, 657a2–5, 181.

[149]Nevertheless, one can still argue that the eyes are *never* affected in the same way and always receive slightly distinct visual species since their positions are distinct. Consequently, the common sense will have to unify these two distinct images.

[150]*CDA* disp. 7, q. 5, n. 9, t. 2, 632. From what has been said above, unlike contemporary research, Suárez does not seem to consider this binocular disparity to be an important source of information about the distance of objects.

[151]For this evaluation of hearing, see the title and the following collection of essays in Burnett, Bend and Gouk (1991).

[152]An exception is the abovementioned French philosopher and mathematician Charles de Bovelles.

works of Galileo Galilei (1564–1642) and Marine Mersenne (1588–1648).[153] Despite the 'visiocentric' approach of the scholastic tradition, the second-class position of hearing, as Suárez's discussion testifies, does not mean that scientific and philosophical enquiries into hearing and sound are entirely derivative and secondary for him. The ontological issue of the character of *sonus* constitutes an important challenge *sui generis* for the Aristotelian theory of perception for at least two reasons. First, more so than with colour, some of Aristotle's formulations concerning the ontological status of sound seem to suggest a reductionist view of sound. Accordingly, the proper sensible of hearing is to be reduced to a common sensible, namely to the motion of air.[154] Second, in *On the Soul* Aristotle makes clear that sound *in actu* resides as in its subject not in the sensibles that produce sound by mutual percussion, but only in air.[155] In the sensibles sound exists only *in potentia*. This statement can be taken as standing in direct tension with other Aristotelian claims according to which the affective qualities of the external senses *in actu* indwell in material substances whose accidents they are.[156] Suárez takes up these two challenges in *CDA* disp. 7, q. 6, where his ontological theory of sound and its subject is systematically elaborated. As we shall see below, in these questions Suárez defends what can be called a version of 'Sonic Realism'.[157]

In the first part, I will consider these two challenges related to issues of the nature, cause, and subject of sound. In the second part, the manner in which sound is propagated *in medio* will be treated. More precisely, I will focus on the topic of the exact co-ordination of natural and intentional *modi propagandi* and *multiplicandi* of sound in the medium. In the third part, I will complete my presentation with an outline of Suárez's view of the organ of hearing.[158]

---

[153]See Burnett et al. (1991, 2). The name of this new science was coined by Samuel Reyher (1635–1714) in his *De natura et iure auditus et soni* (1693). For this, see Wittmann (1987, 271). Of course, it cannot be denied that the mathematical study of harmony and research into architectural acoustics goes back to ancient times with names such as Pythagoras and Marcus Vitruvius Pollio (ca. 80–70 BC–after ca. 15 BC). Yet, acoustics as a complex psycho-physiological enquiry emerged as late as in the period of the Scientific Revolution in the seventeenth century.

[154]Aristotle, *DA* II.8, 420b11–2, 117: '[. . .] sound is a kind of motion of the air'; *On Sense and Sensible Objects* ch. 6, 447a1–2, 269: '[. . .] sound is held to be the motion of something travelling [. . .]'.

[155]Aristotle, *DA* II.8, 419b9–11, 111: 'But the sound actually produced is of something striking against something else in a medium'.

[156]For these challenges, see also Pasnau (1999b, 2000, 2002a, 189).

[157]For an explicit endorsement of Sonic Realism and for a struggle against the 'Tyranny of the Visual' in the recent analytical debate, see O'Callaghan (2007).

[158]For Suárez's theory of sound and hearing, see also Heider (2017c).

4.3.1 Nature, Cause, Subject and Medium of Sound

Suárez opens *CDA* disp. 7, q. 6 'What sound is and how and in which subject it comes to be' with the following concise definition of sound: '(1) Sound is a sensible quality (2) coming forth from a violent percussion or from the division of a body (3) that is apt to receive it [this quality]'.[159] Distinguishing three parts in this conclusion, in (1) Suárez shows that sound is a *qualitas passibilis*, an accident of quality that can be 'suffered' and received by (in) the pertinent sense organ and the corresponding power. As we have seen, perceptual acts are distinguished by their proper sensible objects; these acts then specify and per se *primo* affect the particular senses. This is the reason why sound cannot be reduced to the common sensible of motion. If it were, it could not differentiate the acts and the powers since it could be sensed not only by hearing but also, for example, by sight and touch. This would violate the ontological status of hearing as a distinct power, which would in turn result in a revisionist account of the number of the external senses. As shown in 4.1, this is something that Suárez does not accept. Although Suárez is aware of Aristotle's statements suggestive of this reductionist account of sound, the threat of the revision of the common view is the reason why he rejects it. Sound cannot be reduced to the local motion of air since the 'ratio' of motion is separable from the 'ratio' of sound. If one, namely motion, can exist without the other, namely sound, the former cannot be essentially constitutive of the essence of the latter. In the context of his dismissal of the Neoplatonic heavenly music or 'music of the spheres',[160] he exemplifies this separability by means of the example of soundlessly moving planets. *Sonus* can be a quality that somehow falls upon (*accidere*) or accrues to the motion of air.

Having arrived at 'the genus' of sound, in (2) Suárez approaches the issue of the 'specific difference'. Since a noncircular definition of the specific difference 'in unico verbo' is impossible, Suárez provides the reader with what can be called a causal definition. In the causal context he mentions two models that have been formulated in the history of philosophy and in the history of science.[161] First, sound is caused by the violent percussion of two bodies, one striking (*percutiens*), the other one struck (*percussum*), in a suitable medium. According to this model, the thing struck and the medium are two really distinct things. One can think of the example of the clapping of hands or the ringing of a bell. According to the second model, sound is produced only by the division of a single object. In this case what is struck and the

[159] *CDA* disp. 6, s. 6, n. 1, t. 2, 634: 'Sonus est qualitas sensibilis proveniens ex violenta percussione vel divisione in corpore apto ad recipiendum illum'.

[160] For this rejection, see *CDA* disp. 9, q. 5, n. 1, t. 3, 170.

[161] *CDA* disp. 7, q. 6, n. 3, t. 2, 638. Contrary to Aristotle, who in *DA* II.8 explicitly mentions only a three-member model, Suárez (probably following Avicenna) also adds a two-member model. For this innovation of Avicenna's, see Burnett (1991, 52). Nonetheless, Suárez refers to Aristotle (*DA* II.8, 420b8–9, 115–7) who says that the whistle and the lyre have perfect sound similar to the human voice even though their sound is caused only by virtue of the percussion of air.

medium are not distinct *realiter* but only conceptually. As examples of this second model it is possible to mention the tearing of a paper and the production of a voice caused by a sole percussion of air to the trachea.[162] In line with Aristotle, Suárez proceeds with a survey of the physical properties that sound-producing objects must possess in order to be able to produce a sound or become sonorous. Beside the necessity of a violent motion (a slow motion cannot bring about a noticeable sound), the most fitting properties are smoothness, hollowness and concavity of body. Being hollow and concave are the most suitable properties for retaining air. These two characteristics make objects truly sonorous.[163]

In the next step in his analysis (2) Suárez addresses more precisely the causal issue of the generation of sound. How do forcible contacts of bodies and motion of air causally concur in the production of the quality of sound? Employing the notion of the ontological hierarchy between motion and sound (another instance of Suárez's multifarious hierarchical ontology, which obtains especially between the vital and the nonvital) he raises the question of how an imperfect thing, such as a motion, can produce a more perfect thing, namely the sensible quality of sound. In analogy to the incommensurability of the inanimate and the animate, Suárez says that the common sensible of motion cannot be regarded as the efficient cause of the proper sensible of sound because motion is a less perfect quality than sound. As we have seen, this 'maxim' was a reason why Suárez endorses the 'harmony of the powers of the soul' that interact not causally but 'a-causally' by means of being rooted in the common soul. Of course, this *explanans* is not applicable here. Instead, however, Suárez employs the *topos* of a latent or occult quality, mocked by early modern authors who wrote in the era of the scientific revolution in the seventeenth century. Referring to the example of a projectile (a stone) and a 'latent' impulsive quality (*qualitas impulsiva*) imprinted to this stone, Suárez applies a similar device in his analysis of the efficient cause of sound. This latent (impulsive) quality is called 'sound-generating quality'. Briefly speaking, only this quality can be the efficient cause of sound. The motion of air can be seen at most as its necessary condition. This is also why this quality does not start to exist in the same moment in which the contact concurs but immediately after it. Despite this mere conditional status of the movement caused by percussing bodies, the motion of air is what makes sound a successive or transient entity 'perduring' in the air through its 'temporal parts'. Unlike colour, sound does not have a fixed and permanent being in a sensible

[162]Following Galen (see *De usu partium corporis humani*, book 10, 208), Suárez compares the production of voice by human or animal vocal chords to a whistle which is caused solely by the percussion of air. The whole trachea is compared to the concave shape of a whistle; the lungs are likened to a mouth inspiring air into the whistle's tube; the tongue of the concave larynx corresponds to the upper parts of the whistle. As inspiring air into the whistle while at the same time closing some of its holes and opening others produces certain tones, so the apertures of the larynx–once dilated, another time narrowed down by means of muscles–result in the emission of diverse sounds. For *vox* and its distinction from *sonus* and *locutio*, see *CDA* disp. 7, q. 7, nn. 1–3, t. 2, 652–8.

[163]*CDA* disp. 7, q. 6, n. 1, t. 2, 634–6.

subject. It lasts only while the motion of air is being produced and conserved by the percussing bodies. As soon as percussing bodies cease to be active, sound ceases to exist too.[164]

Having explicated the first two parts of the 'definition' of sound, in (3) Suárez deals with the subject of sound or, more precisely, the subject as a limit case of medium. What body is apt to receive sound? Is it the sound producing sensible object, or is it air commonly considered to be its medium? Aristotle's view seems to be clear. In the beginning of *On the Soul* II.8 he distinguishes between potential and actual sound. The colliding bodies are the subjects of sound only potentially since they are its efficient causes or, in line with Suárez, the necessary conditions. *In actu* sound inheres in air.[165] Clearly, since the subject of sound must have the property of agitability, the subject must be only air.[166] Also our ordinary language (this argument from ordinary language is not used by Suárez, though) confirms this conclusion: We usually say: 'Sound fills the air', 'Music filled the temple' etc. On the other hand, however, Suárez often says that we have a pre-theoretical intuition for the opposite view, based above all on the more or less exact localizability of sounds. By hearing a sound, we can detect the place from which it came. If we do not know where our cell phone is, we ask someone to give us a ring. Consequently, sound cannot have an indeterminate location in air but at least for a while it must be fixed and located in the object producing it.[167] Suárez substantiates this conclusion with three arguments. (a) Let us think of two different pairs of percussing bodies, both generating exactly the same motion (undulation) in air. We often experience that even though both produce the same motion, the sounds produced by those two pairs of bodies are different. However, if the subject of sound were air, both strokes would necessarily produce the same sound. However, this is not the case. In order to explain this phenomenon, we have to 'fix' the sound in the sound producing objects. Consequently, only these sonorous objects will make a difference in the sounds. (b) If the subject of sound were air, the motion of air would have to be the full *explanans* of its quality, say, of loudness. Accordingly, a stronger percussion of air would have to produce a louder sound, a weaker stroke would always lead to a more silent sound and an equal blow would result in an equal sound. However, this is not necessarily the case. A smaller percussion of the mallets on a cymbal can cause a louder sound than a stronger strike on a cloth. This gives us evidence that their difference can be explained only if the cymbal and the cloth, and not air, are the subjects of sound. (c) As a sound produced by a violent discontinuation of parts of

---

[164]*CDA* disp. 7, q. 6, nn. 1–2, t. 2, 636–8.

[165]Aristotle, *DA* II.8, 419b4ff, 109. For this view, see also Thomas de Aquino, *Sentencia libri secundi De anima*, ch. 16, 136.

[166]*CDA* disp. 7, q. 6, n. 4, t. 2 640.

[167]For this stance and the incoherence of the standard view according to which sound exists in the medium, see Pasnau (1999b, 311ff) and (2000). According to Pasnau, we should adhere to the view that sounds exist *in* the objects by which they were generated, and thus we should understand sound in the manner of colour. For this issue of the localization of sound, see more 4.3.2.

air exists in air, so sound generated by a tearing of paper or by a violent discontinuation of parts of the paper will exist in the paper itself.[168]

Like for most early Jesuits, for Suárez a common view is equal to a true opinion. 'Opinio communis' is often regarded by the Jesuit as an important methodological regulative in his decision-making about the truth and probability of philosophical views. Here the common view is represented by the first opinion. The quality of sound principally and primarily inheres in air since air is the kind of subject that is primarily moved by the percussion of sound-producing bodies.[169] A crucial strategy in his reply to the aforesaid arguments for the opposite view attributing the subject of sound to sound-producing objects, however, is the following: Variation in the quality of sound is not only and primarily due to the adjacent or extrinsic air; it is also due to the shape of the bodies and the *intrinsic* air which is contained in the pores of the percussing, as expressed especially in the first and second arguments. The subject of sound will be not only the surrounding or extrinsic air, but also the air inhering in the sound-producing bodies.[170] Admitting this kind of air will save the phenomena adduced for the view according to which the subject of sound is to be assigned to the sound-making bodies. Regarding (a), Suárez briefly replies that even though two pairs of percussing bodies produce exactly the same motion in air, this extrinsic air is distinct from the intrinsic air which is contained in them. As for (b), it need not always be the case that a stronger percussion will entail a louder sound and a weaker percussion will entail a more silent one. Due to the figure and material disposition of sonorous things in which intrinsic air is reflected, it can often happen that a weaker percussion will cause a stronger sound than a stronger percussion. Concerning the last argument based on the production of sound by scission, Suárez replies that sound is produced by scission due to the surrounding extrinsic air or to the intrinsic air which is included in the divided thing.[171]

However, is air the only subject of sound, or can water and other elements also become its subjects? Are there also other kinds of subject of sound except air? It seems that the reply must be negative. The formation of sound requires a forceful contact of sonorous bodies and an easy divisibility and agitability of the medium. But these two conditions are not met in water. Water is too heavy and is not easily divisible. Moreover, air is explicitly mentioned by Aristotle in many passages of *On the Soul* II.8 as the subject of sound: the air which is violently expelled from the interspace between a striking thing and a thing struck produces a sound; when the air is moved as a continuous whole it causes hearing; sound-producing bodies cause movement in a body of air; sound is a kind of movement of the air.[172] All these formulations with some additions of other statements from *History of Animals*,[173] in

[168] *CDA* disp. 7, q. 6, n. 5, t. 2, 642–4.

[169] Ibid., n. 6, 644.

[170] This, however, implies a substantial qualification of the first view.

[171] *CDA* disp. 7, q. 6, n. 7, t. 2, 644–6.

[172] See Aristotle, *DA* II.8, 419b20ff, 110; 419b33–5, 113; 420a3–4, 113; 420b11–2, 117.

[173] Aristotle, *History of Animals* IV.9, 535b12–24, 77.

which Aristotle says that even fish produce sounds by the attrition of spirits, that is, of air, seem to indicate that for him only air can be regarded as the subject of sound.[174] However, Suárez does not share this view. Unlike, for example, Pedro Hurtado de Mendoza,[175] he does not restrict the subject and medium of sound to air.[176] Not only air, but also water, fire, which is even more tenuous than air, and all 'washy' and 'airy' mixtures, such as various liquids and clouds, can become subjects of sound. Although it is clear that water will be far from being the best subject as it does not enable a quick division or compressibility and requires a more forceful movement to strike properly body against a body, it has several properties, such as being soft and being compressible, that qualify it for becoming a subject of sound. Suárez refers to *On the Soul* II.8 where Aristotle explicitly admits that sound can be heard in water too: 'Sound is heard also in water as well as in air, but less loudly'.[177] At the same time he refers to Aristotle's *History of Animals* to justify this conclusion. He says that frogs can emit sound underwater[178] and that fish swimming at the bottom of the sea hear and flee from oncoming ships and oars.[179] He is sure that although Aristotle usually names only air as the subject of sound, he does not limit it to air. By these frequent references he wants to say that air is *its most fitted* subject.[180]

Suárez's solution to the issue of the subject of sound determines his stance to his query about the medium. Importantly, this solution will be further operative in his reasoning for the existence of the audible species. This medium must be a *plenum* extended between the sound-generating object and the auditory organ. As with sight and colours, vacuum cannot transmit sound.[181] If sound can be produced in water, fire and airy mixtures, the same entities must also be conceived as media in which sound can be propagated.[182] Despite this determination of the kinds of media of sound, Suárez adds two additional analyses. First, although he concedes that water is the subject of sound, he is also aware of Aristotle's textual evidence in *On the Soul* against this claim about the medial character of water. Second, dealing with the issue of the list of the media of sound's propagation, Suárez also asks whether terrestrial and dense bodies (*terrestria et densissima corpora*) can transmit sound too.

In the analysis of the first topic, Suárez refers to the following passage from Aristotle's *On the Soul* II.8:

---

[174]For this view, see also Averroes, *Commentarium magnum in Aristotelis De anima libros*, book 2, text 76, 245, 9–11: 'Dicit [Aristotle] quod medium in voce est aer, non aqua, quia animalia que sunt in aqua non sentiunt, ut michi videtur, nisi per voces cadentes in aere extra aquam'.

[175]Hurtado de Mendoza, *Universa philosophia*, De anima, d. 17, s. 3, n. 24, 678.

[176]*CDA* disp. 7, q. 6, n. 10, t. 2, 648–50.

[177]Aristotle, *DA* II.8, 419b18–9, 111.

[178]Aristotle, *History of Animals* IV.9, 536a8–14, 79.

[179]Ibid., IV.8, 533b4–8, 63.

[180]*CDA* disp. 7, q. 6, n. 11, t. 2, 650.

[181]*CDA* disp. 7, q. 8, n. 5, t. 2, 668.

[182]Ibid., n. 2, t. 2, 662–4.

> The medium in the case of sound is air, but in the case of smell has no name; for air and water have certainly a common characteristic, which is present in both of them, and bears the same relation to that which emits smell as the transparent does to colour.[183]

In this passage Aristotle differentiates between the medium of sound and the medium of smell by saying that while air and water are the kinds of media in which fragrances are spread, air is the only medium for the propagation of sounds. Yet, provided that sound can also be propagated in water, as said in *History of animals*, this 'locus sane difficilis' cannot be explained away by the above-employed strategy saying that the reason why Aristotle mentions only air is to highlight its status as the best medium. In fact, fragrances can be best spread in air too, yet Aristotle mentions water here as well. Nor can it be said that Aristotle speaks about the possibility of water's reception of sound only according to intentional being, and not real being.[184] As stated above, water can be the medium of sound also according to real being. Suárez's solution to this textual unclarity seems to be similar to the theory of Pietro d'Abano (1250–1316), as laid out by him in *Expositio Problematum Aristotelis*. For d'Abano sonus can be heard in water since water has not entered the deeper parts of the percipient's ear.[185] If it did, the sound could not be heard. This also makes a difference from fragrances that can be smelled even though water gets into the intrinsic parts of the nose.[186]

The medial elements and mixtures must be of a high *raritas*. Such 'penetrability' does not exist in earthy objects. However, despite their density, which seems to point to a negative reply to the second question, Suárez after all considers the question of whether dense objects can become mediums of sound. As usual, he sides with the common view, which here is the negative reply. Nevertheless, in his typical 'probabilistic balancing', he is not ignorant of the evidence for the affirmative view either. No matter how thick and nonporous the walls in a prison cell are, a prisoner, if attentive, can still (weakly) hear sounds coming from outside. This seems to lead to the inference that sounds, in principle, can penetrate nonporous walls as well. The relevance of this example can be undermined by saying that there are always latent pores in earthy objects, such as walls, since in general there are no absolutely nonporous objects. Consequently, if a prisoner can hear a sound, it is because of these pores filled with air. Thus the real transmitter of sound will be only the intrinsic air in the pores and not the element of earth. Suárez consider this reasoning probable. Nevertheless, he refuses to take it as a decisive argument for the principal impossibility of sound propagation in earthy bodies. Unless we perform an experiment with

---

[183] Aristotle, *DA* II.7, 419a33–4, 109.

[184] *CDA* disp. 7, q. 8, n. 3, t. 2, 664.

[185] For this view of Pietro d'Abano, see Burnett in Burnett et al. (1991, 55). For another doctrinal affinity with d'Abano concerning the *qualitas dolorifera* as the proper sensible of the sense of touch, see 4.5.2. D'Abano's statement can be found also in Aristotle's *DA* II.8, 420a9–13, 113: 'The air in the ears is lodged deep, so as to be unmoved, in order that it may accurately perceive all differences in motion. That is why we can hear even in water, for the water need not enter as far as the air which forms part of the ear; nor even into the ear itself, because of the convolutions'.

[186] *CDA* disp. 7, q. 8, n. 4, t. 2, 664–6.

objects composed of entirely nonporous materials (of course, this experiment could be realized only at a high level of technology capable of producing entirely sound-proof materials), we are not allowed to infer this conclusion. Accordingly, the theory granting sound penetrability to those objects has a probability of its own, especially if one considers that Aristotle nowhere gives a complete list of all the potential media of sound. If he did, fire would have to be excluded from this account. Yet, Suárez finds the opposite claim more probable: Earthy bodies are not mediums of sound. Dense bodies cannot transmit sound because sound is a *res succesiva*. As a transient being sound is to be conveyed through its medium swiftly. Accordingly, its medium must be of such composition as to enable and transduce this quick propagation. Earthy bodies cannot do that.[187]

4.3.2 Ways of Propagation: Real, Intentional or Both?

Compared to the question regarding the kinds of media of sound, the query concerning the manner of sound propagation was a more controversial issue in medieval and Renaissance philosophy. Does *sonus* affect its medium *realiter*, or does it affect it only *intentionaliter*? Or should we say that in air or in another medium sound is spread in tandem, that is, both as a real quality and as species?[188] If the latter is the case, does the spiritual multiplication of sound occur in the first instant of its generation, or does it begin only after the spreading of sound comes to lose its *impetus*? Are sound and its species propagated in the medium in an instant, as visibles are, or are they propagated in time?

As I have said in 3.1 and 3.2, Suárez embraces the species theory for all kinds of human cognition except the beatific vision. The sensible species is a necessary device to secure a cognitively relevant contact of sensible qualities with sense organs and through them with powers. A sound-producing object does not emit only the real quality of sound which is circularly propagated in the manner of air waves, in a way similar to water waves formed on a water surface when a stone is thrown into the water.[189] Suárez endorses the view according to which the auditive species is

[187] Ibid., n. 5, 666–8.

[188] For a controversiality concerning the very existence of the species of sound in Second scholasticism and a progressive denial of the sensible species among early Jesuit authors, see Leinsle (2016, 107–10); for a denial of the sensible species already in the sixteenth century in the case of Antonius Balduin (1533–1585), a Jesuit professor at the university in Dilingen in 1570–1582, see Leinsle (2006, 388). For an overall denial of both the sensible and the intelligible species in early Jesuit philosophy, cf. Girolamo Dandini (1611), *De corpore animato, De anima*, book 3, 1634: 'Hic velut ultimam huius commentarii clausulam adnotabo, sensus ab obiectis non per intentionales formas, ut plerique inquiunt Latinorum: sed per reales qualitates moveri'. For Dandini's rejection of the species theory *en bloc*, see Tropia (2020).

[189] For the circular propagation of sound, see *CDA* disp. 7, q. 7, n. 4, t. 2, 658. This water wave metaphor in the context of the propagation of sound was employed already by Marcus Vitruvius Pollio. For this, see Wittmann (1987, 77).

produced in both the organ and the power.[190] Hearing could not elicit an auditive act without being informed by the audible species conceived as its partial instrumental cause. A more controversial issue is related to the way sound is multiplied and its relation to natural change in the medium. Suárez starts by presenting two extreme positions. The first one holds that sound is spread in air or in another suitable medium only *realiter*. A sonorous object stirs air up to the organ, where it affects this organ and the power by means of a real alteration. If the subject is attentive, an act of hearing is produced in the power. As already mentioned above, this view can be found in Aristotle who says that 'it is air, when moved as one continuous whole, which causes hearing'.[191] Another version of this view asserts that a real transmutation occurs only in air (and perhaps fire), not in water. Assuming that sound can be heard under water, it holds that since sound as a real quality cannot be spread under water, it can be spread there only intentionally. Since the audible species is a more subtle entity than the real quality of sound, the species is better capable of pervading a less thin and less porous medium, such as water. According to the second view, sound is reverberated in air only intentionally. Sound cannot be transmitted *realiter* since (and this is the assumption of the view) it exists *realiter* as in its subject only in sonorous bodies. Since Suárez does not share this assumption, it is not surprising that he declines this view. Considering Suárez's general view of the emanation of sensible species from sensibles it is not surprising to see him dismiss the first view too.[192]

In line with this general stance, Suárez in his first conclusion asserts that sound can be multiplied intentionally in both air and water. He presents three arguments for this view. (1) The first reasoning can be termed 'an argument from hearing distant sounds/voices'. Sounds or voices can be heard at a great distance, say, over two distant valleys by two shepherds. It would be difficult to envisage that these voices could be heard from such a distance, if they travelled from one valley to another valley only *realiter*, that is, if there were nothing else than the violent alteration of the medium by means of a circular commotion of the air. It is more probable to assume that in this case the act of hearing is secured through the multiplication of audible species. This is similar to vision where visual species (being less perfect entities than light) are better disposed to such long distance propagation.[193] Suárez

[190] *CDA* disp. 7, q. 8, n. 6, t. 2, 668.

[191] Aristotle, *DA* II.8, 419b34–5, 113; 420a–420b5, 113–4.

[192] *CDA* disp. 7, q. 8, n. 6, t. 2, 668–70.

[193] For Suárez this multiplication of the species of voices, which is in principle separable from the spreading of the real quality of voices, is, again, an important device in the explanation of the communication among the blessed in the heaven. Since the medium of the highest heaven or heavenly sphere is not air, the propagation of voices in this communication can be based only on intentional multiplication. Clearly, this multiplication requires the previous production of a real quality of voice which must be caused by the emission of the internal air that must be part of the glorified body. In his explanation of how the production of voices in glorified bodies can be possible Suárez says that the internal air circulates in the whole body (the air that is expelled from the lungs will enter the rest of the body through other ways) and enables the emission of air that will

employs an identical argument in his analysis of the acoustic phenomena of echo. He does not flatly reject the account according to which a reflected sound arises through its rebounding from the surface of a concave body back to its source in the manner of water waves that hit a dam and go back to the water's centre. But he leaves no doubt that there are cases in which such real reflection is not possible despite a clear occurrence of the auditory experience of echo. Considering the analogy with the reflection of visual species in a mirror through which we see the reflected objects positioned behind the perceiver, in echo, if standing in a properly disposed concave space, we can hear a reflected sound due to the reflection of auditive *species*. Again, he argues from 'remote perception'. Since we can hear echoes over large distances, as is the case in those shepherds who hear each other's voices reflecting off the opposite valleys, we need to assume the existence of audible species as partial instrumental causes in the production of the auditive act.[194] (2) By employing an 'argument from elimination of the versions of the reductionist view', Suárez also argues that the purely natural multiplication endorsed by the first view, in fact, admits a twofold exposition. On the first understanding, sound is transmitted as a numerically single entity that moves as a flying arrow or a small ball through air. According to the second explanation, sound or voice is carried in the medium through a series of successive and replicated parts of sound, similarly to the way species of colours are multiplied. Neither of these two explanations is plausible for Suárez. Let us imagine the situation of two men standing in different places who are listening to someone else's voice. In line with the first reductionist account they can hear her voice only if a numerically single portion of air 'bilocates' in the different places in which the two perceivers are standing. Obviously, in the natural order this is not possible. If the second version of this view is advocated, obviously, they will not be able to hear the original voice issued by that person but only the last voice in the series of replicated voices, namely the voice that touches or is adjacent to the two perceivers' ears.[195] However, this is at variance with our two pre-theoretical intuitions. One is related to our ability to trace a voice back to its original source. The other one involves the fact that two people standing in different places often have the experience of the *same* voice. By way of concluding, Suárez states that there is no other way to meet these pre-theoretical intuitions than to postulate the intentional multiplication of audible species representing the original voice. (3) Considering the basic metaphysical assumption, sound (as other sensibles) is the proper per se *sensibile* primarily immuting the pertinent sense organ. If no obstacle can be

---

constantly stay within the body of the blessed. Consequently, air necessary for the production of voices will never be lacking. For this description of the manner of perception after resurrection, see Suárez, *Expositio et Quaestiones in quinque priores primae secundae divi Thomae Aquinatis quaestiones materiam de beatitudine continentes*, 'Disputatio de quidditate humanae beatitudinis' (forthcoming 2021), q. 2, 136–8.

[194] *CDA* disp. 7, q. 8, n. 7, t. 2, 670; for the analysis of echo, see *CDA* disp. 7, q. 7, n. 4, t. 2, 658–60.

[195] This statement would amount to what is today called a proximal theory of sound which construes sounds as being located only at the bodily surface of the hearer. For this view, see Casati, Dokic, Di Bona (2020).

found in the medium, this object will 'ex se' affect the sense power without being obliged to be in physical contact with it.[196]

Having reached the conclusion about the necessary multiplication of species that can be separated from the real propagation of sound/voice, Suárez raises a question regarding the beginning of intentional multiplication. If one of the main reasons for endorsing audible species is the audibility of sounds/voices at great distances, one can argue that audible species can start to be replicated only when the propagation of the real sound comes to an end. However, for Suárez this view is false. He aims to show that intentional multiplication starts at the same time as real propagation. He again recurs to the 'argument from the localizability of sounds'. If we are to localize the origin (subject) of sound, intentional multiplication must take place *ab initio*. Since at the very outset of its existence *sonus* is the proper sensible object of hearing, from its very beginning it must multiply its species. If it started to multiply its species only in the last part of the air, it would actually cause the species of the sound that exists next to the perceiver's ear, but not that of the original sound. Moreover, this would also mean that at the beginning this sound is not audible. Provided that it is situated in suitable circumstances, sound as a natural agent multiplies its species wherever and whenever it exists.[197]

However, it may be objected that if audible species are multiplied *ab initio*, sound will necessarily be perceived in an instant. As the case of visual species shows, they are multiplied from the visible objects themselves in the perspicuous medium in no time. In this respect Suárez does not uphold the tenet of parity between the two senses. Timeless multiplication of sound is not consistent with our experience. As Aristotle knew, first we see the contact of two bodies and only then we perceive a sound caused by the bodies' violent. First, we see a flash of lightning and only then we hear a thunderclap.[198] Unlike visual species, in the emission of audible species a time lapse is necessary. However, considering this concomitant temporal lapse, it seems that the theory endorsing the multiplication of species only from the point when the real multiplication ends is better after all. It seems that the view that species are propagated from the beginning leads us to conclude that we hear in an instant. This dilemma is historically rooted in Aristotle. On the one hand, in *On Sense and Sensible Objects* 446a22–5 he declares that sound spreads by means of *kinésis* including a time lapse. On the other hand, considering hearing as the paradigm of sensation in *On the Soul* III.2, Aristotle no less clearly speaks about the temporal simultaneity in the actions of a sounding object and the actual hearing:

> The activity of the sensible object and of the sensation is one and the same [. . .] in saying that they are the same, I mean the actual sound and the actual hearing [. . .] when that which has

[196] *CDA* disp. 7, q. 8, n. 7, t. 2, 670–2.

[197] Suárez finds this view also in Aquinas's *STh*. I, q. 78, art. 3, c., 254 where Aquinas says that in the sense of hearing spiritual immutation occurs only 'ex parte organi' and not 'ex parte obiecti' (see 4.1).

[198] For this, see Aristotle, *Meteorologica* II.9, 369b9–10, 227; *On Sense and Sensible Objects* ch. 6, 446a22–5, 265.

> the power of hearing is exercising its power, and that which can sound is sounding, then the active hearing and the active sound occur together [*hama*; "at the same time"; my italics]; we may call them respectively audition and sonance.[199]

How does Suárez unravel this dilemma of temporality and immediacy in sound propagation? How does he sail between the Scylla of instantaneous change, based on the claim of the simultaneous activities of sonorous objects and the actual perception of the power, and the Charybdis of temporally extended multiplication, established in the propagation of the real quality of sound? Before introducing his (how else?) conciliatory conclusion Suárez makes three preliminary remarks summarizing in part what has been said above. (1) When a sound is produced in the medium, the concomitant action of the multiplication of audible species starts too. (2) An important agent in the process of the multiplication of a sensible is the sensible object's size. Obviously, as a voluminous luminary source can be seen better at a greater distance than a small object, so a sound that emits more species can be spread better and further than a lesser one. Its size contributes to the extension of the sphere of its activity. (3) However, it is necessary to distinguish between two kinds of size, namely between intensional and extensional size. A sound can be larger or smaller 'intensionally' depending on the force with which it was produced, and it can be larger or smaller 'extensionally' if it is propagated through a larger or smaller space. Both sizes are important factors which must be considered in the context of reflecting on the radius of a sound's intentional multiplication and real propagation.[200]

In his reply Suárez comes back to the wave model of propagation based on the water wave metaphor. A sound propagates in air in a manner similar to the circular waves made by the ripples on water after a stone has been thrown into it. After the first (highest) circle wave the process of multiplication begins to fade and continues until the wave motion of water levels out to the water surface. In analogy to water waves, the propagation of sound also circumscribes circles of a different dynamic up to the ultimate (flattest) sound wave. In an explicit reply to the dilemma Suárez adopts the hypothesis of a notional borderline described by the circle formed by the ultimate sound wave. Given the intensional size of a sound, we can assign a borderline at a certain point in the uttermost circle within which we can perceive sounds immediately, although from the beginning sound is not propagated in an instant but in time. Such 'timeless' perception, or perception proceeding in an imperceptible amount of time within this circle, will meet the 'litera' of the abovementioned passage from *On the Soul*. Within this sphere the eyes at the same see the contact of two sonorous bodies and the sense of hearing hears the sound. On the other hand, beyond this boundary the successively proceeding

---

[199] Aristotle, *DA* III.2, 425b26–426a1, 147. This passage is interesting because it seems to undermine the claim advocated by Suárez, namely that sensible qualities exist absolutely and independently of percipients. Here Aristotle, as Pasnau argues, comes with a *relational* understanding of the sensible qualities whose actuality is ultimately dependent on their being perceived. For this interpretation of Aristotle's view which makes a certain concession to Democritus, see Pasnau (2017, 51–5).

[200] *CDA* disp. 7, q. 8, n. 9, t. 2, 676–8.

multiplication of sound and audible species proceed with a temporal lapse. In this case, as Aristotle affirms in *Meteorologica* and *On Sense and Sensible Objects*, the medium is noticeably affected before the power of hearing comes to be affected. The multiplication of species will proceed with a perceptible time lapse.[201]

4.3.3 Organ of Hearing

Suárez's physiological treatise of the ear is significantly shorter than the questions dedicated to the cause, subject and propagation of sound. Despite its brevity, however, *CDA* disp. 7, q. 9, titled 'What the organ of hearing is and which the [auditive] power is', again testifies to Suárez's familiarity with matters pertaining to physiology and also to his hermeneutics related to the exegesis of Aristotle.

In line with the analyses concerning the object and medium of hearing, Suárez formulates a predictable statement according to which the organ of hearing is composed of the element of air. This, however, does not mean that the organ of hearing consists *purely* of air. Since a sense organ as such must be nourished and must be sentient, it cannot be composed only of air since air cannot be nourished. Moreover, air as such cannot be informed by the soul. A sense organ must be called 'aery' due to a relative prevalence of air when compared to the composition of the other external senses' organs. However, it must be composed also of the other elements forming its immediate surroundings and instrumental parts. Although in *Parts of Animals* Aristotle says that, with the exception of the organ of touch, the sense organs are uniform and each of them is conjoined to one of the elementary substances[202] and in *De* sensu he adds that each of the sense organs should be adapted to one of the elements,[203] Suárez makes it clear that this formulation cannot be taken in the literal sense according to which each element is exclusive for a given organ. It must be taken only in the sense of its *prevalence*. Only in the sense of prevalence do we say that the organ of sight is watery and the organ of hearing is aery, since as being watery the organ is best disposed for the reception of the forms of colour and as being airy the organ is proportional to the reception of sounds.[204]

In his anatomical description of the human ear Suárez distinguishes between two kinds of ear, namely between the external ear and the internal ear. While the function of the external ear (called today the external acoustic meatus) which is similar to a labyrinth with 'meanders' is to protect the internal ear from damage caused by a sudden gust of air or dirt and to facilitate the reception of species, in the internal ear

[201] Ibid., n. 9, 678.

[202] Aristotle, *Parts of Animals* II.1, 647a4–13, 113–5.

[203] For Aristotle's statement about this distribution of the elements among the sense organs of the external senses and for the airy composition of the auditive organ, see *On Sense and Sensible Objects* ch. 2, 438b17–24, 227.

[204] *CDA* disp. 7, q. 9, n. 1, t. 2, 680–2.

the proper organ of hearing is to be located. Unlike Aristotle who refers to the eardrum (*derma*) only *en passant*,[205] Suárez underlines the significance of this *membranula* in its role of the crucial boundary line between the internal ear and the external ear. What does the internal ear look like? It makes up a sort of a concave space to which the auditory nerves lead from the brain, which is the root of all sensory cognition. These nerves supply the organ with the animal spirits necessary for the power's (vital) reception of species. Although the external air and the internal air filling these two kinds of ear are the same, the internal ear contains 'vital air' and forms a correspondent 'pars animata' or 'pars spirituosa'.[206] The proper organ of hearing is just this 'pars spirituosa' covered by the thin membrane of the eardrum stretched by the auditory ossicles.[207] Although no anatomical evidence for this 'pars spirituosa' is available, Suárez is sure that this should be regarded as the adequate organ of hearing. So the first phase of hearing in the three-stage process of perception proceeds as follows: A species (and sometimes also sound) arrives at the external ear in which it attains the membrane or eardrum. By means of this it produces the species of sound in the auditive *sensorium*, which is located in the internal ear underneath it. The auditive power is located in this inner ear. Although in *CDA* Suárez often follows the medical authorities (Galen, Vesalius, Vallés), at times he also resists their overt naturalism which is at odds with his Aristotelianism. Sometimes he goes beyond or against their exposition to explain the natural phenomena in line with a broad Aristotelian setting. And this is the case here. Although contemporary anatomy had not detected this vital air in the internal ear and this is also why it 'located' the organ in the nerves, Suárez comes to resolve its 'anatomical absence' with the Solomon-like statement that this vital air vanishes at death. In structural analogy to the treatment of the organ of vision, in which he refused to assign the organ of sight to the optic chiasma, by allocating the *sensorium* of audition to the vital air Suárez *ipso facto* refuses to locate it in the auditory nerve.[208] The auditory nerve cannot be the organ since as such it is too earthy. Obviously, such an organ is not in the right elemental proportion to become the proper organ of sound.[209]

As in the issue of the kinds of medium so in the question of the properties of the external air, the exegesis of Aristotle's texts raises an exegetical problem that stands

[205] 'When this [water enters the depth of an ear] does occur, there is no hearing; nor again if the membrane is damaged [. . .]', Aristotle, *DA* II.8, 420a14–5, 113.

[206] In connection with the interior or vital air, which is crucial for Suárez's conception of the auditive organ, our author refers to Themistius. For Themistius, see *Paraphrasis Themistii peripatetici acutissimi, in Aristotelis Posteriora, & Physica, In secundum de anima, In II De anima*, ch. 28: 'De intermedio in auditu', 500–2.

[207] *CDA* disp. 7, q. 9, n. 2, t. 2, 682.

[208] Like in his analysis of the organ of vision, Avicenna is adduced as an author subscribing to the view that the organ of hearing is a nerve. For Avicenna, see *Liber de anima seu sextus de naturalibus*, p. 2, ch. 5, 159.

[209] *CDA* disp. 7, q. 9, n. 3, t. 2, 684.

at the centre of Suárez's attention. As said, all of Suárez's descriptions of anatomical and physiological facts are always related to philosophical issues. In *On the Soul* II.8 Aristotle formulates the following statement (quoted above): 'The air in the ears is lodged, so as to be unmoved, in order that it may accurately perceive all differences of motion'.[210] In this statement Aristotle suggests that the necessary property of the vital air is its immobility. As noted in 3.2, the sense organ must be deprived of the quality it receives. Even though the cognitively relevant reception is 'spiritual' or intentional, the natural change is often collateral to it. In this case this change is characterized by a concomitant local movement. If in motion, the *sensorium* could not perceive sounds objectively and veridically. Its perception would always be misrepresented since the main condition, that is, the right disposition of the organ, would not be met. As the translucent nature of the crystalline humour in the pupil makes the pupil objectively receptive of visual species and as coloured medium, such as a stained-glass window, affects vision, so the immobility of air makes the organ of hearing suitable for the proper reception of audible species and for their objective representation. On the contrary, the mobility of air and permanent fizzing in the inner ear is a cause of many misrepresentations. However, only a few lines below Aristotle comes with another formulation that seems to give a contrary evidence: '[...] the air in the ears always moves with a special motion of its own'.[211] How does Suárez harmonize these two conflicting statements? Of which ear, as a mater of fact, does Aristotle speak in the second passage? In his exposition Suárez is clear that Aristotle has in mind the external ear, which is not the organ but the adjacent medium of the proper organ of hearing. However, he is sure that if the sense of hearing is to function well, even this 'medial air' of the external ear must be *de se* immobile. So, when Aristotle speaks about motion in the ear, he does not mean that this motion is a sign of its 'essence' and its proper functioning. On the contrary, it is a sign of disorder that does not differ from a coating on a patient's tongue affecting her taste. Suárez quotes from a passage from Aristotle's *On the Soul* following the sentence quoted above: '[...] sound [coming from that "special motion of its own"] comes from an outside source, and it is not a property of the ear'.[212] Fizzle or ringing in our ears is not a natural state based on the natural continuous motion of air, whether in the external or the internal ear. It is a symptom of the intrusion of a foreign air coming from outside. This often happens when we press our fingers to our ears.[213]

[210]Aristotle, *DA* II.8, 420a9–12, 113.
[211]Ibid., 420a16–7, 115.
[212]Ibid., 420a17–8, 115.
[213]*CDA* disp. 7, q. 9, n. 4, t. 2, 684–6.

4.4 Smell

In Suárez's ranking of the sensory modalities according to their 'immateriality', the sense of smell comes on the third (middle) position standing between taste and touch on the one side and sight and hearing on the other. It is similar to the 'distal senses' of sight and hearing because it can be affected merely intentionally by an odour. To produce its act, as we will see below, the olfactory power and its organ do not have to be affected naturally. However, at the same time, the sense of smell resembles the 'contact senses', especially taste. Smell is an important sensorial instrument for the detection and probing of nutrition. As we all know, in this respect sight is often misleading. In investigating the edibility and suavity of food rational and nonrational animals employ olfaction as a sense that is most proximate to taste. Since we are more familiar with sapours and our nomenclature of nutritive odours is taken over from sapours and the nutritive substances whose affections they are, both Aristotle's *On Sense and Sensible Objects* and Aquinas's *Commentary on Sense and Sensible Objects* deal with smell *after* taste. The reason is that human taste is much more fine-grained than the taste of nonrational animals and human smell is much less perfect than the smell of nonrationals. Consequently, sapours and their kinds are more known to us than odours and their kinds. This is the reason why the path of investigation should proceed from sapours to odours, that is, from taste to smell. Although he reverses this order of presentation and is led by the ordering of the senses according to their ontological perfection, in *CDA* disp. 7, q. 10, called 'What odour is, and what sapour is, and about their kinds', Suárez underlines this mutual interdependence of both senses by treating the issue of their object in a single question. However, due to other even more palpable agreements between the contact senses of taste and touch I will delve into the treatment of taste and its kinds together with the treatise on touch in the next section.

In line with the abovementioned overall bottom-up procedure established in the preceding two sections, in this section I will start (1) with the issue of the nature and kinds of odours. I aim to highlight Suárez's realism of odours against the background of the deflationary accounts reducing the quality of odour to more fundamental entities. Notwithstanding our ignorance of the kinds of odour, I will present two main kinds of odours. (2) The most important part of Suárez's treatment will again be devoted to the question of the character of the medium's and the organ's affection both through intentional and natural change. (3) Lastly, I will not leave aside the issue of the olfactory organ.[214]

[214]For these issues, see Heider (2016d, 84–91).

4.4.1 Nature and Kinds of Odour

Due to the affinity of the objects of smell and taste Suárez deals with the issues of the nature of odour and the nature of sapour in the same question. The kinship of the two sensibles derives from the fact that both are secondary qualities arising from the blending of first qualities. Both are constituted by a mixture of the moist and the dry with a significant assistance of the hot. Both can be 'stifled' by the prevalence of the cold. Their particular kinds are interrelated.[215] As in the cases of the sensibles treated before, Suárez starts by rejecting the reductionist account of odour advocated not only by pre-Socratics, but also by some representatives of the Aristotelian and medical tradition.[216] According to this interpretation, odour is nothing else than a vapour or a smoky exhalation (today we would say a dispersal of volatile molecules) constituted especially by the elements of air, water and earth. According to this view, odour is not an accident of quality but a substance. As such it spreads from an odoriferous object via the medium up to the organ. This dispersion is actually all we need to get an account of the production of an olfactory act.[217] In his typical anti-reductionist drive Suárez dismisses this view. As with sight and hearing, he argues that no substance can be a proper sensible of any of the external senses because substances are sensibles perceived by the senses only per accidens. An odour cannot be identified with a substance, however airy and subtle. It is an accident of an affective quality, sc. of a quality that can be "suffered" by a percipient, a quality pertaining to the third kind of accidents of quality.[218] Suárez is aware of the Stagirite's wavering on this issue, as he was in the previous senses. In *On Sense and Sensible* (ch. 2) Aristotle, on the one hand, says that 'odour is a kind of smoky vapour' (438b24–5). In the same treatise (ch. 5), however, while reprehending Heraclitus and his view, he states that 'smell is neither of these [vapour and smoky exhalation]' (443a30–1). In his typical conciliatory exegesis Suárez comes with the following distinction, which has been employed already in 4.1 in the cases of taste

[215]No matter how much Suárez highlights the affinities between olfaction and taste, unlike the common contemporary view, he does not take into account what is called the intersensory modality of retronasal olfaction which arises during eating. This kind of smell, distinct from ortonasal olfaction which perceives odours coming from the outside, is today generally regarded as part of the multisensory flavour experience. For this, see Smith (2015, 324). This multisensory flavour experience, of course, arises at the level of the common sense (see Chap. 5).

[216]In *On Sense and Sensible Objects* Aristotle says that all his precursors defended this theory but explicitly names only Heraclitus. Cf. Aristotle, *On Sense and Sensible Objects* ch. 5, 443a24–5, 249. Suárez mentions also Avicenna. In *Liber de anima seu sextus de naturalibus*, p. 2, ch. 4, 150–1 Avicenna identifies odour with vapour: 'Nos autem dicimus possibile esse ut odoratum sit vapor, et possibile est ut vapor permutetur propter odoriferum et fiat odor et sit idem quod vapor quia, quicquid habet partes subtiles et solet diffundi, cum attigerit instrumentum odorandi et offenderit in illud, sive sit vapor sive aer, permutabitur in odorem et per ipsum sentietur odor'. Pietro d'Abano (1985, f. 212va) says that 'Odor est vaporosa diffusio, sive emanatio: unum enim horum videtur aquae competere, reliquum vero aeri'.

[217]*CDA* disp. 7, q. 10, n. 1, t. 2, 688.

[218]Ibid., n. 2, 688.

and touch: The equivalence of odour and smoky exhalation can be interpreted either formally (*formaliter*), or subjectively (*subiective*). Formally speaking, the essence of odour is not a smoky evaporation but an affective quality. However, if the identity is taken in the subjective sense, the equivalence can be regarded as true. Odour is an accident existing in an evaporation as in its proximate subject.[219] Since it holds that no accident can 'travel' without its substance in a space (while sound is subjected in airy waves moving locally, light inheres in the transparent), odour needs to have a subject too. This subject can be separated from an odoriferous thing, such as an apple, a flower or a mineral, and it can be carried away in air to the perceiver. This subject will be a smoky exhalation (in air) or a vapour (in water).[220] An odour's immediate subject is not air but a smoky evaporation. The smoky evaporation with the quality of odour is carried along by air.[221]

The particular elements are odourless, as they are tasteless. Things that would be purely moist and purely dry would not emit an odour. In order to constitute the quality of odour a blending of first qualities is necessary. Much more briefly than Aristotle in *De* sensu, Suárez specifies the genesis of an odourous mixture. An odour arises by the blending of first qualities in which the dry prevails over the moist and the hot is absolutely dominant.[222] The predominance of the two qualities is clear from experience. Like fire, comprising the dry and the hot, odours moves upward. Odours participate in fire's essential motion upward. Our experience also substantiates the absolute prevalence of the hot. In summer we smell better than in winter. If the moist outweighs the dry, things turn to be less fragrant. Sweet meals are the least fragrant since they are the moistest. The fragrances of the driest, for example, oriental, meals are strongest and sharpest. The same 'logic' holds also for olfaction's organ. If it gets moistened, for example, due to rhinitis loosened from the brain moist, it loses the ability to distinguish odours. However, this does not mean that the dry is the only quality in a smellable object. If a thing is too dry (*arida*), it will emit no odour. The enchymous dry must be dissolved in the humid. The dry must be received and absorbed by what is watery, which can occur both in water and air which, as the ancient elemental physics states, is composed of the first qualities of the hot and the moist.[223] The dry must 'cleanse' the moist. The sea, for example, has

[219] Ibid., n. 4, 692.

[220] For the issue of what the proximate subject of an odourous quality *in medio* is, as explicated by late ancient Neoplatonic commentators, see Ellis (1990). Basically, the Neoplatonic commentators presented two main theories. According to the 'tense solution', an accident of odour inheres in a new subject, in air. According to the 'effluence solution', the fragrance rides on a part of, say, an apple, which gets detached from it. The fragnance reaches the sense power without being separated from the detached part. In this case, with a certain qualification, Suárez embraces the second solution.

[221] *CDA* disp. 7, q. 11, n. 2, t. 2, 700.

[222] *CDA* disp. 7, q. 10, n. 3, t. 2, 688.

[223] Ibid., n. 3, 690.

an odour because the dryness of the earth is combined with the moistness of water.[224]

As I have indicated, Suárez distinguishes between two main genera of odours. While the first kind of odours is connected with taste, the other kind applies to nonnutritive odours, such as those of flowers or minerals. In both cases we stipulate the species of odours according to the odoriferous subjects in which they inhere or in accordance with the sapours with which they are connected (the odour of sweet food, the odour of acid food, the odour of a rose etc.). We distinguish between pleasant and unpleasant odours, and between harsh and dull fragrances depending on the speed at which the 'volatile molecules' affect the organ.[225] Suárez is inclined to the view according to which the smell of nonrational animals (unlike that of humans) appreciates only 'nutritive fragrances'. Or if, after all, the nonnutritive odours somehow constitute an objectual part of the sensorial system of nonrationals, their perception is imperfect. It means that nonrationals will not find pleasure in these fragrances. At the same time nonrationals do not find putrid odours repulsive.[226] As it has been said, although their smell is more developed than ours, Suárez qualifies the higher perfection of some nonrationals, such as dogs or vultures, as merely a superiority in a certain respect. Absolutely speaking, human smell is more perfect. The superiority of human smell can be seen precisely in the capacity to appreciate nonnutritive fragrances. Accordingly, the range of the human smell is broader than that of the smell of nonrationals. Not only do not nonnutritive fragrances stimulate the appetite: they often have a contrary effect. They are repulsive to the appetite. If they are added to a meal, they often discourage us from eating it.[227] But why is it so that human animals can appreciate these odours, while nonrationals cannot? Following Aristotle,[228] Suárez seems to think that the main reason is the different physiology of the human brain which is connected with olfaction proceeding only through inhaling.[229] As I will show in 4.4.3, for Suárez the olfactory organ is situated near (*juxta*) to the front part of the brain, even though it is not identical with it. However, our brain differs from the brain of perfect nonrational animals. When compared to their brain, ours is relatively bigger, colder and more humid than theirs. If an odour is not

[224]For this, see Aristotle, *On Sense and Sensible Objects* ch. 5, 443a12–3, 249.

[225]*CDA* disp. 7, q. 10, n. 7, n. 7, t. 2, 696–8.

[226]It must be noted that the view according to which nonrational animals are not capable of perceiving and appreciating nonnutritious odours was not common in Second scholasticism. The Coimbran authors dissented from this view: '[...] bruta animantia non odores tantum alimentitios [...] sed alios etiam percipiant [...] quia videmus canes venaticos florum odores sentire [...]', Collegium Conimbricensis, *In tres libros de anima, In II lib. De anima*, ch. 9, q. 5, art. 2, 300–1. For this, see also 4.1.

[227]As Aristotle (*On Sense and Sensible Objects* ch. 5, 443b30–2, 253), quoting Euripides, says: 'When you make pea-soup don't put perfume in it'.

[228]Ibid., ch. 5, 444a8–25, 253–4.

[229]*CDA* disp. 7, q. 16, n. 4, t. 2, 768–70.

exceedingly hot, by ascending to the brain it can heat the brain in a degree that will be beneficial for its health.[230]

4.4.2 How Odour Affects Medium and Power

Odour can be perceived through a medium at a distance. Since smell is similar to the two higher sensory modalities, its proper sensible object can be sensed at a distance too. Although air is considered to be the best medium for its spreading, it is not the only one. Water can be its medium too. Fish led by olfaction, which is operative not through inhalation (as in humans) but without it, swim from afar for food.[231] Having noted these two kinds of medium, how does a distant odour actually reach the power of smell? Is it diffused only *realiter*, only intentionally, or in both ways? As with hearing, there are two answers to this question that represent two extreme positions. According to the first view odour is spread up to the power only *realiter* as a quality subjected in a smoky exhalation or vapour. According to this 'naturalist' view, the material affection through the reception of this quality in the first stage of a perceptual process is both a necessary and a sufficient condition for the elicitation of an olfactory act. According to the second reply, *res odorifera* emits *ab initio* only odoriferous species; it does not emit a real quality of odour.[232] Before approaching the special issues related to the query of how an odour is diffused *in medio* Suárez premisses two remarks. (1) Smell cannot be diffused *realiter* in water composed of the first qualities of the moist and the cold, since it is not a suitable subject for the reception of odour in which the dry and the hot prevail. (2) Odour is *realiter* diffused (in air) from *res odorifera*. Since it cannot immediately inhere in air it must be subjected in a smoky vapour or in another tenuous substance circularly diffused to its environment. Suárez formulates five arguments for such real diffusion. (a) After the removal of an odoriferous thing, the medium in which the thing had been located conserves its odour for a certain time. This persistent odour could not be caused by retained odoriferous species since species cannot be conserved without the object represented by the species. They are ontologically dependent in their becoming and being on the qualities they represent. Therefore, the medium must contain the odour which exists in a vapoury substance. (b) As the example of incense shows, it often happens that a thing is redolent only if it is heated and gives off smoke. The concomitant activity of both appears to be the reason why odours are better perceived in spring than in winter. (c) While an odoriferous object stored in a sealed

---

[230] *CDA* disp. 7, q. 11, n. 2, t. 2, 702; d. 7, q. 16, n. 4, 770.

[231] For Aristotle's affirmation according to which the phenomenon of smelling occurs also in water, see *On Sense and Sensible Objects* ch. 5, 443a2–3, 249.

[232] *CDA* disp. 7, q. 11, n. 1, t. 2, 698–700. For the first view, see Avicenna, *Liber de anima seu Sextus de naturalibus I–II–III*, p. 2, ch. 4, 148; for the second view, cf. Averroes, *In Aristotelis De anima librum secundum*, 97, 277.

container keeps its odour well, it loses it quickly once it has been opened. This can be explained only by the fact that the odoriferous object in the container emits the substantial evaporation in which the odour inheres. The mere emission of species would not result in the odour's quick evanescence. A red colour on the inner lining of the container emitting only visual species would not get diminished when the container is opened. (d) A hand that touches an odoriferous thing smells of it for some time. This happens neither due to the inherence of odoriferous species nor due to the inherence of an odour in the hand, but because the pores of the hand are filled with the smoky vapour in which an odour inheres. This is corroborated by the fact that if the hand is properly washed, it loses the smell quickly since the tenuous substance contained in the pores gets dissolved and washed away. (e) As said in 4.4.1, since smelling an odour is often beneficial to the human brain because the vapour in which an odour inheres ascends up to the brain, it must be conceded that an odoriferous thing exhales the smoky vapour which is its vehicle.[233]

Having formulated these two assumptions, in *CDA* disp. 7, q. 11 three special difficulties come to occupy Suárez's mind. (1) Does the real evaporation of an odour have to reach the organ of smell in order for an act of olfaction to be produced? (2) Are *species odorifera* multiplied from an odorous thing *ab initio*, or only from a point where the real multiplication has lost its force? (3) Does the real multiplication of an exhalation (at least for a certain distance) constitute a necessary condition for the production of an olfactory operation, or is it not necessary at all?

(1) Following Aquinas, Suárez says that the more probable view is the negative sentiment.[234] An act of olfaction is possible even if no real evaporation in which an odour is subjected reaches the organ of smell. Setting aside the obvious counterexample to the opposite claim, namely the fact of the underwater diffusion of smell where only intentional multiplication seems to be the case since the first qualities of water, sc. the moist and the cold, are not compatible with the composition of odours, Suárez employs the 'argument from perception over long distances', which he employed already in the question devoted to the sense of hearing. Vultures perceive the odour of carcasses over 50 miles. However, it is not probable that their odorous evaporation, no matter how tenuous, can travel over such a long distance. Moreover, even if the fumes of carcasses were carried along by wind over that distance,[235] it would still be difficult to understand how vultures can identify the places where the carcasses are lying. It would be hardly credible to assume that the whole way from the carcass to the vulture is blazed with the evaporation. Species and their multiplication seem to be necessary to 'save the phenomena'. In an analogy to hearing, if an odour had to be multiplied *realiter* up to the power, the species representing it would

---

[233]Ibid., n. 2, 700–2.

[234]See Aquinas's *Sentencia libri secundi De anima*, ch. 20, 152–3. For Suárez's reference to Aquinas, see *CDA* disp. 7, q. 11, n. 3, t. 2, 702–4. For a purely spiritual affection of smell without a corruptible alteration, see Suárez, *Expositio et Quaestiones in quinque priores primae secundae divi Thomae Aquinatis quaestiones materiam de beatitudine continentes* (forthcoming 2021), 'Disputatio de quidditate humanae beatitudinis', q. 2, 139.

[235]Suárez also mentions the example of pestilence which can be smelled over long distances.

be produced only from an odoriferous body lying next to the olfactory organ. Accordingly, the odoriferous species would not represent the odour of a distant thing but merely an odour that is immediately adjacent to the organ. Besides, if the real evaporation up to the organ constituted a necessary condition for the elicitation of an olfactory operation, olfaction would come to be a sort of touch. Accordingly, there would be no external medium for its diffusion. Olfaction would not differ from touch and the quality of fragrance would be reducible to the tactile qualities. As said, Suárez denies this reduction for all non-tactile senses. Finally, in a structural analogy to the quality of sound, if an odour is the primary per se sensible, it necessarily emits its species. The existence of an odour without the multiplication of species is an impossible state of affairs.[236]

(2) As proper sensibles, odours necessarily emit species *ab initio*. Contrary to the 'spiritual' emission, the natural evaporation is only incidental. Suárez raises two objections to this 'intentionalist' concept. (a) If the olfactory species were multiplied from the beginning, olfaction would have to proceed in an instant, which is not the case. Odours, like sound, are diffused *in medio* in time. As said in 4.2.1, only vision proceeds without a time delay. Perception in the other sensory modalities proceeds successively. In his recurrent analogy to hearing, Suárez replies by distinguishing a (circular) field of perception within which a fragrance can be perceived in an imperceptible amount of time. Beyond this notional ambit it can be sensed only in a time. This is so because beyond this notional ambit the multiplication necessarily takes time since the processes of exhalation and heating accompanying the propagation of an odour takes time as well. (b) The perception of an odour cannot take its course by means of the multiplication of species since the odour's diffusion can be easily affected by external influences, such as wind and other fragrances. Unlike vision, a strong wind can affect the perception of odours. Only the real quality of an odour subjected in a smoky substance can be influenced in this way. In his reply Suárez says that although wind can significantly affect our perception of odours, he surprisingly notes that this applies also to the sensible species itself. Although he recurrently calls the sensible species 'spiritual', we know that it is material and divisible. Its 'spirituality' means material subtlety, not literal immateriality, especially when compared to the represented qualities. Consequently, Suárez also seems to admit a certain ontological gradation even at the level of the sensible species. The reason why wind can affect the dispersion of odours, and not that of colours, is that the *species visualis* is more 'spiritual' than the *species odorifera*.[237]

(3) If the smoky exhalation of an odour is only an incidental feature of intentional multiplication, a real evaporation, even for a minimal distance, is not necessary. Intentional multiplication is sufficient for the production of olfaction. As we experience, some cypresses, aromas and resins, such as amber, remain redolent for many

[236] *CDA* disp. 7, q. 11, n. 4, t. 2, 704–6.

[237] *CDA* disp. 7, q. 11, n. 5, t. 2, 706–8. What we have here is a gradation in the 'spirituality' of the sensible species themselves. This gradation is far from new and was advocated already by Albert the Great. For this gradation in Albert, see Knuuttila (2008, 13–4).

years without the diminution of their fragrances. It means nothing else than that they are redolent *without* evaporation.[238] It may be objected that the evaporation of an aroma that is very subtle and tenuous can take a long time without being diminished and consumed. Given this tenuous evaporation, it is not necessary to postulate only the odoriferous species as an *explanans* of the production of an act of olfaction. Suárez considers this reply to be insufficient. It often happens that a percept of an odour is very strong and acute[239] and yet it lasts over a long time. Therefore, the evaporation of no matter how tiny and subtle smoky vapour cannot be a probable and plausible *explanans* of the given phenomenon.

4.4.3 Organ and Act of Olfaction

As in the other *quaestiones* devoted to the organs of other external senses, also in *CDA* disp. 7, q. 12, in which the organ of olfaction is analyzed, Aristotle is not a decisive authority for Suárez. Here it is especially since 'Aristotelis sententia non est clara [. . .]'[240] On the one hand he seems to assume that the nose is the proper organ of smell,[241] while on the other hand he seems to advocate the view that the nose is only a way (*meatus*) of olfaction.[242] This second opinion comes close to Galen's.[243] For Galen the sense of olfaction is to be placed not in the nose but in the two front ventricles of the brain.[244] In a brief anatomical excursion, besides these two cavities, Suárez mentions two olfactory nerves, the so-called mammillary nerves (*mamillares*). They are called mammilary since their endings look like nipples. These nerves lead from the front ventricles to the nasal cavities. They are commonly considered to be the channels through which the animal spirits are conveyed to the olfactory organs. As such they are structurally and functionally similar to optic and

---

[238] *CDA* disp. 7, q. 11, n. 6, t. 2, 708–10.

[239] For a characterization of acute fragrances as odours quickly stimulatig smell, see *CDA* disp. 7, q. 10, n. 7, t. 2, 696.

[240] *CDA* disp. 7, q. 12, n. 2, t. 2, 712.

[241] When speaking about animals' sense organs in the fourth book of *History of Animals* Aristotle distinguishes between animals with nostrils and animals with olfactory passages, such as birds. He seems to assume here that the nose is the organ of smell. Cf. Aristotle, *History of Animals* IV.8, 533a22–4, 63.

[242] In the first book of the same treatise he claims: 'Smelling, too, takes place through the nose [. . .]', Aristotle, *History of Animals* I.11, 492b13, 45. This seems to be, at least, open to the possibility that the nose is only a channel of olfaction but not the *sensorium*. In the same treatise, in II.12 he says that noses are only passages serving the sensory faculties (504a22–3).

[243] For his view, see Galen, *De usu partium corporis humani*, book 8, 244: 'Propter haec igitur omnia, & propter ea, quae debent dici, solus sensum odoratus in ipso cerebro factus est'.

[244] For Suárez's presentation of this anatomical theory of Galen's in connection with the issue of the interior sense's organ and with the question of the organ of smell, see *CDA* disp. 8, q. 2, n. 8, t. 3, 56–8.

auditory nerves.[245] Vesalius located the olfactory organ just in these two olfactory (mammilary) nerves.[246] These and other views placing the power of smell in various parts of the nose, such as bones, cartilages, or special membranes, show that regarding the issue of the olfactory organ there was little agreement in the medical and the Aristotelian tradition. The variety of these views was also a reason for Suárez's evaluation that the only way to reply to the question of the organ of smell is 'probabiliter procedendo'.[247]

Despite the considerable authority of both aforesaid physicians, Suárez is quick to say that Galen's views cannot be probable (this *ipso facto* applies also to Vesalius's view claiming that the *mamillares* are of the brain substance). His critical stance is based on the arguments formulated some years earlier by Vallés. As shown in *CDA* disp. 6, q. 6, for Suárez the brain is not the proper organ of an external sense since it is the universal source (*fons*), principle (*origo*), or root of sensation. It is not a *sensorium* of an external sense but the source that 'meliorates' the sense organs with the animal spirits.[248] There is no better reason to locate the organ of smell in the brain than in any other organ. If we placed it in the brain, why not assign to it also the senses of sight, hearing, taste, and touch? Besides, employing the argument *ad hominem*, Suárez says that Galen himself admits that there is no touch in the brain.[249] But if there is no touch there, can we really say that the organ of smell is there?[250] One of Galen's reasons for his view is the assertion that respiratory animals can smell only by inhaling. For them aspiration is necessary for smelling. By aspiration odours get through up to the brain. Without sniffing the smoky exhalations of odours cannot reach the inner *sensorium* located in the front cavity. Moreover, the brain gets recreated or damaged by the smelling of exhalations and odours. However, if the brain were not the organ of olfaction, this phenomenon of smelling through inhalation could not be explained. Suárez is not impressed by this argument. He provides a different explanation of this phenomenon of smelling by inhaling. This explanation is based on Aristotle's *On the Soul* II.9.[251] There Aristotle says that animals that smell by aspiration have a sort of a lid in the nose called *operculum* by Suárez, which needs to be opened so that fragrances and their species could enter. This lid opens when they breathe in and closes when they breathe out. It has the same function as an eyelid protecting the eye as a sheath against a potential damage. Humans and some other animals having soft eyes possess the eyelids as such a cover. On the other hand, animals that can smell without aspiration, such as aquatics, do not have this *operculum*. Likewise, hard-eyed animals do not have to open their eyes to be able to see. They see directly what appears in front of them without having to

[245] *CDA* disp. 7, q. 12, n. 1, t. 2, 710–2.

[246] For this view of Vesalius's, see his *De humani corporis fabrica*, book 4, ch. 3, 322–3.

[247] *CDA* disp. 7, q. 12, n. 3, t. 2, 712.

[248] See *CDA* disp. 6, q. 6, n. 10, t. 2, 540.

[249] For this view of Galen mentioned by Suárez, see *CDA* disp. 7, q. 14, n. 3, t. 2, 740.

[250] *CDA* disp. 7, q. 12, n. 3, t. 2, 714–6.

[251] See 421b28–422a8, 123–5.

open a cover. In principle, Suárez agrees with this opinion of Aristotle's. The only proviso he makes concerns the existence of this *operculum*. Instead of employing such a lid (again he refers to 'an entity of which physicians know nothing'), Suárez proposes a different explanation of smelling by inhalation. He says that inhalation is necessary to expand the narrow internal cavities in the nose. Without aspiration the nasal tracts would be too narrow for the exhalation and the odoriferous species to go through.[252]

In the positive part of this question Suárez asserts that since the sense organ and the power of olfaction cannot be located in the brain, the organ must be placed in the interior part of the nose. He mentions two arguments for this claim that are taken over from Vallés.[253] First, the organ of sense perception must be located where the pleasure and pain associated with it are sensed. However, they are felt in the nose, not in the brain. Second, the olfactory organ must be situated outside the cranial wall (*calvaria*, the borderline between the nasal cavities and the front ventricle), since if it were not located there, the brain would be inundated with the chaos of various exhalations. As a consequence, the power would have difficulty in discriminating them. In an analogy to the mouth and the tongue in the case of taste, there must be a 'sensorial explorer' situated in front of the brain to discern and prevent potential damage.[254]

4.5 The Contact Senses: Taste and Touch

It has been shown that taste and touch are two distinct sensory modalities. Taste cannot be reduced to touch since they have different formal objects. While the tactile objects are the primary and secondary qualities, such as the hard and the soft, the proper sensible of taste amounts to the secondary quality of sapour which results from a special mixture of the first foundational qualities. The two senses have different organs. While taste is located in a single part of the body, namely in the tongue, touch is spread all over the body with the exception of the inanimate parts, such as bones, teeth and hair. The two senses can be separated in their operations not

[252]For Suárez's presentations of Galen's and Aristotle's accounts, see *CDA* disp. 7, q. 12, n. 7, t. 2, 718.

[253]See Vallés, *Controversiarum medicarum et philosophicarum libri decem*, book 2, ch. 24, 99: 'Tamen non dubito instrumentum odoratus esse in partibus nasi, nam qua parte fit ex iis quae sensibus obiiciuntur, voluptas & dolor: illa fit sensus [...] Sed ego mihi saepe visum sum, & alii etiam ita dictitant, ex admotione rei fragrantis odore, in ipsius nasi parte interiori, voluptate affici: ut contra ex odore iniucundo dolore. Fit ergo odoratio in ipsius nasi tunica. Praeterea consentaneum magis est rationi cum ex halitibus subeuntibus in cerebrum, spiritus animales in illius ventriculis anterioribus generentur, naturam ante illos extra calvariam collocasse sensum illorum exploratorem, ut gustum in ore fecit: quam exploratorem adeo recondere, ut malus vapor discerni ante non possit, quam cerebrum laeserit. Quod cum observatum in ventriculo & gustu sentiamus: cur non perinde observari in cerebro & odoratu putemus?'

[254]*CDA* disp. 7, q. 12, n. 3, t. 2, 714.

only in respect to various kinds of animals (some imperfect animals do not have taste but only touch), but also in a human being. A sick person whose tongue is covered with a bitter coating and cannot taste has not lost the sense of touch. Despite these symptomatic differences, the two powers stand in close mutual proximity, which we have assessed as being even closer than that between taste and smell. They both are 'contact' senses since both need to be in physical touch with the sensibles to produce their operations. They necessarily apprehend only 'tangendo'. Even though the intentional affection of the organs of these two external senses is necessary, this affection must always be accompanied by natural (tactile) change. Without natural change these senses do not apprehend. While the alteration in touch can be manifold depending on the nature of a tangible (it can be cooling, warming, moistening etc.) and thus touch gets mostly affected by the quality of a tangible itself, the tongue does not need to become bitter or sweet. As such, however, it needs to be physically predisposed by creating a moistening saliva that will dissolve food in the mouth and will aid in spreading it on the tongue. Only as moistening a food can the tongue perceive differences between sapours. When Aristotle defines taste as a sort of touch, he has in mind not a specific sensory modality but only 'touch' denoted by the common name, that is, touch that is common to all the senses that can be affected by physical alteration. As Suárez notes, taste can be touch not formally, but only *praesuppositive*.

Due to these crucial agreements between the senses, I will lay out two of the three fundamental issues concerned with them in tandem. These two issues will be as follows: (1) How the tangible and the tasteable affect the powers? and (2) What are their organs? Although there is a distinction between their organs which is conceived as a distinction between part and whole, the structural analogy and wide employment of identical statements and arguments make common treatment inevitable. For this reason I will consider the two modalities together in 4.5.3 and 4.5.4. I am only going to disconnect the queries about the nature and kinds of the two sensibles (4.5.1 and 4.5.2). In the latter section I will raise the important question whether, beside the primary and sensible qualities of the *exteroceptual* tangibles, Suárez admits also so-called *interoceptual* sensibles, such as physical or bodily pain, as something that constitutes a part of the total proper object of touch.[255]

4.5.1 *Nature and Kinds of Sapour*

The kinship of sapour and odour is evident in Suárez's characterization of the *gustabile*. Sapour is a secondary quality resulting from the blending of elemental qualities, which constitute the basic *explanans* with respect to secondary qualities. In line with the exposition of odour, sapour cannot be reduced to substance or common

---

[255] For an exposition of these issues related to these sensory modalities, see Heider (2016d, 91–102). For the problem of bodily pain in relation to touch, see Heider (2018).

sensibles, such as shape, affecting fundamental entities, such as atoms, which come to being only relative to our perception of them.[256] Both reductions would violate Suárez's general teleological ordering between the proper sensible qualities and the external senses. Like odour, sapour can arise only by a blending of the first qualities. Neither the dry without the moist, nor the moist without the dry, nor any other first quality as such has a sapour, as it does not have an odour. Like odour, sapour also requires a combination of the moist and the dry with a significant assistance of the hot.[257] The quality of cold supresses both taste and smell. As with odour, the dominant quality in sapour is the hot. The main difference between these two qualities lies in the fact that in sapour the moist gains dominance over the dry. Accordingly, the moist must be conceived as a prevalent quality in the organ of taste. The tongue must be moistened by saliva to be capable of perceiving sapours. If not, as happens with the sick, sapours cannot be perceived or are not apprehended objectively. Unlike smell, in which the fiery element is prevalent,[258] the dominant quality in sapour is the moist. If the hot attains an inordinate degree, the moist evaporates and the dry prevails. Consequently, an odour is produced. By contrast, if a *res odorifera* is moistened, its odour is diminished.[259] However, this does not mean that the dry is not an important ingredient in the blending out of which sapour originates. On the contrary, since flavours are always connected with nutrition, the quality of the terrene dry is crucial. Being the affection of nutrition sapour must be an affection of the nutritive earth.

In an analogy to the kinds of odour, the different types of sapour result from different mixtures of the elemental qualities. Unlike odours, there are no nonnutritive sapours. In respect to the mixtures, there is an analogy between the kinds of sapours and colours. In line with the two extreme colours, white and black, there are two extreme sapours, that is, the sweet (*dulcis*) and the bitter (*amarus*). While the sweet is formed by a mixture of the hot and the moist, the bitter is produced by a strong hot with a strong presence of the dry. This is also why ashes or a burned food taste bitter. Albeit these two kinds of sapour are not extreme opposites, since the sweet and the bitter differ only in the degree of the dry and the moist, not in the hot common to both, they are contraries with respect to taste. In an analogy to the 'medial' colours, all other intermediate sapours are characterized by a proportional approximation to or recession from the extremes. Unlike odours, however, these kinds have special

---

[256] This subjective existence of colours and sapours is expressed in Democritus's famous fragment: 'By convention sweet and by convention bitter, by convention hot, by convention cold, by convention color; but in reality atoms and void'. (Taylor, 1999, 9). Some atoms that are jagged and tear the tongue create bitter sensations in the perceiver; some that are smooth and roll easily over the tongue produce the sensation of sweetness. These qualities do not exist intrinsically in the sensible objects but only dependently on our modes of observation. For a critique of Democritus's reduction of flavours to the atoms' shapes, see Aristotle, *On Sense and Sensible Objects* ch. 4, 442b10–2, 247.

[257] *CDA* disp. 7, q. 10, n. 5, t. 2, 692.

[258] *CDA* disp. 7, q. 12, n. 5, t. 2, 716. See also *On Sense and Sensible Objects*, ch. 2, 438b21–2, 227.

[259] *CDA* disp. 7, q. 10, n. 5, t. 2, 692–4.

names. They have names since our gustatory organ is much more perfect than our organ of smell.[260] Although there are different taxonomies related to the number and kinds of sapours depending on this or that author, it is not surprising that Suárez follows Aristotle's classification. In the fourth section of *On Sense and Sensible Objects*, Aristotle distinguishes eight kinds of sapours[261]: Beside the two extreme kinds, namely the sweet and the bitter, two similar sapours are mentioned, namely the rich-tasting sapour (*pinguis*), which is almost the same as the sweet, and the salty sapour (*salsus*), which is almost the same as the bitter. Beside these four there are the following four 'middle' sapours: the pungent (*ponticus*), which is called the sapour with a tang, the harsh (*acerbus*), that is, vinegary sapour, the astringent (*austerus*) and the sharp (*acutus*).[262]

4.5.2 Object of Touch: The Problem of Corporeal Pain

In the exposition of the proper sensible object of touch Suárez is even briefer than in his disquisition of the object of taste, as if he were suggesting that the reply to this question is more than clear. It must be said, however, that it is anything but that. In *CDA* disp. 7, q. 13 he asserts that the object of touch is a set of primary qualities and other qualities resultant from them.[263] What does he mean by these resultant qualities? Obviously, they are not sapours or odours. Although they are secondary qualities, *as such* they are not tangible. In the second book of *De generatione et corruptione* (*GC*), disp. 4, q. 1 Suárez mentions five pairs of non-basic tactile qualities, which all constitute binary contraries.[264] They are the heavy and the lightweight, the hard and the soft, the viscous and the brittle, the rough and the smooth, and the coarse and the fine. In line with the mainstream scholastic tradition he deems them to be secondary qualities. Except for the opposition of the heavy and the lightweight, all the other pairs are grounded in the primary qualities. Since the primary qualities are foundational, the touch which perceives them is the fundamental sense power. As Suárez says, *tactus* is 'quasi sensus universalis'. Its universality is due to the fact that its organ is spread all over the body and it can feel almost all the proper sensible objects of the other four external senses even though it cannot perceive them *as such*. This is also why the natural action of sensibles in most cases accompanies the intentional action of sensible objects. This concomitant action is a reason why the organs of all the sense modalities can be injured. Natural

[260] See *CDA* disp. 7, q. 16, n. 6, t. 2, 772.

[261] Aristotle, *On Sense and Sensible Objects* ch. 4, 442a13–442a29, 245.

[262] *CDA* disp. 7, q. 10, nn. 6–7, t. 2, 696. For this taxonomy of flavours, see also Thomas de Aquino, *Sentencia libri De* sensu *et sensato*, tr. 1, ch. 10, 59–61.

[263] *CDA* disp. 7, q. 13, n. 1, t. 2, 720.

[264] For this, see Francisco Suárez, *In Aristotelis De generatione et corruptione*, disp. 4, q. 1, 25–6.

alteration can destroy their appropriate mixture of the first qualities, which is necessary for the organs' proper functioning.[265]

So far, so good. However, what about other qualities called today 'interoceptual', that is, qualities perceived by bodily self-awareness, such as fevers, swells, itches and corporeal pains and pleasures, which all are characterized by the special phenomenal quality of 'ownness' and 'bodily self-awareness'? Are these qualities a part of the total proper sensible object of touch, too? Does Suárez hold, as Aquinas claimed,[266] that only the exteroceptual tangibles constitute the total tactile sensible object and corporeal pain (*dolor*) and other interoceptual qualities are relegated to the sensory appetite? *Or* does he, in a linkup to Avicenna, the medical and the Augustinian tradition, extend the scope of its proper sensible object in the direction of these interoceptual qualities? To put it tersely, is pain a sensible object perceived by the tactile power, *or* is it a feeling and emotion?[267]

Truly, Aristotelian psychology pays much more attention to the external tangibles than to the internal ones. The main reason is that the tactile power, being passive as all the external senses are, could not be actuated by itself but only by an external object. It needs to be reduced to act only by something external that is already in act. Basically, all the sense modalities including touch are directed outside, namely

---

[265]For the possibility of a damage of the organs through the natural action of a sensible object, see *CDA* disp. 5, q. 2, n. 31, t. 2, 336–8.

[266]Thomas de Aquino, *STh.* I-IIae, q. 35, art. 1, 240: '[. . .] ita etiam ad dolorem duo requiruntur, scilicet coniunctio alicuius mali [. . .] et perceptio huiusmodi coniunctionis. Quidquid autem coniungitur, si non habeat, respectu eius cui coniungitur, rationem boni vel mali, non potest causare delectationem vel dolorem. Ex quo patet quod aliquid sub ratione boni vel mali, est obiectum delectationis et doloris. Bonum autem et malum, inquantum huiusmodi, sunt obiecta appetitus. Unde patet quod delectatio et dolor ad appetitum pertinent [. . .] Cum igitur delectatio et dolor praesupponant in eodem subiecto sensum vel apprehensionem aliquam, manifestum est quod dolor, sicut et delectatio, est in appetitu intellectivo vel sensitivo. Omnis autem motus appetitus sensitivi dicitur passio [. . .] Unde *dolor, secundum quod est in appetitu sensitivo, propriissime dicitur passio animae*'. [my italics]

[267]This doctrinal split associated with the different replies to the issue is not restricted to the medieval and post-medieval scholastic debate. In recent scholarship the core of the philosophical controversy about the (paradoxical) nature of pain oscillates between two main positions which bear a striking resemblance to their scholastic counterparts. While one view focuses on the objective aspects of pain, the other underscores the complex subjective elements in our experience of pain. The former understands pain as an attribute of a physical object, commonly identified with the tissue damage of a part of our body. Accordingly, pain is tightly associated with a physical damage objectively located in a bodily part. Thus pain is something that can be perceived in the same way as the sensibles of the other sense modalities, such as a red colour. This perceptualist concept is contrasted with the approach underlining the subjective and non-perceptual aspects inherent in our experience of corporeal pain. In this approach, more in line with the contemporary scientific approach, our pain experience is, phenomenologically speaking, of a much richer composition than the advocates of the purely perceptualist view are prepared and willing to concede. Far more than an object or a cause, 'being in pain' is a multidimensional matter involving evaluative, affective, emotional, motivational and imperative elements (the view that better explain phantom pains). For these two basic approaches to pain, see Hardcastle (2015), Aydede (2019) and Grahek (1991).

toward the proper extramental sensibles. In defiance of this mainstream scholastic position, however, there were not infrequent alternatives in the Middle Ages. Olivi, who is sometimes mentioned as a significant source of Suárez's psychology, was one of the authors who extended the field of the tangible to include the interoceptual qualities. Pietro d'Abano, to name another author, broadened the field of the tangible to comprise pain (*dolor*) as a proper sensible of touch.[268]

It must be said that even though it has been shown in 2.5 that Suárez seems to be open to interoception, in *CDA* disp. 7 I have not found an explicit endorsement of tactual interoception. Although Suárez (as I will further show in 4.5.3) rejects the Aristotelian-Thomistic view according to which the sensible species of the proper sensibles of all the external senses must be transmitted through a medium, and albeit he rejects the all-inclusive validity of the Aristotelian dictum that 'a sensible placed on a sense impedes sensation' (*sensibile positum supra sensum impedit sensationem*), which brings him near to an affirmation of bodily self-awareness, he is explicit that touch and taste do not perceive qualities *inherent* in the pertinent organs, but only qualities of external things that are *tangent* to the tactile organs.[269]

But how does Suárez argue for the non-perceptibility of qualities inherent in the tactile organ?[270] He says that the argument is clear in the case of taste. From everyday experience we know that our gustatory sense does not perceive a sapour inherent in the tongue but a sapour pertaining to a tasted food. We say that this meal tastes better than another.[271] As regards the tangible, Suárez formulates four arguments for his conclusion about the non-perception of inherent qualities. (1) Considering the argument that can be called 'from parity with the other senses' he says that

---

[268]As regards Pietro d'Abano's extension of the external proper sensible of *dolor*, see his *Conciliator controversiarum quae inter philosophosos et medicos versantur*, differentia LXXVII: 'Utrum dolor sentiatur, nec ne', f. 118va: 'Rursus dictum est antea per Avicennam quod dolor sit contrarium impressorem repente sentire. Adhuc illud, quod est signum, sentiatur [. . .] Amplius sciendum quod Aristoteles ponit ibidem dolorem fore unum tactus sensibilium inquiens'. For Olivi's broadening of the scope of touch by referring to various inner states of the body such as feverish heats or swellings, see Olivi, *Quaestiones in secundum librum Sententiarum*, vol. II, q. LXI, 574. For both authors, see Yrjönsuuri (2008, 105–6), (2006, 158–61) and Toivanen (2013a, 372–3). As regards Olivi's conception of touch as a power of bodily self-perception, contrasted with the Aristotelian approach according to which touch perceives only qualities that are external to us, see Toivanen (2013b, 285ff). For the sentiment that physical pain is sensed directly in Avicenna, Aquinas, d'Abano and Olivi, see Knuuttila (2011, 909–10). Declaring the up-to-date importance of this position, Olivi is also mentioned by contemporary philosophers of perception. For this reference to Olivi, see De Vignemont and Massin (2015, 296).

[269]See *CDA* disp. 7, q. 13, n. 3, t. 2, 722; q. 14, n. 6, t. 2, 744–6. Suárez raises a question similar to that posed in contemporary phenomenology of perception, namely what is it like to feel. Suppose that we approach fire with our hand. What do we feel? Do we feel a heat inherent in our hands, or a heat of the fire itself? Suárez's reply here seems to be 'externalist': We primarily perceive the heat of the fire. For an overview of the divergent positions on the issue 'what it is like to feel', see Mattens (2009).

[270]Of course, he does not mean a situation when one hand touches the other hand. This is a case not of interoception but of exteroception.

[271]*CDA* disp. 7, q. 13, n. 3, t. 2, 722.

it generally holds that the other senses including taste cannot perceive qualities inherent in the organs (not those that are contiguous with them!). (2) If touch perceived inherent qualities, it would have to perceive them continuously. Some first qualities always inhere in the organ. However, such continuous perception is at odds with our dynamic and changeable tactile experience. (3) *Tactus* often perceives a greater heat than that which inheres in the hand. A greater heat can be a heat of fire, rather than a heat inherent in the hand. If I bring my hand to a fire, I will soon feel the fire's heat. It would be false to say that at that moment I perceive a heat inherent in my hand. In the short time in which I come to perceive the heat of the fire the heat of the fire could not produce such a heat in my hand. (4) Not only the primary qualities but also the non-basic tactile qualities, such as the hard and the soft, constitute the proper sensible object of touch. The perception of these secondary qualities shows even more clearly than the perception of the first qualities that we perceive the qualities of *external* objects. At least in short term my hand does not become physically soft when touching soft sensibles or hard when touching hard objects. (Admittedly, in the long term we can distinguish between the hands of an intellectual and those of a manual worker.) Therefore, I do not perceive the softness of my hand but the softness of an external object.[272]

This exclusion or, to be more charitable to Suárez, this non-inclusion of corporeal pain in the total proper sensible of touch results in the conclusion that bodily pain, properly speaking, is not an object of touch but an act of the sensory appetitive power. Accordingly, the affection of *dolor* will not be dissimilar to the passion of distress (*tristia*). In order to explicate what this statement about pain being an emotion precisely means I have to now outline some aspects of Suárez's theory of emotions.[273]

Like the mainstream scholastic tradition, Suárez distinguishes between two main kinds of appetite, sc. sensory and rational. These faculties are really distinct from each other and differ *realiter* from the other (sc. cognitive) powers and from the soul. In the case of elicited appetites, which–contrary to natural appetitive powers–are triggered by a preceding cognition, the two appetites differ in the character of the antecedent cognitive element. While the rational appetite, sc. the will, starts from and comes after intellectual cognition, the sensory appetite follows sensory cognition. Both kinds of cognition yield the necessary evaluative judgment that concerns the convenience/inconvenience of a perceived object, which is conceived as good (*sub ratione boni*) and conceived as bad (*sub ratione mali*). This judgment then elicits the

[272]Ibid., n. 4, 722–8.

[273]For Suárez's theory of emotions and his ontological reduction of the concupiscible and the irascible appetitive powers (conceived by Aquinas and Thomism as two really distinct powers) to a single power, see King (2002, 238–44) and Heider (2016a). For a recent translation of Suárez's *De actibus qui vocantur passiones*, his crucial treatise on emotions, into Portuguese and Spanish, see Francisco Suárez (2014).

affective elements of pleasure, or displeasure.[274] Importantly, in many places of *CDA* disp. 10 and 11 Suárez suggests that the sensory appetite follows the cognition of the *interior* sense power, and not the apprehension of the external senses, since only the internal sense is capable of this estimation.[275] Since for Suárez, as we shall see, there is only a single interior sense generating various kinds of operations (Aquinas had ascribed these to really distinct senses), there is also a single sensory appetite accomplishing the various kinds of affective acts, that is, concupiscible and irascible ones. The manifold of these acts is structured both by means of the nature of the appetible objects and by the manner in which these appetitive acts are related to their objects.[276]

According to the well-known deductive taxonomy of emotions advocated by Aquinas in his *Treatise of Passions* from *Summa theologiae*, which Suárez also endorses (albeit only for pragmatic or pedagogical reasons[277]), there are two main kinds of adequate objects, namely the sensible good and the sensible evil. Speaking only about concupiscible acts, the good and the evil can be approached in a threefold way. The sensible good can be followed and the sensible evil can be avoided either (a) as the good/the evil in an unqualified sense (absolutely or without a temporal index), (b) as the good/the evil *qua* present, or (c) as the good/the evil *qua* future. The timeline of the past is irrelevant since as such it does not contain the feature of the good and the evil. The good that is vitally and absolutely followed by an appetent is equivalent to the emotion of love (*amor*), the evil vitally avoided by the appetitive power amounts to the passion of hate (*odium*). While on the level of affective acts love is a temporally unqualified movement of the soul toward a beloved thing (person), hate is the inverse movement of avoidance of an odious thing (person). While the *pondus* of the passion of desire (*desiderium*) is related to a future good, the emotion of avoidance (*fuga*) concerns the evasion of an oncoming evil. While the affection of joy (*delectatio*) applies to an acquired good, the affect of distress (*tristia*) relates to a present evil.

At first sight it is clear that a present evil vitally avoided by the sensory appetite can be not only an intentional distress or sadness, but also a crude bodily pain (as we know from our own psychosomatic experience, both often go together). Considering Suárez's non-inclusion of pain in the tactile object jointly with his statement that only the interior sense proposes an object to the appetite seem to lead to the conclusion that corporeal pain is nothing else than an act of the sensory appetite. Considering his theory of the sympathy or harmony of the cognitive and the appetitive powers, Suárez stresses that an appetitive act, strictly speaking, is not

---

[274] *CDA* disp. 10, q. 1, n. 3, t. 3, 286–90. The adequate object of appetite is the good. In the case of sensory appetite it is the sensible good. See *CDA* disp. 10, q. 2, n. 3, t. 3, 292 and disp. 11, q. 1, n. 4, t. 3, 330.

[275] *CDA* disp. 11, q. 1, n. 2, t. 3, 326.

[276] Ibid., n. 6, 330–2.

[277] For this taxonomy, see Francisco Suárez, *De actibus qui vocantur passiones*, disp. 1, s. 12, 474–7.

caused but only conditioned by a precedent cognition of the interior sense. As elsewhere, Suárez rejects causal efficacy between cognitive objects or acts and the operations of the really distinct affective powers. His claim that the organic root of sensory cognition is the heart and the principle of sense cognition is the brain does not pose obstacles to this mediation. He makes it clear that if in respect to the elicitation of an affective act an apprehended object does not generate proper causal efficacy but only metaphorical motion, the two powers need not be located next to each other. The soul 'a-causally' mediates their contact.[278]

In harmony with *CDA* disp. 7, q. 13, Suárez embraces the view that *dolor*, like *tristia*, is to be ranked among the passions. Bodily pain is necessarily perceived under the characteristic of a present evil. However, as said, the evil *qua* present is an object of the sensory appetite. Accordingly, pain is an emotion whose main characteristic is active bitterness (*amaritudo*) and active avoidance of the evil object.[279] As such corporeal pain forms a 'conceptual tandem' with the emotion of distress and consequently stands in contrary opposition to the affection of pleasure or delight (*voluptas*).

Having placed *dolor* in the set of the concupiscible passions, the next question is about *the cause* of this emotion. In *CDA* disp. 11, q. 2, n. 5 Suárez says that despite the affinity of the passions of distress and pain, these two emotions are not the same. He explains that while *tristia* concerns a present evil, which does not harm the body, bodily pain applies to a present evil that does harm to it. Our experience shows that we suffer corporeal pain in the very part of the body that hurts.[280] When a doctor asks where it hurts, we can commonly more or less localize our pain. Following the early Aquinas of the *Commentary on the Sentences*[281] (in this text the Angelic Doctor takes a different position from the stance advocated in *Summa theologiae* I-IIae) Suárez declares that the cause of *dolor* is perceived by touch. However, admitting this type of apprehension does not imply that Suárez supplements the cognition of the external sense of touch with some corresponding particular sensory appetite that would be localized in the part that hurts. On this view, which states that each external sense is to be paired with the corresponding sensory appetite[282] *dolor* is

---

[278]For this metaphorical efficacy, see *CDA* disp. 10, q. 3, n. 8, t. 2, 310; see also 3.4.

[279]*CDA* disp. 9, q. 2, n. 5, t. 3, 340.

[280]*CDA* disp. 11, q. 2, n. 5, t. 3, 340.

[281]Thomas de Aquino, *Scriptum super Sententiis*, book 3, d. 15, q. 2, art. 3, qc. 2 c.: 'Secundo quantum ad perceptionem: quae quidem in dolore semper est secundum sensum tactus, ut dictum est, in tristitia autem secundum apprehensivam interiorem. Tertio quantum ad ordinem istorum duorum: quia dolor incipit in laesione, et terminatur in perceptione sensus, ibi enim completur ratio doloris; sed ratio tristitiae incipit in apprehensione, et terminatur in affectione; unde dolor est in sensu sicut in subjecto, sed tristitia in appetitu'.

[282]This theory was defended several decades after Suárez's composition of *CDA* by the Spanish Jesuit Pedro Hurtado de Mendoza, 1578–1641. See Hurtado de Mendoza, *Universa philosophia*, De anima, disp. 17, s. 11, 685: 'Opinor diversis sensibus externis diversos respondere appetitus [...] bonum apprehensum est obiectum appetitus: sed bonum visum, auditum, etc. est bonum apprehensum, ergo est obiectum alicuius appetitus: sed non appetitus communis, ergo particularis'.

not an act of the common sensory appetite but of a particular appetitive power, namely the tactile one, which is preceded by tactile perception. Suárez does not seem to show much understanding for this theory as it redundantly multiplies sensory appetites. Two appetites, the common sensory appetite and the rational appetite, sc. the will, are sufficient.

However, the following difficulty arises: In the *Topics* Aristotle explicitly claims that the *cause* of pain is a separation of conjoined parts accompanied by violence.[283] Accordingly, *dolor* is nothing else than the separation or dissolution of a conjoined whole. However, as such it cannot be part of the proper sensible object of touch. Referring to Averroes, Suárez rejects this view by denying that the disruption of naturally conjoined parts is a proximate cause of pain. The main reason is that the disjoined parts cannot be regarded as a proper sensible object of touch since they are common sensibles, rather than proper sensibles. As such they can be perceived by more than one external sense, at least by touch and sight. At most we may say that cutting and the resulting plurality of parts is one of the remote causes of the emotion of pain. The proximate cause, however, must be some other quality that is produced by this scission or by another remote cause, such as calefaction or commotion. In his gradual elimination of the possible candidates, Suárez presents a second view (advocated by Averroes), according to which this proximate cause of corporeal pain is an *intemperies* of the bodily primary qualities, that is, a lack of temperateness caused by an excess of a primary quality. Accordingly, such an excess of the hot or an excess of the cold in the body violating the overall equilibrium of the *humores* of the primary qualities is to be taken as the proximate cause of pain. With reference to our experience, however, Suárez insists that we often perceive pain without recognizing whether the pain comes from an excess of the hot or from a superfluity of the cold. So it cannot be said that this *intemperies* caused by such an excess is the proximate cause, which as a proper sensible is apprehended by the sense of touch.[284] Suárez is clear that the hot, the cold, a scission, or a commotion cannot be regarded as proximate causes of pain. The immediate cause of pain can only be a quality that results from a lack of temperateness in the primary qualities. So the lack of moderateness can at most be conceived as a middle link in the causal chain of remote and proximate causes. It is a part of our experience that our hand hurts even more long after it has been removed from the fire and when, for example, it has already been cooled in cold water. The same holds also for an injury caused by cutting. Sometimes a pain caused by cutting arises long after the act of scission has been terminated. Regarding the role of an excess in a primary quality as a middle link, it may be said that a scission causes a humour, which brings about an excess in one of the primary qualities, from which a new quality arises. And precisely this new quality, which is a result of the ratio of the primary qualities, is the secondary quality

---

[283] Aristotle, *Topica* VI.6, 145b13–4, 603: 'For the separation of the conjoined parts is not pain, but a cause of pain [. . .]'

[284] *CDA* disp. 11, q. 2, n. 6, t. 2, 342–4.

for which Suárez is searching. As such it is based on the first qualities.[285] And since in his *GC* Suárez says that secondary qualities are innumerable *(innumerae)*,[286] it may be surmised that there is no a priori obstacle to include this (new) *dolorogenic* quality in the overall list of the proper sensibles of touch together with the other secondary qualities. There is no textual obstacle to include this quality in the total ambit of the proper sensible of touch.

Concluding, Suárez shows that *dolor*, formally speaking, is equivalent to an act of the sensory appetite. This conclusion is confirmed also in *De angelis*, book VIII, chapter XIII 'Whether fire tortures demons by apprehension alone or by an intentional action, or also by a natural action' where he rejects the theory according to which pain as such lies primarily in the cognitive power, that is, that pain as such is perceived by touch. The damage of the flesh and the perception of *qualitas dolorifera* are only the first two steps concurring in the production of the full mental experience of pain. Suárez explains that this full experience is nothing else than 'dissensio animae a tali laesione', which amounts to escape and dissatisfaction, viz. operations whose proximate subject is the sensory appetite.[287]

However, assuming that in *CDA* disp. 11, q. 2 Suárez embraced the dolorogenic quality, why did he not include it in the list of the proper sensibles of touch in *CDA* disp. 7, q. 13, where he *ex professo* delved into the issue of its proper sensible object? Charitably speaking, with reference to *GC* disp. 4, q. 1 where the innumerable character of secondary (tactile) qualities is mentioned, we may say that in *CDA* disp. 7, 13 Suárez did not aim to present a comprehensive list of tactile sensibles. On the other hand, considering his proverbial conciliatorism, it may be said that Suárez's reasoning sometimes seems to border on incoherence.[288] He is well aware that this proximate cause of *dolor*, namely the quality inherent in the body, is not regarded as a proper sensible object in the Aristotelian tradition. At the same time he is also conscious of the view wide-spread in the medical tradition that pain is a quality directly perceived by touch. As I have shown, throughout *CDA* this tradition constitutes an important referential point alongside the broad Aristotelian tradition.

---

[285] Ibid., n. 7, t. 2, 346.

[286] Suárez, *In De generatione et corruptione*, disp. IV, q. 1, 25.

[287] Francisco Suárez, *De angelis*, book 8, cap. 13, n. 9, 1025: '[...] dolorem formaliter esse actum cognoscendi, falsum est, ut in tractatu de Anima inferius ostendam, et tradit div. Thomas, 1, 2, quaest. 35; nec Augustinus supra dixit, sed agens de dolore sensibili ait, *Doloris, qui dicitur carnis, animae esse in carne, et ex carne* [...] Tria ergo secundum Augustinum, et veritatem concurrunt ad sensibilem dolorem, scilicet laesio carnis, perceptio laesionis, et dissensio animi a tali laesione, quae dissensio non est actus cognoscendi, sed est fuga quaedam, et displicentia appetitus; et in hac consistit formalis actus doloris, laesio autem carnis, eique disconveniens per doloriferam qualitatem est objectum doloris: perceptio autem, seu cognitio laesionis est applicatio, seu propositio objecti doloriferi'.

[288] He borders on incoherence, especially when we consider the post-Suarezian Jesuit Francisco de Oviedo (1602–1651) who *explicitly* includes the dolorogenic quality and the quality incurring pleasure (*qualitas voluptuosa*) in the list of the total sensible object of touch. See Francisco de Oviedo, *Cursus philosophiae, De anima*, Controversia IV: De potentiis materialibus tam externis, quam internis, Punctum primum, s. 4: De gustu, et tactu, n. 20, 51.

From this point of view, it is therefore not difficult to see why Suárez's view wavers as far as the two abovementioned textual blocks of *CDA* disp. 7 and *CDA* disp. 11 are compared.

4.5.3 Taste and Touch: Only Real, or also Intentional Affection?

While Suárez believes that an odorous quality can intentionally affect the power without natural alteration, the opposite does not hold. As far as the dispensability of natural change of the organs is concerned, the situation with the contact senses of taste and touch is different. A natural affection of the organ is necessary for both. Some authors go even so far as to say that it is not only necessary but also sufficient to produce a pertinent sensation. Accordingly, in their explanation of the production of these powers' acts intentional immutation is fully dispensable. The title of *CDA* disp. 7, q. 13, called 'Whether *gustabilia* and *tangibilia* affect the corresponding senses only *realiter*, or also *intentionaliter*', shows that Suárez (as he was in the preceding senses) is well aware of this naturalist account which he associates especially with representatives of the medical tradition.[289] He introduces three arguments for this position according to which the tangibles and the tasteables concur immediately, that is, without the mediation of sensible species, in the formation of the proportional sensorial operations.

(1) It is part of our experience that if these two powers are actualized by sensibles, their organs always undergo a material alteration. When we touch fire, our hand gets warm. By touching snow, it gets cold. When we put our hand in water it becomes wet. While tasting ice cream our tongue comes to be penetrated with a humid liquid through its pores etc. Why should we then be obliged to introduce the tangible species or the saporific species as the necessary instrumental co-causes of tactile and gustatory operations? Why not conceive the causal concurrence of these qualities in a purely naturalized manner, that is, through the organs' literal affection?[290] The answer to these questions seems to be *a fortiori* affirmative, if we consider the general conditions of the introduction of the intentional species, as formulated by Suárez himself:

> The intelligible species is to be posited either in order for an intelligible object in potency to become an intelligible in act, or [and this is applicable also to the sensible species] in order

[289] In this context he names Vallés. For his view, see *Controversiarum medicarum et philosophicarum libri decem*, book 2, ch. 29, 108: 'An, quo sensus fiat, non satis est, proprium instrumentum alterari, & facultatem sentire alterationem? At vero, ut hoc fiat: non opus est medio, neque speciebus, praecipue in tactu'.

[290] While discussing Aristotle's theory, Cynthia Freeland (1995, 231) calls the theory, according to which being actually and literally affected by the relevant qualities suffices for eliciting a perceptual operation, a 'literalist position'. No doubt, this theory is more in line with Ockham's razor.

> that an object that was separated or distant and not proportional [to the power] could be connected or proportioned to the power.[291]

Symptomatically, just this essentially contact-based nature of these two 'lowest' senses was the reason why a number of post-Suarezian Jesuits denied the existence of the gustatory and the tangible species as the necessary principles of these senses' operations. They argued that if neither distance nor disproportion is extant between touch and taste on the one side and their objects on the other, the gustatory and the tangible species are fully dispensable.[292]

(2) If touch apprehended its object through species, it would have to be able to perceive a distinct object's heat of four degrees (not Centigrade; it is Suárez's example) even if a heat of the same degree inhered in its organ. Since there is no contradiction in the co-existence of the real quality of heat and its species, touch would have to be capable of receiving the sensible species of a distant object's heat of four degrees. However, as our experience shows, this is not the case.

(3) If these two senses produced their operations through the reception of species, they would have to perceive their objects in an instant since species (as the general argument against the existence of sensible species always assumes) are emitted only instantaneously. However, this is false since a successive approximation of fire to a hand determines an increasing intension in the perception of its heat.[293]

Suárez's denies this 'species-less theory' of touch and taste. This denial comes directly from his four-argument criticism of the theory according to which we (can) sense the qualities inhering in the tactile and gustatory organs, as presented in the previous section. He is sure that since touch perceives a heat produced in the surrounding medium by a fire (and not that inherent in the body), it necessarily needs to receive its species. It has been said in 3.2 that, as any other affective quality, heat generates not only physical action but also intentional action.[294] Suárez's claim about the necessity of the intentional species applies also to the two lowest senses and their proper sensibles. These two senses are not excluded from his general claim about the necessary existence of the sensible species.

But why cannot the power of touch perceive an external heat of fire through a heat inherent in the organ? Clearly, if this were the case, Suárez's ontology would be more parsimonious. In his main argument Suárez introduces the example that has already been presented in the previous section and argues for the impossibility of perceiving qualities inherent in the body. If perception of inherent qualities were possible, the tangible species would be redundant. According to the argument it holds that we often feel a heat that is of greater temperature than that inhering in our body. A hand approximated to fire actually feels a stronger heat of the fire's immediate thermal zone than that which was immediately produced in

---

[291] *DM* disp. 35, s. 4, n. 18, t. 2, 464. For this passage, see also 3.1.

[292] See, e.g., Hurtado de Mendoza, *Universa philosophia*, De anima, d. 12, s. 1, n 10, 611. For Hurtado, see also Heider (2019a, 129–31).

[293] *CDA* disp. 7, q. 13, n. 1, t. 2, 720.

[294] See *CDA* disp. 6, q. 2, n. 6, t. 2, 476.

it. Moreover, the hand perceives this greater heat very quickly. In this short time, however, this perceived heat could not have been produced in the hand. Although, as said, it holds that the intentional action of the fire's heat must be preceded by the fire's natural action, sc. by heating, it is often not the case that this heating arrives at the body of the perceiver in the same degree as it exists in the medium or in the immediate thermal zone of fire. Therefore, it is impossible that the heat inhering in the hand, and not the tactile species of the external heat, is a 'ratio sentiendi' through which the heat of the fire is sensed. This would be even more improbable in the case of the perception of the non-basic tactile qualities, such as the hard and the soft. These qualities do not produce similar real qualities in the tactile organs, at least in short term. Yet, they are perceived. Therefore, their sensation can run only by means of their species's reception in the power. Accordingly, starting from the assumption of a single sensory modality of touch, if there is a single sense of touch, there will also be a single way of the sensation of its various proper sensibles. But if the sensation of hardness and softness runs through the sensible species, the same must hold also for the sensation of the first qualities.[295]

Coming back to the arguments for the naturalist view according to which natural affection is all that is needed for the production of sensation, why cannot touch, in which the heat of four degrees inheres, perceive another heat with the same degree, if the tactile species of the external heat can coexist with the real quality? Why cannot a heat inhering in the tactile organ produce its species in this organ, if it can do that in a distinct subject? Suárez's answer is based on a specific feature of the tactile senses' immutation according to which their intentional action is necessarily connected with natural action. There is a natural *ordo* between the intentional and natural activities of these sensibles. An intentional action of the sensible is always accompanied and preceded by its natural action. Our hand simply cannot feel the heat of another object without being somehow warmed at the same time. The reason for this connection is the imperfect and least noble character of *tangibilia* and *gustabilia*.[296] Unlike the distal senses, in the natural order the tangible can intentionally affect the organ, only if it undergoes a material alteration. However, this alteration must proceed only by a deflection from a midpoint *secundum excessum*, that is, only by means of the

---

[295] *CDA* disp. 7, q. 13, nn. 4–5, t. 2, 724–30. Suárez also introduces a special argument regarding the tasteable species. Since this sense perceives the qualities of external objects, we taste a quality of the food, not a quality of our tongue. Therefore, for the production of a gustatory operation we need to receive the species of sapours. The non-inherence of sensible qualities implies sensation through sensible species. See *CDA* disp. 7, q. 13, n. 3, t. 2, 722; and n. 5, 728.

[296] Interestingly, even though Suárez assumes this connection *in statu isto* as necessary, he seems to reserve the possibility of the separation of these two actions in the glorified body. Since glorified bodies will be impassible, they will not be able to suffer material alterations. They will have to be immuted purely intentionally. Material alterations of glorified bodies would also have to be equivalent to the necessary existence of vegetative operations. However, these will not be part of the resurrected body. According to Suárez *in patria* there will be tasting without digestion. See Suárez, *Expositio et Quaestiones in quinque priores primae secundae divi Thomae Aquinatis quaestiones materiam de beatitudine continentes* (forthcoming 2021), 'Disputatio de quidditate humanae beatitudinis', q. 2, 139–40.

application of an external quality that is registrable by the tactile organ, which must be appropriately 'tuned' by the elemental 'medial disposition' of the first qualities.[297] Unlike the organs of the other senses (the organ of eye is colourless, the organ of ear is motionless etc.), touch cannot be deprived of the qualities it receives. It cannot be without its cold or its heat. This is why it must be well 'tuned' or placed in a mean state between the contrary tangibles to be able to objectively register oncoming tactile qualities. A too cold or hot hand will not objectively judge the temperature of external objects.[298] This is the reason why the organ of touch cannot be affected by a quality of an equal degree. Only if the balance of the primary qualities of its organ becomes noticeably 'deflected' can a tangible object be perceived at all. But such 'deflection' can come only from an external sensible. Only an external tangible is able to bring about such an excess. And since natural action proceeds only successively (this is the reply to the third argument for the 'naturalist' view), the multiplication of these species can run only succesively according to particular parts and degrees at which warming successively arrives.[299]

4.5.4 *Organ of Taste and Touch: The Problem of Medium*

While one of the most controversial issues related to the topic of touch is the question about its organ, the organ of taste does not pose a problem for Suárez. In line with Galen[300] he says that the organ is the inner part of the tongue, namely the lingual nerve, while the outer part of the tongue is its medium.[301] However, since these external parts of the tongue are porous, if food is sufficiently chewed and humidified by saliva, it can penetrate through its pores to the internal part of the tongue and can become tangent to the proper organ. Accordingly, even though it in a way possesses an intrinsic medium, touch can sense the *gustabile* by being in immediate contact with it.[302]

The issue of the organ of touch is more dubious (*maius dubium*). There are several competing views. On the first sentiment, attributed to Aristotle's *De sensu*, its organ is the heart; the other parts, such as the skin and the flesh, are merely the

---

[297]For the organ of touch as a sort of "mean" between the relevant sensible extremes, see Aristotle, *DA* II.11, 424a4–6, 135.

[298]For this mean state of touch situated between contrary tangibles as being in potency to its objects, see also Thomas de Aquino, *Sentencia libri secundi De anima*, ch. 2, 167.

[299]*CDA* disp. 7, q. 13, n. 6, t. 2, 730–2.

[300]For Galen's view, see his *De usu partium corporis humani*, book 16, 450–1.

[301]As for the composition and functions of the human tongue, Suárez (as usual) refers to Vesalius's *De humani corporis fabrica*. For the anatomy of the tongue, see book 2, ch. 19, 252–5.

[302]*CDA* disp. 7, q. 14, n. 1, t. 2, 734.

medium.[303] According to the second tenet, the organ of touch is the nerves that abound immediately under the flesh and the skin. As a web they are spread all over the body. Also, here the flesh is only the medium of tactual perception. Although Aristotle did not know much about nerves and the nervous system,[304] Suárez states that in *On the Soul* he comes close to this view by assigning the flesh as the medium of the tangible.[305] Some other authors, such as Avicenna, advocate this view with a qualification. For them the organ is not the nerves but rather a combination of the nerves and the flesh, namely the nervous flesh.[306] According to yet another opinion espoused by Galen, the organ of touch is the most external part, namely the skin. The skin, unlike the flesh which is too hot since it is too sanguine and contrary to the nerves which are too cold because they are bloodless, is best tempered for perception of the first tactile qualities. On the last view, which in a way synthetizes the preceding ones, the organ is the whole body or all the parts except for the hard and earthy parts, such as bones, fingernails, teeth and hair that seem to be not animated by the soul.[307]

Although in the conclusion he embraces the last view, Suárez formulates it synecdochically. He says that the organ of touch is the flesh. But by presenting the flesh as *the* tactile organ he is, however, far from limiting the organ to it. It is also the skin that is to be adopted even as the best tactile organ. In his reasoning Suárez starts from experience. If we are pricked, our flesh and skin feel a stab. Even when the skin is separated from the flesh due to injury, still this 'naked' flesh, for example, in an ulcer, can feel a stab. Second, there is tactile perception in parts in which no nerves exist, such as the stomach, and in which we cannot distinguish between the flesh as the medium and something else as the organ. Third, the organ of any sensory modality is to be searched for in the place where the most appropriate *temperamentum* for this sense is to be found. We have seen that for the organ of sight it was colourless crystalline humour in the pupil, for the organ of hearing it was the motionless animated air in the inner ear etc. The best *temperamentum* for the reception of the first qualities can be found in the skin. Only in a lesser degree can it be found in the flesh, and even in a lesser degree does it occur in the nerves as these

[303] Aristotle says that a part near the heart is not only the organ of touch but also of taste. See Aristotle, *On Sense and Sensible Objects* ch. 2, 439a1–2, 229: '[. . .] the sense organ of both taste and touch is near the heart'.

[304] 'The nerves' for him were rather sinews that as ropes move the limbs. As Friedrich Solmsen (1961) notes, the theory of the *pneuma* and the specific channels of their flow was still in the process of evolving in Aristotle's time. It was not until post-Aristotelian medicine, exemplified by authors such as Herophilus (335–280 BC) and Erasistratus (ca. 304–ca. 250 BC), when *pneuma* currents identified as *neura* (both sensory and motor nerves) were discovered.

[305] Aristotle, *DA* II.11, 423b23–7, 133–5: 'From this it is clear that that which is perceptive of what is touched is within [. . .] the medium of the tangible is flesh'.

[306] See Avicenna, *Liber de anima seu Sextus de naturalibus I–II–III*, p. 2, ch. 3, 138–40.

[307] *CDA* disp. 7, q. 14, n. 1, t. 2, 734–6. This claim also raises the issue of the 'limits of body'. For this issue in the theories of early Jesuits, see Des Chene (2000, 196–9). For this issue according to early Jesuits, especially in connection with the query about the animation of blood connected with the theological topic of Eucharist, see Sander (2017).

are too cold and dry. Since, unlike the other senses, touch cannot be deprived of the qualities it perceives, it always has some degree of the hot etc., it must be in potency to its sensibles by being set into a well-balanced blending of these qualities. This well-balanced mixture can at best be found in the skin. And the human skin is the best tactile *sensorium* in the animal kingdom at all. Fourth, the organ of touch must be placed in the peripheral parts of skin and flesh because *tactus* was instituted for the protection of animals against external dangers. Touch is most needed for the preservation of life and defence against external harms. Teleologically speaking, the main function of the external 'soft tissues' is to protect the 'hard parts', such as bones, that by nature are established to carry the whole weight of the entire body. This is also why these hardest parts are unsuitable for feeling.[308] Although Suárez refers to the flesh and the skin as to the proper organs of touch, he inclines to the last abovementioned view by conceding that the other bodily parts can also, in a greater or lesser degree, 'participate' in the ideal disposition of the skin and the flesh and thus they can become organs of touch. He is well aware of Aristotle's claim that not only the flesh but also 'what is proportional to it' can be the tactile organ.[309]

If the skin is the best 'tuned' organ of touch, the first two abovementioned views that consider the skin and the flesh to be the medium of tactile perception must be false. We feel heat in our hands, not in the heart. We can speak about touches of the heart either metaphorically, or in relation to emotions. As I have made clear, for Suárez the heart is neither the organ of tactile perception nor the *radix sentiendi*, but the organ of the sensory appetite. The second view is also problematic. It cannot be denied that Aristotle comes close to it by saying that:

> In a general sense we may say that as air and water are related to vision, hearing and smell, so is the relation of the flesh and the tongue to the sense organ in the case of touch.[310]

In this quote Aristotle is clear that the difference between the three higher senses and the two lower senses is that the former ones have an *external* medium (e.g. air, water), while the latter ones are operative through an *internal* medium, such as the uppermost part of the flesh. He shows that the difference between distal and contact senses is not that the higher senses have a medium, while the lower do not. In fact, *all* the senses essentially operate through a medium since (as Aristotle says) 'That what is placed on the sense organ should be imperceptible is common to all senses'. In general, there is no sensation without a medium. The medium must be assumed as a necessary factor in the first stage of a perceptual process in general.[311]

Throughout *CDA* Suárez denies this claim. Unlike the impossibility of the perception of qualities inherent in the body, qualities that are contiguous to the body do not constitute a problem for him. But how does he assess Aristotle's general

---

[308] *CDA* disp. 7, q. 14, n. 2, t. 2, 736–8.

[309] Ibid., n. 3, 738. Aristotle, *Parts of Animals* II.1, 647a20–1, 115: 'So the sense-organ which deals with these [sc. hot and cold, solid and fluid and the like] viz. the flesh, or its *counterpart* [...]'[my italics].

[310] Aristotle, *DA* II.11, 423b17–20, 133.

[311] Ibid., II.9, 421b16–8, 123.

claim? At first he says that although his conclusion is not in line with *De anima*, it can be authorized by Aristotle's *Parva naturalia*, such as *De generatione animalium*, *De historia animalium*. Since these texts are later than *De anima*, it is more secure to follow them than the former.[312] Second and more importantly, Suárez agrees neither with the universal validity of the proposition 'sensibile supra sensum non facit sensationem' nor with the statement according to which the difference between the visible, the audible and the smellable on the one side and the tangible and the tasteable on the other side is due to the fact that while the former senses operate through an external medium, the latter operate through an internal medium. In his very elaborate reasoning he shows that one cannot start to explain the role of medium in the external senses according to the model of the visibles. A visiocentrism based on the impossibility of the perception of colours that touch the eye is detrimental to a true understanding of the function of medium, especially in the lower senses.

What are the general reasons for embracing the axiom 'Sensible supra sensum non facit sensationem'? One of the main reasons for adopting it, advocated especially by Thomists, is the essential *spiritualizing* role of the medium. Since the sensible species immediately emitted by the sensible qualities are spiritual, which for Suárez means more subtle than the represented sensible objects, they need to be 'spiritualized' through a medium intervening between the power and the object. The species must come to be spiritual to become proportional to the 'spiritual' character of the sense powers. The medium, so to say, fulfills the role of the agent sense together with a higher 'elevator'. From what we have seen especially in 3.2 it is clear that Suárez rejects this view and this argument. Not only that Suárez refuses an ontological elevation of the sensible species through the participation in the activity of the separated forms, he declines also a 'spiritualizing' role of the medium. The 'spiritual' species, which is more subtle than the represented sensible, is produced by the sensible quality. At the same time it cannot be said that this denial of Aristotle's axiom implies the denial of the existence of the sensible species. The contiguity of the sensible with the power does not make the sensible species redundant. The contiguity of the power with an object does not impede its sensation either.[313]

Aristotle's famous distinction between the senses operating through an external medium and the senses operating through an internal medium is a mere fiction. In his inductive reasoning Suárez shows that not only is no external and internal

[312] *CDA* disp. 7, q. 14, n. 3, t. 2, 738.

[313] Ibid., n. 6, 746. For a dispensability of the sensible's intentional action from the tangible and the tasteable through the medium in Suárez's theory and its evaluation as non-traditional, see also Knuuttila (2015, 207).

medium necessary for the senses of touch and taste, but that the absence of a medium can *also* be the case in the affections of sound and odours. The senses of hearing and smell can be both 'hit' by a smoky evaporation carrying an affective quality of odour and by a compressed wave of the movement of air which is the subject of sound (of course, with the concomitant intentional change) without being prevented from the sensation. The contiguous existence of an object with a sense organ seems to make sensation impossible only in a single case, sc. in the vision of colours. As it has been said in 4.2.3, colours need to be illuminated to become proximately visible and a coloured object placed on the eye cannot be seen. However, the case seems to be different with light which can touch the pupil and yet (through the species) be seen. It is not true that the tangible and the tasteable can be perceived only through an internal medium, either. As it has been said, both sensibles can be perceived while they are contiguous to the sense powers. While the skin or open flesh can feel the first qualities or the hardness of an object, if food is appropriately humified and chewed, a sapour can penetrate through the pores up to the lingual nerve, which is the adequate organ of taste. Moreover, it is not correct to claim that only the first three senses can sense through an external medium, such as air. Touch can be affected by the heat of a distant fire through the air too. In sum, Suárez considers Aristotle's aforesaid clear-cut distinction between two groups of external senses based on operability through external and internal medium to be implausible.[314]

Given Suárez's heuristic principle of charity applied in his exegesis of Aristotle and other authorities, especially Aquinas, how does he assess the validity of the proposition 'That what is placed on the sense organ is imperceptible to all the senses' that is typical of Aritotelian-Thomistic philosophy? Is this proposition plausible for him in some sense at all? In his reply, Suárez distinguishes between two possible meanings of the sentence. First, this sentence can be interpreted in the sense of inherence. If understood in this way, it is true. Taking into account Suárez's overall approach to interoceptive perceptions in *CDA* disp. 7 (not his exposition in *CDA* disp. 11), the external senses cannot sense inherent qualities. Second, the aforesaid phrase can be interpreted in the sense of 'contiguity'. If understood in this way, the sentence is valid only in the case of sight. For a coloured object to be visible it must first be illuminated and consequently its medium (but not necessarily) must be lit too. If a visible colour is superimposed on the eyes, the eyelids are either closed or can get damaged. Yet, in the other senses this proposition is not true.[315]

---

[314] *CDA* disp. 7, q. 14, nn. 4–7, t. 2, 740–8.-

[315] *CDA* disp. 7, q. 14, n. 6, t. 2, 744–6.

Chapter 5
Internal Senses

Our perceptual experience is far from being completed by the operations of the external senses. It is not a simple process through which we perceive isolated 'sense data' or discrete psychological atoms, such as colours, sounds, odours and sapours. Beside the external senses, which perceive the proper sensibles, and derivatively also the common and *in concreto* incidental sensibles, there are several post-sensory operations contributing significantly to the formation, retention and combination of our unified percept. We perceive the sensibles belonging to different modalities in a combined way as belonging to one and the same sensible object. At the level of sensory perception, we recognize agreements and distinctions not only within the sensible qualities of a particular sensory modality such as white and black in the case of sight but also among the sensibles belonging to the different external senses. We perceptually judge that white is not sweet or that what is white, for example, milk is also sweet. Moreover, we often apprehend these combined perceptual wholes with an additional aspect that is expressed in terms of the object's convenience or inconvenience. We do not perceive objects 'neutrally' or from a purely theoretical point of view. In fact, only when perceived as convenient or inconvenient can sensible objects stimulate operations of the sensory appetite and be 'supervised' and commanded by the higher appetite, namely the will, and followed by a behavioural response executed by the locomotive power.[1] Accordingly, our perception is frequently 'tinted' by practical intentions that are not detected as being separate from the perceived sensible wholes. Such separation can only be performed by the intellect.[2] Concrete sensible wholes are apprehended also according to other aspects. We perceive this white man as Thaddeus and this small girl as Mary.

[1]For Suárez the *virtus motiva* is to be understood as a power really distinct from the appetite and irreducible to a mere passive obedience of the appetite. For this, see *CDA* disp. 13, q. unica, 424–43.
[2]See *CDA* disp. 9, q. 3, n. 16, t. 3, 136. Only the intellect can consider friendship or enmity as such.

D. Heider, *Aristotelian Subjectivism: Francisco Suárez's Philosophy of Perception*, Studies in the History of Philosophy of Mind 28,
https://doi.org/10.1007/978-3-030-67341-3_5

Moreover, not only do we perceive present objects, we are also aware of absent or even impossible objects that do not and cannot exist extramentally.

In this chapter, I aim to lay out Suárez's doctrine of these post-sensory operations and cognitive functions through which our perceptual experience is processed and unified. At the same time, beside the presentation of the acts that *are* implemented by the internal sense and contribute to this aim, I will highlight also the operations that according to Suárez are *not* accomplished by the internal sense.[3] As I have already partly shown and will show more below, there are other powers such as the external senses and the intellect, and indirectly also the soul, that in Suárez's account take over several functions that had commonly been attributed to the interior senses. Against the background of the pluralist views of the internal senses endorsed especially by Avicenna and Aquinas, in 5.2 I will show that Suárez's emphasis on experiential unity (despite his general theory of really distinct powers) in perception can be observed above all in his reductionist view of the number of the internal senses. In 5.3, in the context of comparison with the traditional view regarding the cogitative power (*vis cogitativa*) and recollection (*reminiscentia*) represented by Aquinas, I will claim that the numeral reduction of the interior senses undertaken by Suárez is accompanied by a substantial limitation in the internal senses' *radius operandi*. As I will indicate, both reductions correspond to Suárez's downplaying of the interior senses, which is mirrored already in the space devoted to this topic of the interior senses. In a striking contrast to the preceding disputation 7, *CDA* disp. 8 devoted to the subject matter of the interior senses consists only of *two* questions.

Before presenting these reductions in 5.2 and 5.3, I will first lay out Suárez's views of the interior senses' objects and the way the internal senses' species are produced. As we have seen, all the cognitive powers including the internal senses require the existence of intentional species conceived as instrumental causes causally concurring with the powers and the soul in the production of cognitive operations. The second topic of the origin of the imaginative species or the species of the interior senses also testifies to Suárez's aim to explain the unified character of perception. In Sect. 5.1 I will leave aside the issues that have been dealt with in Chap. 3, that is, those that are common to all the cognitive powers, such as the theory of the principles of a cognitive act, the tenet of its terminus and the issue of its nature. Following Suárez's approach 'in particulari', initiated in disp. 7, I will focus only on those topics that are specific to the interior senses.

[3] Although Suárez advocates a theory of a single interior sense, I have sometimes applied and will at times employ the term in plural, 'the internal senses', since Suárez himself assumes that there are conceptual distinctions within this single power and uses the phrase in plural too. These conceptual distinctions are based on a comparison with its various acts (*CDA* disp. 8, q. 1, n. 24, 44–6). See also 5.2.

5.1 Objects and Sensible Species of Internal Senses

Unlike the external senses, whose proper sensible object is limited to a single generic type, the proper object of the interior sense, at its own level, is all-embracing. The object of these (as they are called) 'senses of the brain' are *all* the sensibles (*omne sensibile*) perceived by the external senses.[4] Scholastics commonly divided this *omne sensibile* into two main groups. While the one group concerns extramental sensibles, the other one applies to intramental items.[5]

Considering the first group, Suárez is explicit that, in fact, the only function of the common sense, which is compared to a spring from which the five senses as five rivulets flow,[6] is to distinguish between the sensibles of the five sensory modalities.[7] It can perform this discriminative function, only if it can perceive the objects of all the senses. As we have seen, it is called common not because it apprehends the common sensibles in a way that would go beyond the cognition of the external senses, but because it apprehends the sensible objects perceived by all the external senses. It is the power that (*in actu exercito*) judges that a colour is not a sound. While sight can make intrasensory judgements about white and black colour, it cannot distinguish between white colour and sound, and taste cannot discern between sapour and odour. Only the common sense can distinguish the sweet-sensible from the hard-sensible. Only a higher and single capacity whose *radius operandi* extends to the objects of all the external senses can simultaneously perceive them all. Albeit not explicitly, Suárez indicates that the common sense *ipso facto* unifies the atomic data perceived by the external senses. Even a non-rational animal, such as a dog, can see an object that looks like a bone, and upon approaching and tasting it can find that it is only a dummy bone. Accordingly, it will know that it is not a real bone, but only a fake bone. This can occur only if it perceives the bone's colour and the bone's taste as combined and belonging to one and the same sensible object. Beside the proper sensibles, the common sense's apprehension covers common sensibles too. Since these common sensibles' manner of representation amounts to modifications of the external senses' species representing the proper sensibles, the common sense, rooted in the same soul as the external senses, will apprehend these in the same way. In the same way it will also apprehend incidental sensibles, such as material substances. It will apprehend them concretely or confusedly, that is, as part of an accidental aggregate or concrete

---

[4] *CDA* disp. 8, q. 2, n. 1, t. 3, 46: 'Obiectum ergo istius sensus est omne sensibile quod sensu externo percipitur'.

[5] For this division, see, e.g. Mastrius and Belluto, *In libros De anima*, disp. 5, q. 8, art. 2, § 242, 117: 'Quae possunt obiici interno sensui ad duo membra videntur reducibilia, vel enim sunt sensationes ipsae, vel res ad extra'.

[6] *CDA* disp. 8, q. 1, n. 3, t. 3, 16. For this comparison, see Augustine, *De Genesi ad Litteram libri duodecim*, book 12, ch. 16, n. 32.

[7] For this function of the common sense, see ibid. n. 2, 14–6.

sensible whole composed of substance and accidents, of which only the latter are per se sensibles objects.[8]

The interior sense apprehension not only follows the external senses, but according to some (being a higher power) it also upgrades the external senses' cognition in that it reveals modifications that have not been detected by the external senses. We will see that (at least according to one possible interpretation) this is not precisely the view that Suárez seems to be proposing. For him, in a direct analogy to the modifications of the external senses' species representing the proper sensibles through the representations of the common sensibles and the external senses' acts, the interior senses perceive the same modifications that Avicenna called 'intentions'. These intentions are crucial for an animal's well-being and life preservation. As a classical example has it, a sheep runs away from a wolf because it apprehends the intention of its enmity. Such intentions can be perceived not only in objects perceived *hic et nunc* but also in those that are imagined or remembered.[9] A sleeping dog often barks because it is dreaming about objects and persons perceived in light of convenience or inconvenience.[10]

Beside *real* external objects the interior sense also conceives impossible objects. Due to its capacity to combine various items the human phantasy can connect, for example, the notion of gold and the notion of a mountain to contrive the composite notion of a golden mountain. It can put together the notion of a man and the notion of a lion to produce the composite notion of a chimaera. Albeit the phantasy can create these compositions and deliver material for the constitution of fictitious notions, it is not, properly speaking, the capacity that forms beings of reason as such (*entia rationis*). Only the intellect can conceive what is *not* a being *as if* it were a being. If not corrected by a higher instance, sc. the intellect, the interior sense always conceives these objects as true and real beings.[11] Apart from these composed beings, Suárez also admits another category of nonbeings which he assigns to the total objectual ambit of the interior sense. These are *privations*, such as darkness and silence. In darkness we perceive what we cannot see; in silence we perceive what we cannot hear. We do not apprehend these privations positively, that is, through a sensible species representing them, since such species obviously have no efficient cause. Light cannot produce a visual species of darkness and sound cannot emit an

[8]For the perception of incidental sensibles, which according to Aquinas are *distinctly* detected by the *vis cogitativa*, and for Suárez's critique of this view, see 5.3.1.

[9]*CDA* disp. 8, q. 2, n. 1, t. 3, 46–8.

[10]*CDA* disp. 8, q. 1, n. 11, t. 3, 24.

[11]*DM* disp. 54, s. 2, n. 18, t. 2, 1023–4. If not corrected by the intellect, the phantasy often falls into illusions since sensory powers can consider their objects only *with* reference to existence whether present or past. This is also why, properly speaking, they cannot conceive something as nonbeing. Unlike the intellect, they cannot abstract from existence and consider the pure essence of an object. If the phantasy were left as it were to itself, i.e. without the supervision of the intellect, it would fall into incorrect judgements about these impossible beings. It would always judge them to be extramentally real. This often happens in dreams when the intellect is 'turned off'. For this see *CDA* disp. 6, q. 5, n. 9, t. 2, 528.

audible species of silence. These items can be perceived only privatively, sc. through the absence of the pertinent acts. In the dark sight is not affected and the common sense apprehends the absence of light *through the comparison with the species of light* that is retained in the memory.[12]

Beside this group of extramental sensibles in a broad sense, including fictitious beings and privations, the second set of the interior sense's proper objects involves intramental items, namely perceptual acts. Clearly, intellectual acts are ruled out from the objectual ambit of the internal sense. A lower faculty cannot cognize the acts of a higher capacity. A material power cannot apprehend the operations of an immaterial power. However, for a number of scholastics including Aquinas (in some texts), Scotus[13] and Olivi,[14] the operations of the external senses constituted the common sense's intentional proper object. For them the common sense as a higher power can be aware of the lower-order acts of the external senses in the way that it receives the *proper* species of these acts. This capacity to sense these acts, which those authors attributed to the common sense, is the main reason why we remember. For Scotus and Olivi these acts seem to even constitute the *primary* objects. It is a fact that we often remember our acts, rather than the objects of these acts. We remember *that* we heard or saw something, but we do not recall *what* exactly we heard and saw. Nonhuman animals are endowed with this capacity too. They search for a place where they have found food or shelter. In 3.7 I showed that Suárez rejects the view that the common sense directly and *intentionally* cognizes the external senses' operation as its object. I also suggested that the soul, through the interior sense *qua* the common sense, produces the species' additional mode representing awareness of the external senses' act on the basis of the exemplar causality of the lower external senses' acts.[15] I also put forward that according to Suárez the common sense does not bring an additional interpretative value to the perceptual process of the external senses. No act of the interior sense can perceive an act of the external senses by means of a species representing it as its own object. This kind of self-awareness exceeds the capacity of the material powers including that of the interior sense and it can occur only at the level of intellectual operations, which can

[12]See *CDA* disp. 5, q. 2, n. 26, t. 2, 328–30 and disp. 6, q. 1, n. 11, t. 2, 468.

[13]Scotus, *Ord.* IV, d. 45, q. 3, n. 95 (Vat. 14: 167). For Scotus, see also Cross (2014, 27–32).

[14]For Olivi, see Toivanen (2013b, 275–81).

[15]It may be objected that this modularistic isolation of the common sense's act from the operation of the external senses necessarily leads to the fragmentation of our perceptual experience, which is at odds with the requirement of a unified percept. The senses perceive at first the proper objects and concomitantly their acts and only then the common sense's (unified) species modified by the external senses' awareness of their operations is produced and its act is brought about. I agree that in this respect Suárez's common sense can be understood as being detached from the external sensorium of the particular senses. However, I do not think that on the phenomenological level this fragmentation must be a necessary consequence. One ought not to forget that for Suárez this sequence is not temporal. The external senses' and the common sense's operations mediated by the soul take place at the same time.

reflexively inspect both the acts of the internal sense and their own intellectual acts.[16] The common sense can perceive the external senses' acts only through the species representing the external sensibles modified by the external senses' acts.[17] The memory retains both the species representing an external object and the modification representing a past perceptual act. The only possible concomitant (non-intentional) self-awareness is caused primarily by the external senses whose operations are triggered by the *soul's substantial attention*. Since these acts are vital and attentive, they are not only concomitantly adjudicative but also virtually aware of themselves. According to this same-order theory of consciousness the perceiver apprehending a sensible is simultaneously aware not only of the sensible object but also of her own operation. From this point of view, when compared to the theories of Aquinas, Scotus, Olivi, among others, Suárez's view of the common sense's self-awareness can be regarded as that which restricts the functional scope of the common sense.

But how precisely do this modification and other modifications of the common sense's species arise? How do these modes accrue to this species? In order to explain their origin, I first have to focus on the theory of the production of the simple (unmodified) sensed species of the interior sense, or the species that represents qualities that have already been represented by the external senses. Due to a structural similarity with the question of the origin of the intelligible species, I refer to this theory too. However, before that, for completeness's sake, I must briefly mention Suárez's reasoning for the necessary existence of the *species phantastica* that was introduced by the Jesuit already in *CDA*, disp. 5.

As in the case of the external senses' species, Suárez claims that the interior sense cannot bring about its acts without the reception of sensible species called *species phantasticae*. He argues that we experience the formation of notions and memories in our phantasy, which cannot occur without the impressed species. If no species remained in the phantasy, no imaginative and memorative acts would be possible. The sensible species of the internal sense must be retained and activated by one and the same power, if these acts are to be possible.[18] These species, or *phantasmata*, as Suárez sometimes (to wit, inconsistently) says,[19] cannot be substantial corpuscules or animals spirits. They must be accidents belonging to the category of quality. If they were corporeal substances, they would have to occupy a space in the brain. Not being aware of 'a universe' in the brain in the form of the minute sizes and gigantic

[16]For Suárez's theory of the intellect's knowledge of its own acts conceived as qualities by means of the proper intelligible species, which is a view different from that of Aquinas (who does without them in this kind of intellectual self-awareness), see *CDA* disp. 9, q. 5, n. 6, t. 3, 174. See also Rode (2015, 373–5) and Perler (2014a, 277–9).

[17]For this, see *CDA* disp. 8, q. 2, n. 5, t. 3, 50.

[18]*CDA* disp. 5, q. 1, n. 5, t. 2, 290.

[19]In most places of *CDA* disp. 5 and disp. 9 Suárez employs the term 'phantasma' in the sense of the internal sense's species, which (as we know) has only virtual representation. On the other hand, there are other places in which he takes it as a second act or mental representation. One of these is the passage in which he says that a phantasm is an exemplary cause of the agent intellect's production of the intelligible species (for this, see *CDA* disp. 9, q. 2, n. 12, t. 3, 96).

numbers of neurons and their synapses with chemical reactions, Suárez argues that since our cerebral cavities are small and since we have memories of myriads of individual things, the imaginative species cannot be substantial corpuscules.[20]

As in 3.2, where I discussed issues of the origin of the external senses' species, so also in his theory of the production of the internal sense's species Suárez strives to avoid the problematic case of ascendant causality.[21] As said, the reception of this causal axiom based on the impossibility of the production of the higher by the lower made Suárez, in part, a follower of Augustinian philosophy. In the context of the *species phantastica*, however, this axiom is employed also in the context of the causal relationship between distinct powers. First of all, he makes it clear that imaginative species cannot be efficiently caused by species of the external senses since the former are more perfect than the latter and since *no* direct causal communication between two really distinct powers, including the external senses and the internal sense, seems to be possible. The *species phantasticae* are more perfect because the interior sense of phantasy with its operations is more perfect than the external senses. Unlike the species of the external senses, the species of the phantasy are not bound to objects existent here and now. While the being and becoming of the sensible species of the external senses naturally depend on the extramental objects, this is not the case with the imaginative species.[22] If the visual species of Peter caused the imaginative species representing him in the phantasy, then whenever an external sense received a species of Peter, the phantasy would have to receive the same species too, which is not the case. Throughout *CDA* Suárez refers to the phenomena of 'dead' reception of species, such as the reception of species in the states of ecstasy or apoplexy. Regardless of how strongly the external senses are affected by the species, the cognitive powers whose perceptions occur in those mental states do not produce cognitive acts at all. This also seems to be the reason why Suárez denies that we can remember objects whose species we have received *unconsciously*. He is sure that there must be at least a marginal attention of the external senses to guarantee the production of the interior sense's species. We cannot have memories of an object to which we did not attend at least with a liminal attention. If the external senses are not reduced to second act and if they remain only in first act as determined and affected by the species, the phantasy will remain intentionally unaffected too. Without a reduction to second act, which only is the formal representation of the object, we cannot have a notion of the object co-caused by the imaginative species. Suárez is sure that experience supports this view. We receive the imaginative species in the interior sense only if the object has been previously cognized by an external sense or if the species were reduced to second act

[20] *CDA* disp. 5, q. 2, n. 3, t. 2, 298.

[21] For Suárez's view of the origin of the species of the interior sense, see Heider (2019b, 176–8).

[22] *CDA* disp. 6, q. 5, n. 7, t. 2, 524.

by an external sense.[23] Typically of Suárez, it is the same power that receives this species and elicits a cognitive act of the represented object.[24]

If the sensed species (*species sensata*) of the internal sense cannot be produced by the sensible species emitted by a sensible object, the other option that Suárez takes into account is that they are co-caused by the external senses through their acts. This view seems to be *prima facie* consistent with Suárez's theory of the end-product of the cognitive act of the external senses or the *species expressa* conceived as a quality. This quality can well produce another quality, namely the *species phantastica*, because it can produce a habit. A repeated cumulation of these cognitive qualities can lead to the formation of a habit, such as the knowledge of a language or a science. This probable tenet can be confirmed by phenomenological experience: The more we pay attention to the affected external senses and aspects of sensible things, the more clearly and lastingly the sensible objects imprint their species on the internal sense.[25] No matter how convincing this view is it must face the following objection: It is not only the case that the imaginative species are ontologically higher than the *species sensibilis* of the external senses (while the former do not depend on its object in its becoming and being, the latter do), but the same holds also about the imaginative species and the perceptual acts of the external senses. As 'located' in a higher faculty, the species will be more perfect than the perceptual acts of the external senses. Considering the impossibility of ascendant causality, the actual vision of Peter cannot cause imaginative species of him, because then the being and conservation of the imaginative species would be dependent on external sensation. The phantasy would be limited to the perception of sensibles *hic et nunc*. In his reply Suárez agrees that if we compare the first acts of the powers with each other, the sensible species of the internal sense are more perfect than those of the external senses. However, if we compare the first act of the internal sense with the second act of the external senses, then the act of an external sense as a vital act will be more perfect than the *species phantastica*. But will not then the *species phantastica* be dependent in its conservation on external sensation? In the context of this view, Suárez does not think so. He shows that this dependence is determined rather by the nature of the subject, which in this case are the external senses. Accordingly, it will be external sensation that imprints a perfect likeness on the internal sense and precisely this perfect likeness is caused in the interior sense whose distinct cognitive disposition makes it possible to conserve the new species even in the absence of the represented sensibles.[26]

No matter how compelling this theory seems to be, it is not Suárez's final word. In the last conclusion he makes a noncausal claim that seems to be more in line with the theory of the powers' harmony: 'It is probable that these interior species result in the internal sense through its own activity, and not through the efficiency of something

[23] *CDA* disp. 6, q. 2, n. 9, t. 2, 478–80.
[24] *CDA* disp. 6, q. 5, n. 8, t. 2, 526.
[25] *CDA* disp. 6, q. 2, n. 10, t. 2, 482.
[26] Ibid., n. 12, 484.

exterior'.[27] Again, with a reference to the immanent character of cognitive acts Suárez claims that an act of cognition cannot transitively cause a distinct quality that would be 'localized' in a (really) distinct faculty. What does it mean that the *species phantastica* arise through the activity of the interior sense? In answering these questions Suárez again refers to the *topos* of the attentive soul, which is responsible both for the production of the external senses' acts and the indirect self-awareness of the external senses' perceptual acts. He is explicit that even though an act of cognition produced by the lower faculty cannot directly cause a quality in the higher power, this mediation can proceed *indirectly*. Although he endorses the view that perception is an act of the sensory power as really distinct from the soul, as we have seen in his theory of the efficient causes of cognitive acts he states that the principal agency of a cognitive act belongs to the attentive soul. The soul is aware of its operations and as the common root (*radix*) of all its vital operations is the principle of their unity. This general awareness allows it to coordinate and harmonize all the vital activities occurring in the different powers of the animated subject at the same time. The soul not only perceives these activities but also produces the sensible species in the interior sense. Unlike the sensible species of the external senses which are produced from outside by the sensibles (their production cannot be an instance of upward causality since they are only the traces of sensibles) the sensible species of the internal sense are caused by the *soul* through the phantasy.[28]

Given it is the soul that produces the imaginative species, what is the role of external sensation? In line with his treatment of the causes of intellection in which he says that phantasm alias the interior sense's cognitive act constitute a quasi-exemplar for the intellect's 'depiction' (*depingere*) of the same thing,[29] the same is to be said also about the imaginative species. Also in this context the external senses' operations are not efficient causes but only quasi-exemplary causes.[30] In the context of the production of the intelligible species Suárez explicitly refers to his theory of the production of the imaginative species and states that the *same* procedure and theory should be followed. The substantial influx of the soul in the powers and their

[27]Ibid., n. 13, 486: 'Probabile est huiusmodi species interiores resultare in interiori sensu propria vi et activitate illius, et non per efficientiam alicuius extrinseci'. Even though in *CDA* disp. 6, q. 2, n. 13 Suárez does not explicitly say that this view is more probable (he says only that it is 'probabile'), the overall context of *CDA* disp. 5–6 and especially disp. 9, q. 2 devoted to the issue of the production of the intelligible species and his explicit reference to the theory of the harmony of powers and their acts make it, in my opinion, more probable than the previous conclusion.

[28]Ibid., n. 13, 486.

[29]For Suárez's theory of the origin of the intelligible species where phantasms are only quasi-exemplary or occasional causes, see *CDA* disp. 9, q. 2, n. 12, t. 3, 94–8. For this theory, see Spruit (1995, 301–7) and South (2002).

[30]See also *CDA* disp. 2, q. 3, n. 45, t.1, 228: '[...] operatio unius potentiae potest dependere ab aliqua materiali potentia [...] tamquam a proponente obiectum eiusdem ordinis et rationis, ut pendet [...] operatio phantasiae a sensibus externis [...]' For Suárez's theory of cognitive acts (formal concepts) conceived as exemplary causes in connection with his tenet of the endpoint of a cognitive act, see also 3.5.

operations in the form of the overall awareness seems to be a sufficient guarantee of what Suárez calls the harmony or sympathy of the distinct powers' operation. If the soul perceives Peter by means of the visual power, the same soul produces a likeness of Peter in the phantasy and the intellect.[31] Therefore, the intellect and the interior sense perceive the same object, namely a material singular. Of course, this does not mean that the intellect cannot add any other cognitive operation to the interior sense. As an immaterial capacity of 'intus-legere', of course, it can generate a number of operations that go well beyond those of the internal sense, such as universalizing abstraction, discursive and inferential disclosures of material substances, souls, powers, habits and intentional species, judgement by composing and dividing, intentional self-awareness etc.

Accordingly, in his conclusion Suárez formulates a strict asymmetry between the production of the species of the external senses and the interior sense. Somewhat obscurely he states that, unlike in the external senses, in the internal sense we have to posit the agent sense (*sensus agens*).[32] In the case of the imaginative species we need a 'lifter' of the sensible species of the external senses. Against the background of the theory of John of Jandun, Adriaan Pattin refers to this position with the pejorative label 'eclectic'.[33] He finds that while Suárez rejects the agent sense in the case of the production of the sensible species of the external senses, he admits it when analysing the production of the imaginative species. Two reasons should make us reluctant to embrace Pattin's negative connotation of this evaluation unreservedly. First, provided that the theory of the *sensus agens* represented by Jandun's teaching (see 3.4) is essentially connected with the real distinction between the passive and the agent sense, this doctrine of the agent sense can hardly be ascribed to Suárez. As in the treatise on the nature of the distinction between the agent and the potential intellect, in which Suárez approvingly refers to Nifo,[34] the two senses are only conceptually distinct for Suárez. It is one and the same power, or rather the soul that through one and the same power produces this species, receives it and produces the pertinent act. *Expressis verbis* Suárez denies that there are two distinct powers, of which one would be receptive of the sensible species and the other cognitive of absent

[31]This is also why Suárez holds that the primary objects of intellectual cognition are material and sensible singulars (substances aggregated with per se sensible accidents) and not universal quiddities. For this theory, see *CDA* disp. 9, q. 3, n. 15, t. 3, 130–2. For this statement about the intellectual cognition of material singulars considered as a significant point of departure of Suárez's epistemology in general, see Rinaldi (1998). I will show in 5.3.1 that this immediate intellection of material singulars is also one of the main reasons why Suárez refuses to attribute the apprehension of the so-called particular intentions to the *vis cogitativa*.

[32]*CDA* disp. 6, q. 2, n. 16, t. 2, 492.

[33]Pattin (1988, 420).

[34]For Suárez's view of the real identity of the agent and the passive intellect, which parallels the real identity of the agent and the passive sense in the interior senses, see *CDA* disp. 9, q. 8, n. 18, t. 3, 234–6 and disp. 9, q. 10, n. 8, t. 3, 272, where Suárez 'inserts' only the conceptual distinction between them. For the same view of Agostino Nifo, see 'De sensu agente', in *In librum Destructio Destructionum Averrois commentationes*, 129a.

objects.[35] Second, the agent sense in the context of the internal sense does not seem to be a power really distinct from the soul. Rather, the agent sense is much more an abbreviation for the soul's mediating and coordinating activity which is not an ad-hoc and eclectic feature of Suárez's system but, quite on the contrary, its 'basso ostinato'.

Beside the imaginative cognition of simple sensibles co-caused through simple sensible species (*species sensata*) representing simple and basic objects, as said, there is another operation that combines these simple species and phantasms and creates impossible objects. In fact, these composed objects can be formed in two ways. (1) The interior sense focuses on the species of gold and a mountain that have been retained in the memory from a previous sensory cognition and by means of an original creative act that has these two as intentional objects it forms the composite notion of a golden mountain. Only *after* the production of this first act a new (composed) species of a golden mountain, which can later be re-activated, is formed. (2) This cognition of a golden mountain is at the beginning produced by a single species already representing a golden mountain. This composite species originates through the merging of the species of gold and a mountain in the memory *before* the very first production of the imaginative act representing a golden mountain. While the production of imaginative acts about impossible objects in (1) is based on a *deliberate* imaginative process; the merging of the two species in (2) represents a rather *spontaneous* and *involuntary* process.[36] Suárez is explicit that the two views are compatible. When the phantasy cognizes a golden mountain for the first time, it employs the plurality of the species representing the discrete sensibles, as stated in (1). It produces a single act that is representative of a golden mountain since it employs these two species together in the manner of a single object (*per modum unius*). When cognizing the composite object for the first time there is no other source from which the phantasy could acquire this composed species as a principle of its act. The only way to produce these composite notions for the first time is to employ the species of the two sensibles. A new sensible species representing the composite sensible of a golden mountain can be available only after an act representing it has been elicited. The main function of this new species, then, will be to facilitate another cognition of the object. As Suárez says, this cognitive processing has its direct counterpart in the production of the intelligible species representing (metaphysical) universals. Having abstracted the universal from the individual 'haecceities' for the first time, the intellect will later more easily conceive it without recurring to the memory of the singulars from which the universal has been abstracted. This shows that after the first intellectual (universal) precision there remains the intelligible species of that universal in the intellectual memory. As stated in *CDA* disp. 9, q. 3, precisely the *same* happens after the first formative act of impossible objects, such as a golden mountain.[37]

---

[35]For this, see *CDA* disp. 6, q. 5, n. 8, t. 2, 526.

[36]For this distinction presented also by Olivi, see Toivanen (2013b, 305–6).

[37]*CDA* disp. 9, q. 3, n. 14, t. 3, 128–30.

But what about the sensible species representing, for example, the above-mentioned 'practical' intentions of enmity? When we have seen how the simple and composed species are produced, the remaining question to be answered is how these species representing these 'non-sensed' features, for example, of utility or enmity are formed. Clearly, they have to be perceived through the sensible species or somehow be part of it since the intentional species is the necessary principle in all kinds of cognition. Suárez starts his exposition of the production of the principle of these intentions by introducing the famous theory advocated by Avicenna[38] and Aquinas,[39] according to which these intentions must be represented by a species that is really distinct from that representing the sensible aspects of the objects. Accordingly, there are two distinct kinds of sensible species, one of which represent things under their external features that have been before apprehended by the external senses (*sub extrinseca ratione*) and the other under a more latent characteristic (*sub alia magis abscondita ratione*) that was not the object of the external senses and is revealed only by the internal sense. Since these two features are substantially distinct they cannot be represented by one and the same sensible species. These species must be distinct also because it is possible to have a representation of one, sc. the sensible features perceived by the external senses, without a representation of the other, namely the intention of enmity. A wolf can have the species of another wolf that represents it according to its sensible features, such as its greyness, ovalness and howling. without having the other species that represents it according to its enmity. On this view, the non-sensed intentions seem to be similar to incidental sensibles somehow 'hidden' behind the per se sensibles. Importantly, these two kinds of species would have to be the principles of two really distinct interior senses.[40]

In line with his reduction of the cognitive functions of the interior senses Suárez rejects this real distinction and dismisses the view that an interior sense, whether the estimative or the cogitative power, detects these 'hidden' intentions. He states that these non-sensed species are not and cannot be distinct in this way from the sensed species.[41] Suárez introduces two arguments for this claim. (1) These distinct non-sensed species are *redundant*. A sheep apprehends the wolf's hostility through the same species through which it apprehends the wolf under its sensible qualities. Led by the principle of parsimony he stresses that there is no necessity to distinguish between these two sensible species since through one and the same species the sheep

[38]For Avicenna, see his *Liber de anima seu Sextus de naturalibus*, p. 1, ch. 5, 86.

[39]For Aquinas see, e.g., *De veritate*, q. 25, art. 2, c., 733: 'Sicut vis imaginativa competit animae sensibili secundum propriam rationem, quia in ea reservantur formae per sensum acceptae; sed vis aestimativa, per quam animal apprehendit intentiones non acceptas per sensum, ut amicitiam vel inimicitiam, inest animae sensitivae secundum quod participat aliquid rationis [. . .]'

[40]*CDA* disp. 6, q. 2, n. 15, t. 2, 490.

[41]For a famous critique of these intentions as distinct objects, see Scotus, *Ord.* I, d. 3, p. 1, qq. 1–2, n. 62 (Vat. 3: 43–4). Scotus says that a lamb would flee from a sheep miraculously changed to a wolf with all of its sensible qualities. However, it would not do that, if it had an estimation of the non-sensed agreeability of the object.

(in actu *exercito*) judges 'This is a wolf' and 'This should be evaded'. He shows the redundancy of such a species also on a different example. Birds that have never seen a rain can apprehend a future rain from a cognition of a wind. But who could believe that from this perception of wind they somehow infer and acquire a non-sensed species representing a future rain? (2) These distinct species are also *impossible* items. However, as really distinct they would have to be able to exist without each other. But a sensible species cannot represent enmity as being prescinded from its subject. As I have said above, only the intellect can perform such formal abstraction. It is not possible for a sensible species to represent a wolf under the aspect of enmity without representing it according to its colour, shape, odour etc. Concluding, in analogy to the acts of the external senses, these non-sensed intentions can be represented only as *modes* of the sensed species.[42]

But how precisely does this modification of the species representing sensible qualities or this 'modal increment' occur? Suárez's reply is brief: '[...] from what has been said it is clear how it is produced'.[43] What does he mean by this easy-going 'see above'? In any case, he seems to make clear that this modification cannot be an invariable and absolute modification that would not take the cognizing subject into account. If it were an absolute modification, a perception of a wolf as such would have to always include the intention of enmity in all potential cognitive subjects, including other wolves. This intention is relational in the way that it is produced only in sheep, and other animals and humans whose instincts in this regard are substantially different from those of wolves or tigers. The distinct natures of sheep and wolves are the ultimate reason why on seeing a wolf a sheep produces an imaginative species modified by the intention of enmity and wolves do not. In an analogy to the modification of the *species phantasticae* through acts of the external senses Suárez conceives these intentions also as modifications of the representations. On seeing a wolf the sheep produces a visual act of seeing a wolf. This act as an *exemplum* gives an occasion to the soul's production of the *species phantasticae*. However, this production can be understood in two ways. (a) This modification, as said, can be *already* a part of the sheep's visual act which as an exemplary cause gives rise to the production of a species modified by this intention (the same happens with the common sense's species representing an act of an external sense). (b) This modification will not be implemented *until* the imaginative species is produced by the soul through the internal sense. According to this understanding the sheep's visual act will not include the intention of enmity. It will be intention-free. This intention will be added only with the soul through the interior sense.

So, which explanation does Suárez favour? It is not easy to decide. *CDA* gives good reasons for both accounts. On the one hand, in a structural analogy to the concomitant self-awareness of the external senses' acts, these practical modes seem to be virtually included already in the external senses' acts. As modified by these intentions, these acts of the external senses will serve as exemplary causes for the

[42]*CDA* disp. 6, q. 2, n. 15, t. 2, 492.

[43]Ibid.: '[...] ex dictis patet, quomodo fiat'.

formation of the interior sense's species. On the other hand, however, Suárez is explicit that sensible objects in terms of convenience and inconvenience are recognized *only* by the internal sense.[44] As it has been said, an emotional reaction is triggered by an estimative act of the phantasy, and not by the external senses. If pushed to decide which of these accounts is more likely, I would be in favour of explanation (b). While the argument for exposition (a) is only indirect, based on a structural analogy with the self-awareness of perceptual operations, account (b) is supported by Suárez's *ex professo* affirmation.

5.2 Number of Internal Senses

The denial of the real distinction between the sensed and the non-sensed species and the emphasis on the perceptual interconnectedness of the sensible qualities and the (practical or action-oriented) intentions lead Suárez directly to his reductionist account of the number of the internal senses. This question about the number of the internal senses can be regarded as one of the most important scholastic issues in the field of the interior senses. A growing reductionism in the issue of the number of the internal senses in Jesuit philosophy at the end of the sixteenth century and the beginning of the seventeenth century is an obvious fact. All Jesuit philosophers of the first generations, such as Francisco de Toledo, Manuel de Góis or Pedro de Fonseca, advocated theories reducing the number of five or four internal senses to a lesser number. All these thinkers contribute to the progressive dissolution of the Pluralist View of the interior senses and, in a way, to the abandonment of faculty psychology in early modern philosophy.[45]

Before I present his arguments for the existence of a single internal sense three versions of the so-called Pluralist View, according to which there is a plurality of really distinct internal faculties, must be introduced. Despite differences in the descriptions of the particular senses' functions and in face of the different answers

---

[44]See, e.g. *CDA* disp. 11, q. 1, n. 2, t. 3, 326: 'Hi [the external senses] enim non movent appetitum immediate; sumendum est ergo est ex interiori sensu'.

[45]For their theories, see Heider (2017a); for Suárez and Góis, see also Silva (2020). Considering Suárez's theory of the single internal sense which is in charge of the manifold operations I cannot share Harry Wolfson's (1935, 126) assessment according to which this reduction of the internal senses to the unique sense of phantasy took place as late as with Eustachius a S. Paulo (ca. 1573–1640) and that Eustachius set the tone for the early modern debates. For Eustachius's view, see *Tertia pars Summae philosophicae*, tomus posterior, tract. III., disp. III, q. 1, 391–4. Moreover, leaving aside the fact that Eustachius's *Summa* was nothing but a brief excerpt from Jesuit authors, especially from Suárez and the Coimbrans, there were also several other authors in medieval philosophy, such as Olivi, who embraces the theory of a single internal sense. For Olivi, see Toivanen (2013b, 247–65).

to the issue of the number of the internal senses in medieval philosophy,[46] ranging from one to six, the theoretical approach to the scholastic topic of post-sensory powers especially in Second scholasticism was largely determined by the theories of three authors: Avicenna, Aquinas[47] and Galen. In fact, any possible 'deduction' of the internal senses in medieval philosophy is to be based on principles formulated by Ibn-Sīnā. Generically speaking, his criterion was twofold. The first was physiological or anatomical. According to this standard, based on a modification of Galen's anatomical knowledge, interior senses are to be multiplied according to their localizability in different ventricles. The allocation of a particular interior sense is correlated with a particular qualitative disposition occurring in a particular sinus. The second principle of this multiplication was philosophical: Interior senses are to be differentiated according to their distinct sensible objects and according to the different ways of their apprehending. Interestingly, the theories of the three aforesaid authors can be ranked according to the number of interior senses endorsed: five (Avicenna), four (Aquinas) and three (Galen). These theories constitute a fundamental background against which Suárez elaborated his view of the single internal sense.

5.2.1 Criteria for the Multiplication of Internal Senses

In medieval debates regarding internal senses the common point of departure was Avicenna's theory. In his *Kitāb al-shifā'*: *De anima*, first part, chapter 5, Avicenna formulates three epistemological principles, resulting in his fivefold conception of post-sensory faculties.[48]

1. For every different type of sensible object there must be a distinct internal sense to apprehend this object. If we get typologically distinct objects, these objects must be attributed to really distinct powers.
2. Receptive powers differ from retentive powers. Reception substantially differs from retention.
3. Active powers differ from passive powers. Active powers and passive powers are mutually exclusive capacities.

---

[46]Deborah Black (2000) concludes that the cognitive operations and functions of particular internal senses in the theories of representatives of medieval scholasticism differ to the extent that it is difficult to detect common features in them.

[47]For Aquinas's view as the backdrop of Suárez's theory, see South (2001b).

[48]For these criteria, see Avicenna, *Liber de anima seu Sextus de naturalibus I–II–III*, p. 1, ch. 5, 85–90. Describing these principles is a set agenda in the secondary literature devoted to the topic of the internal senses. See Black (2000, 59–60); Toivanen (2013b, 231–45); Hasse (2014, 308–9); Simon Kemp and Fletcher (1993, 561–5); Casini (2006, 100–2); Harvey (1975, 43–6), and many others.

In applying the first principle Avicenna distinguishes between the two kinds of sensibles introduced in the previous section. The first object corresponds to the sensible forms perceived by the external senses. Various sensible aspects of the objects, apprehended by the visual, auditory, olfactory, gustatory and tactile faculties, are processed by internal apprehension. As we have seen in the previous section, the principle, recruited from the external senses, that triggers the corresponding operations was later named 'the sensed species' by Aquinas.[49] Avicenna introduces the above-mentioned example of a sheep perceiving a greyish oval spot emitting horrible noises, namely a wolf. The wolf is apprehended by the sheep by means of its visual, auditory (and perhaps also olfactory) sensory modalities. The second kind of object is completely different. Unlike the object cognized by the sensed species, this object is, sort of, 'hidden' behind the images. Although it is latent, it can still be detected by a corporeal power. In order to explain the sheep's reaction and behaviour on encountering the wolf, its fear and flight, a special kind of object, called 'intention' by Avicenna (*ma'ânî*), translated as 'connotational attribute', must be posited.[50] This connotational attribute of the wolf's enmity cannot be perceived by any of the external senses. It can be detected only by a power making instinctual judgements, which is more perfect than a faculty restricted to the apprehension of sensible forms. Avicenna calls this capacity 'the estimative faculty'.

In line with the second criterion he asserts that a power cognizant of an object *in praesentia* and one knowing its object *in absentia* must be two separate faculties. Obviously, a power cognizing its object abstractively, in its absence, is more perfect than one apprehending it intuitively, in its presence. The capacities to receive and to retain require distinct material dispositions. What is good for reception is usually not convenient for retention. While water, as a malleable subject, is suitable for the reception of an impression made by a signet ring (this is also why the pupil consists of the crystalline humour), it is entirely unfitting for retaining it. On the contrary, what is not fitting for reception is usually fitted for retention. While a stone, as a stable substrate, is well suited for retaining that impression, it is not suitable for receiving it. This organic difference goes back to a difference in the prevalence of the quality of moisture, or the dominance of dryness. While moisture is the quality convenient for reception, the quality of dryness is good for retention.[51] For Avicenna

---

[49] See Thomas de Aquino, *STh*. I, q. 78, art. 4, 255–7.

[50] Accordingly, this 'intention' cannot be understood in the sense we know from logic, philosophy of mind, or theory of action. Neither can it be identified with logical intention, nor with a conceptual content of a mental state, nor with an ethical aim of an agent. Although it is immaterial, it exists only as connected with a sense-perceived form. This is also why it is properly translated as 'connotational attribute'. For this translation, see Hasse (2000, 127–32).

[51] In Chap. 4 we have seen that this logic of the mixture based on the first qualities, namely on the difference in the prevalence of the moist over the dry and the dry over the moist, was also the reason for the constitution of the different mixtures giving rise to the different sensible qualities of sapour and odour.

(less explicitly so for Aquinas[52]) the layering of moisture and dryness is connected with Galen's ventricular theory according to which the brain is divided into three ventricles, that is, the front, the middle and the rear ventricle.

According to Galen, these cavities are the seats of *three* interior senses, which are the phantasy *sive* common sense, the cogitative power *sive* intellect (Galen conceived the intellect as a corporeal power) and memory. The main rationale for this localization lies in Galen's experimental medical knowledge concerning the correlation of the physical damages of particular cavities and corresponding cognitive disorders. Lesion of the front ventricle causes error in apprehension; injury in the middle ventricle results in incorrect judgement; and damage of the rear ventricle eventuates in bad memory. For Galen the interior powers are distinct because they are located in distinct cavities.[53] All the ventricles are interconnected by means of the fluid animal spirits (*pneuma*) leading from the organs of the external senses to the front part and through a narrow passage called the vermis to the middle ventricle and then up to the back occipital cell. Adding two bisections in the front and in the middle ventricles to Galen's division Avicenna arrives at a fivefold division of the brain. The common sense, apprehending its object intuitively, and the retentive phantasy, cognizing its object abstractively, are located in the front ventricle. The back ventricle, possessing the driest disposition, became the seat of the memory. Unlike the retentive phantasy, its function was to conserve non-sensed intentions.

Besides the first two criteria, employed also by Aquinas, Avicenna additionally adopted a third criterion based on the mutual exclusivity of active and passive powers. According to this criterion it is necessary to distinguish between the retentive phantasy and the compositive (active) phantasy. The compositive phantasy, composing and dividing both forms and intentions, cannot be identical with the retentive phantasy since this power is nothing more than a storehouse of sensible forms. Unlike nonhumans, the compositive phantasy in human animals can be controlled and harnessed by the intellect. If it is controlled, it is called the cogitative faculty by Avicenna. The incessant activity of the compositive phantasy is the reason why it differs from the estimative power, which is substantially passive. Accordingly, this distinction is the reason why the middle ventricle is to be 'bisected' too.

In sum, contrary to Galen who advocates a theory of three interior senses, Avicenna endorses five post-sensory faculties. Two of them are in the front ventricle,

---

[52]Despite Aquinas's reluctance to explicitly correlate the internal senses with the individual ventricles, in his *Opera omnia* we can find the following statements from *Quaestiones disputatae de veritate*, q. 10, art. 5, c., 309 locating the cogitative power in the middle ventricle: '[Et] sic singularibus se immiscet mediante ratione particulari, quae est potentia quaedam sensitivae partis componens et dividens intentiones individuales, quae alio nomine dicitur cogitativa, et habet determinatum organum in corpore, scilicet mediam cellulam capitis'; see also Thomas de Aquino, *STh.* I, q. 78, art. 4, c., 256: 'Et ideo quae in aliis animalibus dicitur aestimativa naturalis, in homine dicitur cogitativa [...] Unde etiam dicitur ratio particularis, cui medici assignant determinatum organum, scilicet mediam partem capitis [...]'

[53]See Galen, *De symptomatum differentiis*, ch. 3, 225–7. For Suárez's presentation of this view of Galen's, see *CDA* disp. 8, q. 2, n. 8, t. 3, 56.

the receptive common sense and the retentive phantasy. The two are situated in qualitatively distinct parts of this ventricle. In the middle ventricle the active cogitative faculty, or active compositive phantasy, is located. Besides, this middle ventricle contains the passive estimative power, which is receptive of non-sensed intentions. The memory is placed in the back (not bisected) ventricle, called cerebellum. Unlike the retentive phantasy, a treasury for images received by the common sense, the memory is a storehouse for both intentions received by the estimative power and complex forms composed of intentions and the sensible forms produced by the compositive phantasy.

No matter how closely Aquinas follows the Avicennian model, inspired also by Averroes the Angelic Doctor makes two important modifications which, in a way, foreshadow Suárez's reductionist approach.[54] First, and for our purpose more relevantly, Aquinas denies the validity of the above-mentioned third criterion based on the distinction between passive and active powers. The retentive phantasy and the compositional phantasy are one and the same power called *imaginatio* alias *phantasia*. Albeit brief in his justification of this identification, Aquinas gives us a clue in *STh.* I, q. 78, art. 4, c. Unlike Avicenna, Aquinas shows that the compositional phantasy, called by him simply the phantasy, occurs only in humans and not in nonhuman animals. However, if it holds that we do not have to posit the cogitative power (the so-called particular reason that deals discursively with individual intentions) and reminiscence (the memory that syllogistically seeks for a recollection of the past by individual intentions) as two additional capacities to the instinctive estimative power and the memory of perfect non-human animals, we do not have to posit a new capacity of phantasy in humans either. The higher perfection of the cogitative power, memory and phantasy is not a reason for introducing new faculties. This higher perfection can be explained by their participation in the intellect.[55] Second, in line with Averroes, Aquinas explicitly associates abstract intentions with individual intentions that can be perceived through incidental perception. In this way Aquinas substantially revises the cognitive function of Avicenna's cogitative faculty. In his rendering it becomes the more perfect human counterpart of the nonhuman estimative power. He also designates this power as 'ratio particularis' since it is capable of distinct cognition of incidental sensibles and of proto-syllogistical reasoning with particular objects as terms.[56] In sum, Aquinas advocates a theory of *four* human internal senses, which are the common sense, the cogitative

---

[54]For these modifications influenced by Averroes, see Black (2000, 66–8).

[55]Aquinas, *STh* I, q. 78, art. 4, c., 256: 'Avicenna vero ponit quintam potentiam mediam inter aestimativam et imaginativam, quae componit et dividit formas imaginatas, ut patet cum ex forma imaginata auri et forma imaginata montis componimus unam formam montis aurei, quem nunquam vidimus. Sed ista operatio non apparet in aliis animalibus ab homine, in quo ad hoc sufficit virtus imaginativa'.

[56]As regards the crucial role of this cogitative faculty in Aquinas's system of internal sense apprehension, see Lisska (2016, 237–72). For Suárez's critique of the *vis cogitativa* conceived as 'ratio particularis', see 5.3.1.

power, the phantasy alias imagination, and the memory alias reminiscence proceeding discursively from the known to the yet unknown.

5.2.2 *One Internal Sense and One Organ*

Although Suárez regards Aquinas's theory of four senses as more probable than the other views (except his own, of course),[57] for him the most likely is the view of a single interior sense. Before rejecting the two above-mentioned standards for the multiplication of internal senses and for the correlation of the tripartite division of the brain with the individual senses, in *CDA* disp. 8, q. 1 Suárez premises two main *fundamenta* of his reductionist procedure. (1) Senses are not to be multiplied if one sense can perform several functions by itself. They are to be differentiated only if one cannot exercise the act of another; that is, only if operations are incompatible in such a way that they cannot be attributed to a single power. As Suárez shows, this criterion can well be applied to the distinction between the external and the internal senses. The external senses can do what the internal senses cannot do, namely they can be immediately intentionally affected by external sensible objects. Moreover, this can be applied also to the external senses themselves. A particular sensory modality cannot be immuted by a sensible of the other modalities. This is why the five external senses cannot be reduced to a lesser number.[58] (2) An equality in the number of human and nonhuman senses is to be defended. Perfect nonhuman animals are endowed with a sensory nature having a perfection that is proper to their order. If there is an operation in humans that is more perfect than that of perfect nonhumans, this is not a reason to add a new interior sense in nonrational animals. The higher perfection of this operation in humans comes from the powers' 'rootedness' in the more perfect (to wit, rational) soul. As it was for Aquinas, this second *fundamentum* is the reason why in the overall equipment of human internal apprehension Suárez refuses to embrace the active (compositional) phantasy productive of fictitious notions, such as golden mountains or chimeras as an additional internal sense power.[59]

The second *fundamentum* is important also because it inhibits a general argument for the Pluralist View of the internal senses that is based on the powers' separate existence. According to this argument, if powers are mutually separable, they must be really distinct, according to the definition of *distinctio realis*.[60] Since some animals have only touch and perhaps taste but no other external sense, these two external senses are to be conceived as really distinct from the three remaining ones.

---

[57] *CDA* disp. 8, q. 1, n. 14, t. 3, 28.

[58] Ibid., n. 15, 32.

[59] Ibid., n. 16, 32–4.

[60] For the definition of real distinction based on the mutual separability of the extremes, see *DM* disp. 7, s. 2, n. 2, 261.

In the middle part of *CDA* disp. 8, q. 2, devoted to the issue of the subject of the interior senses,[61] Suárez states that all animals including imperfect ones possess an interior sense that is capable of perceiving external objects in terms of their usefulness or harmfulness. Since imperfect animals have the sensory appetite through which they strive for nutrition, they must be able to perceive sensibles from this point of view. At the same time Suárez is explicit that it is neither true that all nonrational animals can exercise the same operations, nor are their operations related to the same adequate objects.[62] In this Suárez follows what Aristotle says at the beginning of his *Metaphysics*:

> Now animals are by nature born with the power of sensation, and from this some acquire the faculty of memory, whereas others do not. Accordingly the former are more intelligent and capable of learning than those which cannot remember.[63]

Aristotle does not say which animals have and which do not have memory. However, in *DM* disp. 1, s. 6 Suárez notes that the animals that have memory are those that can move over long distances.[64] These animals are those that remember the places where they have found shelter or nutrition. Without memory they would not be able to return to those places. But assuming that there are animals with a basal phantasy that can perceive only what is useful and harmful but do not have traces of memory, should we not conclude that the power of phantasy or estimation is a distinct power from the power of memory? For Suárez this is not the case. He does not consider this separability to be a cogent reason for postulating two really distinct powers, namely the memory and the phantasy. In the same way as there are differences in perfection and scope in the acts of the interior senses of rational and nonrational animals, so there are also differences in the perfection and number of operations in perfect and imperfect animals (i.e. those with memory and those without memory). As it is not necessary to postulate a plurality of interior senses in rational animals due to their capacity to produce compositive imaginative acts, so when compared with perfect animals it is not necessary to envisage a lesser number of interior senses in imperfect animals.

In his reasoning for his first conclusion, in which the identity of the common sense and the phantasy and that of the estimative power and the memory are stated, Suárez is guided by the first above-mentioned premise. As he argues, this identity claim holds for all powers generating intuitive and abstractive apprehension.[65] As for this first identification, Suárez shows that a power cognizant in the absence of its object can and must beforehand apprehend an object in its presence. Abstractive cognition is not incompatible with intuitive cognition. On the contrary, the latter precedes the former. There cannot be a division of cognitive labour in the sense that

---

[61] For this segment, see *CDA* disp. 8, q. 2, nn. 3–6, t. 3, 48–54.

[62] Ibid., n. 3, 48.

[63] Aristotle, *Metaphysics* I.1, 980a28–33, 3.

[64] *DM* disp. 1, q. 6, n. 17, t. 1, 58.

[65] *CDA* disp. 8, q. 1, n. 17, t. 3, 34.

one faculty cognizes only intuitively and a distinct faculty can cognize only abstractively. The fact that one power apprehends intuitively and another abstractively *plus* intuitively is not in itself a reason for positing two distinct powers. A power that can render more perfect cognition, that is, abstractive, can exercise also a less perfect one which is an operation bound to the cognition of objects existing here and now.[66] Once the external senses as active potencies cognizing intuitively are affected by an extramental object, the interior sense gets *simultaneously* affected by the same object too. It makes no sense to say that in order for the phantasy to receive the sensible species some time must elapse between the external senses' and the interior sense's affections. This time could make no difference in disposing the phantasy better to the affection than it was disposed before. Since there is a substantial psychological unity between these two powers, it holds that when one (attentive) power gets affected, the other will be intentionally immuted as well. The external senses and the phantasy do not differ in the manner that the former cognize only in the presence of sensibles and the latter only in their absence. The difference lies in the fact that the phantasy can apprehend objects in their presence *and*, additionally, in their absence.[67] As is usual for Suárez, he tries to corroborate his own views with Aristotle's formulations. In this regard he refers to *On Memory and Recollection* where Aristotle speaks about the common sense and phantasy as a single power. Suárez refers to this claim: '[. . .] the image (*phantasma*) is an affection of the common sense'.[68] If the image, commonly regarded as the affection of the phantasy, is said to be the affection of the common sense, the phantasy and the

[66]The structurally same argument but in a reversed order is applicable in Suárez's reasoning for direct intellectual cognition of material singulars. If a less perfect capacity, such as the senses, can directly apprehend singulars, then the intellect, being a higher power, must be capable of doing it too. For this, see *CDA* disp. 9, q. 3, n. 9, t. 3, 120.

[67]For this, see *CDA* disp. 8, q. 1, n. 4, t. 3, 20.

[68]Aristotle, *On Memory and Recollection*, ch. 1, 450a10–1, 293. Just this claim of Suárez's became the target of Anthony Lisska's critique: 'If this distinction between these two internal sense faculties [sc. the common sense and the phantasy] is denied, then it is easy to have a phantasm become the object of both imagination (Fancy) and the *sensus communis* [. . .] The phantasm becomes the *tertium quid*, which is the direct object of knowledge in the *sensus communis*. While Suárez may have held this position on the epistemological placement of the phantasm, Aquinas did not. To render a phantasm the direct object of knowledge in the *sensus communis* in Aquinas is to render Aquinas's theory into the account possibly put forward by Suárez'. (Lisska, 297). Elsewhere he says: 'Deely notes that Suárez's position on sensation was structurally similar to Locke's–i.e. an image is required for sensation'. (ibid., 185, footnote 29); 'Suárez: The sensible qualities are images (phantasms) formed by the sense faculties through the causal interaction of the thing with the sense faculty. This is epistemological representationalism'. (ibid., 231); 'Considering the influence of Suárez on these philosophy-of-mind matters, Deely has written: 'Locke's (and Descartes's) position is rather that of Suárez, who held that external sense, no less than internal sense and understanding, required the formation of *species expressa* (i.e. an image or an idea) to produce its object of apprehension''. (ibid., 235); 'Suárez possibly adopted an 'image position' for his account of phantasm'. (ibid., 298). For Lisska the presence of a phantasm in the common sense is the result of Suárez's excessive reductionist account of the internal senses. Unlike Aquinas, Suárez conflated the external *sensorium* of which the common sense is part with the internal *sensorium* of the phantasy which is productive of images conceived as that in which the objects are cognized. That

common sense have to be considered as a single power. As I have repeatedly highlighted above, in Suárez's *CDA* the mediation between less perfect and more perfect capacities is justified by his theory of the sympathy of powers according to which once the lower power is operative, the higher is active too. Concluding, the phantasy and the common sense must constitute a single power. The same holds also for the memory and the estimative power. Before apprehending abstractively the same power must cognize intuitively.[69]

Suárez adopts a critical stance also to the criterion (for the distinction of powers cognizant intuitively and abstractively) grounded in the anatomical theory of qualitative distinctions in the brain ventricles. He is sure that one and the same material, such as lead, can be both receptive (if hot it is well malleable) and retentive (if cold).[70] Suárez points out that affection by sensible is not a purely corporeal process but first of all an intentional reception. No matter how the organs' dispositions contribute to the reception of sensible species, these are not received in the organs but in the powers.[71] Moreover, the distribution of the qualities of moisture and dryness in the brain chambers is not as clear as the older scholastics and physicians supposed to be.[72] Therefore, the tripartite cerebral division conceived as a criterion for determining the number of internal senses should not be regarded as the defining one. As said above, Suárez agrees that the brain is the organ of the interior sense. However, he differs from Galen's view as presented in his *De differentiis symptomatum*. There are for him, at least, two reasons to deny Galen's view. (1) There are animals that have the memory, the phantasy and other interior senses, or more in line with Suárez the cognitive operations that are attributed to these supposedly really distinct powers, but do not have the distinct ventricles. (2) Galen is not consistent. In *De usu partium* he proposes a different view from that in *De differentiis symptomatum*.[73] The physiological theory from the latter treatise, with a certain modification, complies better with Suárez's view of the single interior sense. He makes it clear that the only function of the front ventricle (which, as we know, is the organ of the olfactory power for Galen[74]) is to produce the animal spirits by which the organs of the external senses are 'irrigated'[75] (the ancient physician regarded this as the second main function of this front sinus). The middle

---

this view of the expressed species is not that of Suárez has been sufficiently shown above in 3.5. For this, see also South (2001a).

[69]See *CDA* disp. 8, q. 1, n. 17, t. 3, 34–6.

[70]For the same rejection of this physiological criterion, see Olivi in Toivanen (2013b, 252–3).

[71]For this, see *CDA* disp. 5, q. 1, n. 6, t. 2, 292–4 (see also 3.1).

[72]*CDA* disp. 8, q. 1, n. 18, t. 3, 36. For the doctrinal variability related to the number and the function of the particular brain cavities, see also Mastrius and Belluto, *In libros de anima*, disp. 5, q. 8, art. 5, § 282, 124: 'Nec minor dissensio est inter hac sententias [concerning the number of the interior senses], ut assignent organa uniuscujusque potentiae; hinc cellulas cerebri ad libitum assignant'.

[73]Galen, *De usu partium corporis humani*, book 8, ch. 10–11, 479–84.

[74]For Suárez's denial of Galen's view, see 4.4.3.

[75]See *CDA* disp. 6, q. 6, n. 10, t. 2, 540.

cavity then serves only as a passage (*meatus*) through which the animal spirits with the sensible species are conveyed to the third (ultimate) ventricle. And, finally, in this back chamber the organ of the interior sense is to be located. With a sceptical view on the philosophical relevance of medical experiments Suárez notes that various kinds of disorders can be explained by different kinds of lesion in the very same organ. One and the same ventricle and its *temperamentum* can be damaged in manifold and unknown ways that can cause a wide range of disfunctions in one and not in another cognitive operation.[76]

Aside from denying the need to multiply powers based on the criterion of intuitive and abstractive cognition, we have seen in 5.1 that Suárez radically declines also the criterion based on the essential difference between sensed and non-sensed species. The rejection of this distinction based on the arguments from the redundancy and the impossibility of non-sensed species leads him to identify the estimative power and the memory with the phantasy.[77] If the former two powers were really distinct from the last one, one could not satisfactorily explain the phenomenological fact of perceptual unity between 'speculative' and 'practical' perception. In addition to *CDA* disp. 6, q. 2, in *CDA* disp. 8, q. 1 Suárez also rejects a division of labour in the processing of information in the internal senses – employing again the example of sheep and wolf – according to which the sheep's apprehension of the wolf as such and the judgement 'This wolf is hostile' would lead to attributing these acts to two distinct powers, that is, to the common sense and the estimative power. He denies this modular procession of information because the judgement 'This wolf is hostile' is intrinsically included in the cognition of what is hostile. Apprehensive and adjudicative acts cannot be acts pertaining to two distinct powers. Practical judgement could not be exercised without a single power that is both cognizant and judicative of both extremes at the same time. This is why the common sense and the *aestimativa* have to be taken as one and the same interior sense. But if the common sense is deemed to be equivalent to the phantasy, the estimative power and also the memory must be identical with the phantasy.[78] Suárez again refers to Aristotle who assigns estimative acts to the phantasy:

> But now, about movement in space, we must consider what it is that causes the living animal to exhibit a travelling movement. Obviously it is not the nutritive faculty; for this movement always has an object in view, and is combined with imagination and appetite (432b13–7) [...] Imagination, too, when it starts movement, never does so without appetite (433a20–1) [...] but it is not capable of appetite without imagination [...] (433b29).

The phantasy is presented as the power that triggers the sensory appetite. This view can be confirmed by an analogy with the intellect. In *CDA* disp. 9, q. 9 Suárez denies the theory according to which the theoretical and the practical intellect are to be

---

[76] *CDA* disp. 8, q. 2, n. 8, t. 3, 56–8.

[77] For a similar identification of the estimative faculty and the common sense based on the interconnectedness of the psychological functions of perception and estimation in Olivi, see Toivanen (2013b, 332–9).

[78] *CDA* disp. 8, q. 1, n. 19, t. 3, 38.

conceived as two distinct intellects. With reference to the interior sense Suárez embraces the theory of a single intellect which is plural only conceptually, that is, according to the various ways of consideration. Theoretical and practical intellect cannot be really distinct since speculative acts are closely linked to practical ones. The former are the foundations of the latter.[79]

If the estimative power and the memory do not differ from the phantasy, considering what has been said above, there must be *realiter* only a single internal sense called the *phantasy*. In this case Suárez does not even apply the Scotistic formal distinction, even though one could argue for it on the basis of the distinct definitions of various operations: While the common sense is the power that judges about sensibles only in their presence; the phantasy judges about them in their absence etc. Suárez leaves no doubt that *realiter* and also *formaliter* there is only the single sense of phantasy. Since there is only one formal and adequate object (the sensible *in commune*), there must be formally a single internal sense. Interior sense apprehension can be conceived as plural only conceptually, that is, according to our inadequate concepts that are based as on their *fundamenta* that are the various acts generated by this single capacity.[80]

5.3 Reduction of the Interior Sense's Cognitive Functions

Suárez's reduction of the plurality of the interior senses to the single power of phantasy has its counterpart in his reductionist account of the scope of the cognitive functions attributed to the interior senses by his predecessors headed by Aquinas. It is not the case, as one might assume, that Suárez's single internal sense is in charge of all the operations attributed by Aquinas to four distinct powers. In his recent book on Aquinas's theory of perception, Anthony Lisska argues that Aquinas's philosophy of human perception would be an embarrassment without the *vis cogitativa* and the sensory memory conceived as a storehouse of the intentions that are acquired by the cogitative power and as a faculty capable of seeking sylogistically for a recollection of the past.[81] George Klubertanz claimed that the human cogitative power is concerned only with the singulars of action (*operabilia*) as standing under the intelligible light of reason. For him Aquinas's *cogitativa* is operative only in the order of practical knowledge and is restricted to discerning the useful from the harmful. Its function is limited to action determination, which is to be undertaken here and now by comparing individual intentions standing under the universal

[79]Ibid., n. 20, 40. The same reasoning can be seen also in Suárez's identification of the concupiscible power and the irascible power. See *CDA* disp. 11, q. 1, t. 3, 320–33 (cf. also 4.5.2). For this, see also Heider (2016a).

[80]*CDA* disp. 8, q. 1, n. 24, t. 3, 44–6.

[81]Lisska (2016, 327). For sensory memory, see ibid., 265–9. As regards this broad account of Aquinas's cogitative power, which is seen as 'the real center of our interior life', see also Fabro (1938, 352–3).

principle of practical reason.[82] For Lisska, however, the *vis cogitativa* '[…] at its core, has more than the practical function Klubertanz puts forward'[83] and '[it] is far more important than Klubertanz acknowledged'.[84] Lisska highlights that apart from detecting the latent 'practical' intentions, such as hostility or usefulness, the *vis cogitativa* as a comparative and discursive power is for Aquinas also the faculty by means of which we are distinctly aware of the per accidens objects of sensation, that is, primary substances as instances of natural kinds. This cogitative power which detects the non-sensed particular intentions of accidental sensibles and structures the corresponding phantasms makes a crucial preparatory tool for the agent intellect's abstractive (univerzalizing) process that leads to the abstraction of the quiddities of material substances which are then received in the really distinct passive intellect. By this empowered *cogitativa* Aquinas bridges the gap between the material powers that apprehends *only* material singulars and the spiritual power of the intellect whose object are *only* the *quiddities* of material singulars.[85]

In the following two parts I aim to show that this is anything but Suárez's view. I provide evidence that in *CDA* the Jesuit is sceptical about attributing proto-intellectual operations to the human, let alone the nonhuman,[86] phantasy. These operations can be found only at the level of intellectual workings. In 3.7, 3.8 and 5.1 I have shown that, compared to some formulations of Aristotle and theories of authors such as Aquinas, Alhacen and Scotus, Suárez undertakes this functional reduction also in the operations of the common sense. The common sense is neither the power that apprehends the common sensibles that cannot be perceived by the external senses, nor it is the capacity through which a second-order (intentional) self-awareness comes. The common sense is not the first sensory faculty that apprehends the acts of the external senses since already at the level of the particular sense modalities we are not only aware of the proper sensibles but are concomitantly perceptive of the acts through which we apprehend them. Equally, the common sense does not exercise a more perfect self-awareness that would be mediated by the impressed species of the external sense operation. There are no special species of the external senses' acts and there is no intentional apprehension of these acts, let alone a perfect reflexive act that would be part of the common sense's cognitive scope. The common sense apprehends the acts of the external senses only through a modification of the sensible species representing the extramental sensibles that have been produced by the soul looking at *exempla* of the operations of the external

---

[82]Klubertanz (1952, 202–3, 205, 256–7, and elsewhere).

[83]Lisska (2016, 245).

[84]Ibid., 317.

[85]Lisska (2016, 258–61, 317, and *passim*).

[86]Adam Wodeham (1298–1358), among others, went so far as to attribute such 'proto-intellectual' operations to the interior senses of nonrational animals as well. For this, see Tachau (1993). Attributing these operations to perfect animals is motivated by the observation of remarkably prudent and providential kinds of behaviour among brutes. Many of these phenomena can be found in Aristotle's *History of Animals* IX.6, 611b33ff, 245ff.

senses. As we will see in this section, this reduction, in fact, applies also to other operations attributed by Aquinas to the cogitative and the memorative power.[87]

5.3.1 Cogitative Power

As Aquinas scholars largely agree,[88] the central internal sense for Aquinas is the *vis cogitativa*. In line with what in 2.5 I have called the Principle of Continuity in the *Scala Naturae*,[89] exemplified by the hierarchically ordered powers of the human soul (the highest part of the lower faculty 'touches' on the lowest part of the higher faculty, whereby the lower faculty gets epistemologically elevated in its operations[90]), through its proximity to the intellect this internal sense can be elevated to such a degree as to be able to exercise judicial, inferential, discursive (syllogistic) and deliberative acts that are similar to intellectual operations. We have seen that this principle of participation of the lower in the higher has a wide application among Thomists in a number of issues, which derives from Aquinas's theory about the natural flowing of the lower powers from the higher ones (2.7) and from his doctrine of the eminent containment of the soul's lower parts in the higher soul (2.5). Given this metaphysical principle, it is not surprising that Aquinas calls this sense the particular *reason* (*ratio particularis*). Like the *vis aestimativa* of nonhuman animals, it detects the non-sensed intentions of convenience and inconvenience by means of the *species non sensatae*. Nevertheless, it also discovers incidental sensibles, viz. the sensibles that are accidentally conjoined to the proper and the common sensibles. While the common sense apprehends this or that bundle of proper and common sensibles, the cogitative power goes beyond this bundle or deeper by adding an important interpretative moment to our perceptual experience. Due to this power, as Lisska notes, we can sensorily discover individual substances (just this has been denied by a significant part of the empirical tradition in modern philosophy for which material singulars and even the self are equivalent to bundles of qualities or perceptions). The cogitative power's capacity of 'interpretation' consists in detecting particular intentions, such as this or that individual substance. The *cogitativa* is operative when I see this white object as my daughter Mary. Importantly, being

[87]For Suárez's systematic downplaying of progressive cognitive roles in internal sense apprehension and for the view according to which the internal sense is the power of mere replication and synthesizing of the information received from the external senses, see South (2001b) and Heider (2017d).

[88]For the centrality of the cogitative power in Aquinas's cognitive and affective theory, see in addition to Lisska and Fabro, the recent contribution by De Haan (2014).

[89]For this principle, see the above-quoted Forigari (1968, 325–6) and Heider (2015b). As regards Aquinas's employment of this principle, see Montagnes (1968).

[90]For this principle in Dionysius the Aeropagite, see his above-quoted *On the Divine Names and the Mystical Theology*, ch. 7, § 3, 79. For Pseudo-Dionysius's (and also Proclus's) influence on medieval and Renaissance philosophy in respect to this *scala naturae*, see Mahoney (2000).

elevated by the power of the intellect, this particular reason apprehends individual substances as instances of certain *kinds*. It subordinates individuals under kinds. It apprehends this white as 'this *human'*. This is also why this power prepares phantasms for the oncoming universalizing operation of the agent intellect. Moreover, unlike the estimative power of nonrational animals led by natural instinct according to which animals of the same kind are determined to behave always in the same way, the proximity of the *cogitativa* to the intellect capacitates it to compose and divide the detected particular intentions and, consequently, enables our sensory apparatus to behave more variably. After apprehending these intentions, the cogitative power can generate judgements, such as 'Peter is not Paul', or it can infer the practical conclusion 'Do this!' in a practical syllogism. All in all, as Aquinas says, while it operates discursively in respect to these particular intentions, with universal intentions the intellect works discursively. In order to apprehend individuals, the intellect needs to return to the phantasms in the interior sense to get a cognition of material singulars. It is not surprising that, as a consequence, the sensitive appetite, whether it is the concupiscible or the irascible power, which is triggered by the operations of this *vis cogitativa* can be regarded as, in a way, a voluntary capacity which is 'free' and indifferent in such a way that, as Cajetan explicitly says, it can even become the subject of some moral virtues, such as fortitude and temperance, or can commit a pardonable sin (*peccatum veniale*). As such these sensory appetites, next to the proper rational appetite, can be regarded as more or less independent 'wills' housed in a single person.[91]

[91]In many *loci* in the works of Aquinas, Cajetan and other Thomists these operations are granted to the cogitative power. The most frequently quoted passage by Aquinas is *STh*. I, q. 78, art. 4, c., 256: 'Sed quantum ad intentiones praedictas [intentiones quae per sensum non accipiuntur], differentia est: nam alia animalia percipiunt huiusmodi intentiones solum naturali quodam instinctu, homo autem etiam per quandam collationem. Et ideo quae in aliis animalibus dicitur aestimativa naturalis, in homine dicitur *cogitativa*, quae per collationem quandam huiusmodi intentiones adinvenit. Unde etiam dicitur *ratio particularis*, cui medici assignant determinatum organum, scilicet mediam partem capitis: est enim collativa intentionum individualium, sicut ratio intellectiva intentionum universalium'; Thomas de Aquino, *Sentencia libri de anima*, book 2, l. 13, 191–7, 121: 'Si vero apprehendatur in singulari, utputa cum video coloratum, percipio hunc hominem vel hoc animal, huiusmodi quidem apprehensio in homine fit per vim cogitativam, quae dicitur etiam ratio particularis eo quod est collativa intentionum individualium, sicut ratio universalis est collativa rationum universalium [...]'; Thomas de Aquino, *Summa contra gentiles*, book 2, cap. 73, 460: 'Virtus cogitativa non habet ordinem ad intellectum possibilem, quo intelligit homo, nisi per suum actum quo praeparantur phantasmata ut per intellectum agentem fiant intelligibilia actu et perficientia intellectum possibilem'; Cajetan In *STh* I-IIa, q. 74, art. 4, IV, 39: 'Vires namque animae in parte sensitiva, sicut in cognoscendo sunt secundum se excellentes totam naturam brutorum, ut patet in cogitativa et reminiscentia habentibus collativam vim particularium; ita in appetendo sunt secundum se excellentes totam naturam brutorum, per hoc quod appetunt cum quadam libertate. Ita quod non solum ex hoc quod actualiter moventur vel non moventur a voluntati; sed seipsis, quia sunt tales, quod superius vocavimus *quia sunt volentis*, sunt quodammodo liberae. Dico autem *quoddamodo*, quia non habent tantum libertatis ut possint in actum deliberatum; sed vestigium quoddam libertatis sortitae sunt ex elevatione in naturam rationalem, a qua fluunt. Eadem autem vires appetitivae sensitivae, ut actualiter motae et moventes a parte superiori, tantum libertatis participant ut etiam in deliberatos actus possint [...] Quo fit ut, quia sensualitas secundum se habet

Suárez is far from likening the cogitative power to the intellect and far from attributing the above-mentioned proto-intellectual operations to the interior sense. At least in *CDA*, in its *radius operandi* this power seems to amount to nothing significantly above the *vis aestimativa* of nonhuman animals. Its main function is to apprehend sensibles in terms of their usefulness and harmfulness. What Aquinas calls *ratio particularis* Suárez prefers to liken to a power that cognizes the aspects of convenience and inconvenience and is led by a natural instinct. Indeed, in humans such sensory cognition and behaviour is often regulated by the intellect and the will. The will has a 'political', viz. not a despotical *dominium* over the sensory appetite. The will cannot immediately move the sensory appetite. It can do this only by means of the intellect and the interior sense. It can hamper a movement of the sensory appetite by moving the attention of the phantasy to another object.[92] But this does not entail that the estimative power as the proximate ontological subject can perform the above-mentioned operations of judgement and discourse, which both Aquinas and Cajetan attributed to the cogitative power, and that the sensory appetite can be the subject of virtues such as fortitude and temperance[93] and of venial sins. Why is that so? How does Suárez deprive the phantasy comprising the function of the estimative power of the operations that Aquinas and his disciples attributed to the *cogitativa*? Suárez's main strategy is clear: All these operations are to be relegated to the intellect whose first object is the same as it was for the sensory powers.

Suárez is sure that incidental sensibles such as material substances are objects that, unlike God who is known only discursively and *a posteriori* from his created sensible effects, are sensed. They are immediately cognized confusedly or concretely, that is, with their per se sensibles of which these substances are the subject. As such these sensibles per accidens constitute a part of per se sensibles. They are not represented *in abstracto*, that is, they are not represented as separated from the subject. They are represented *in concreto*, sc., as 'grown together' with the subject.[94] And since, as said, the representation of the intelligible species is determined by the representation of the interior sense's species our intellect primarily does not have a direct access to material substances. In the first stage, the intellect does not produce a

---

tantum libertatis ut sufficiat ad peccatum veniale, ideo peccatum ei tribuitur: quia vero non habet tantum rationalitatis ut possit in peccatum mortale, quia non est eius respicere bonum rationis ut finem, ideo ei non attribuitur peccatum mortale'. See also Thomas de Aquino, *STh.* I-IIae, q. 74, art. 3, 32. For other Thomists than Cajetan, see Domingo Báñez, *Scholastica commentaria in primam partem Angelici Doctoris D. Thomae: a quaestione LXV vsque ad CXI*, q. 78, art. 4, dub. 2, 273–7. For Suárez's critique of Cajetan's view according to which some cardinal virtues, namely temperance and fortitude, are subjected in the sensory appetite, see Heider and Machula (2019).

[92]While according to the despotical subordination view the higher power, namely the will, forces the lower power, sc., the sensory appetite, in such a way that the lower cannot decline the command of the higher, according to the political subordination view it governs it in such a way that it can decline. For this, see *CDA* disp. 11, q. 3, t. 3, 356–67; see also Suárez, *De voluntario et involuntario*, disp. 10, s. 2, 272–3. For Aristotle's claim, see his *Politics* (1984) I.5, 1254b3–6, 1990.

[93]For the claim that fortitude and temperance are virtues of the irrational parts of the soul, see Aristotle (1984), *Nicomachean Ethics* III.10, 1117b24–5, 1764.

[94]*CDA* disp. 6, q. 1, nn. 2–3, t. 2, 456.

proper and distinct concept of a particular substance.[95] If we had this access we would have to be able to grasp the new substance of the body and the blood of Christ after Eucharistic consecration. This would recall its character of the mystery.[96] Considering the importance of incidental sensation in our everyday sensory cognition, which, concisely speaking, can be expressed as 'seeing *as*' (we see this white object as Peter), it would be odd if Suárez did not concede this kind of cognition to the senses too.[97] In fact, throughout *CDA* he speaks about the phantasm of Peter which serves as an *exemplar* for the formation of the intelligible species that represents an individual that has previously been represented by a phantasm. So, clearly, Suárez admits that we have an immediate proper concept of the substance of Peter which can be applied solely to Peter.[98] In this sense, he denies that this concept is a common concept (*conceptus communis*). At the same time, he is explicit about the difference between the intellect and the senses, which remain purely within the ambit of 'externorum accidentium sensibilium'. He is sure that the only proper and distinct concept of material substance we acquire is through the additional discursive operation of the intellect.[99] Considering these two dialectical statements it can be said that there are for Suárez, in fact, only *two* kinds of cognition of incidental sensibles. While the external senses (and then the internal sense and the intellect in its first stage) grasp them confusedly, the only more perfect (distinct) knowledge is accomplished only through the discursive operations of the intellect. Only this power can discover a hidden 'invariant' behind the temporal variations of accidents. Only after the quidditative detection of this invariant we form the intelligible species that represents it. We do not have such a species before. In the same way we come to know our soul. We do not have immediate access to it. We can cognize it only *a posteriori*, that is, from its own effects, that is, from its operations with which we are directly and primarily familiar.[100] No other third and middle (preparatory) cognition reserved for an interior sense, such as the cogitative power, is necessary and possible.[101] This claim can be confirmed by Suárez's above-mentioned denial of a

[95] *CDA* disp. 9, q. 4, nn. 1–4, t. 3, 152–8.

[96] *CDA* disp. 9, q. 4, n. 5, t. 3, 158–60: 'Et confirmatur utraque conclusio [i.e., the intellect does not primarily cognize the incidental sensibles through their own intelligible species and it does not produce the proper and the distinct concept of these sensibles] ex mysterio sacramenti Eucharistiae, nam ibi eodem modo sentiuntur accidentia panis quo aliae res sensibiles; et in aliis rebus non magis sentitur subectum accidentium quam ibi; sed ibi ex phantasmate panis consecrati non producitur in intellectu species propria substantiae panis nec proprius illius conceptus, quia ibi vere non est talis substantia, ergo neque in aliis rebus sensibilibus acquiritur illa species vel formatur talis conceptus, alias posset intellectus diiudicare an sub his accidentibus esset substantia necne'.

[97] It should be said that the recognition of an object as *X* already requires the involvement of memory. In perceiving an object as Peter I assign my perceptual experience to one of the objects that are conserved in my memory.

[98] *CDA* disp. 9, q. 4, n. 3, t. 3, 156.

[99] Ibid. n. 7, t. 3, 162.

[100] *CDA* disp. 9, q. 5, n. 4, t. 3, 172–4.

[101] I cannot agree with the argument according to which the emergentist origin of the species of the interior sense (not caused efficiently by an act of the external senses) is the reason why already the

distinct *species non sensata*, which for Aquinas is representative also of incidental sensibles. As I stressed, for Suárez there are no such sensible species functioning as principles in the interior sense apprehension.[102] In addition to the information brought by the external senses, further cognitive value comes with the discursive operation of the intellect. Only intellectual discourse that at first is cognitively related to the various species of accidents can lead to the elicitation of an intelligible species representing their substratum, viz. an individual substance. Accidents are what our intellect apprehends first, not a substance or a universal essence. Considering the overall parallelism between metaphysics and epistemology in Suárez's philosophical system it can be said that if he does not explain the individuality of accidents in terms of their subject, or, in other words, if the accidents of quantity and quality conceived as *res*, which do not have to inhere in its subject but can be through the divine power conserved in their existence without its bearer, are individuated by themselves (by their *entitas tota*), they can well be intellectually grasped at the beginning of concept formation without a significant cognitive involvement of substance. True, this would be much more difficult if we claimed that their principle of individuation is their substance.[103] Concluding, only the intellect, not the interior sense, can infer a more perfect (distinct) cognition of a subject from its discursive elaboration of its accidents.

In his long critique of Aquinas's attribution of judicial and discursive operations to the *cogitativa*, Suárez premises the division of two kinds of judgements that have already been introduced in 3.8. While Suárez concedes judgement in the broad sense to all the senses (as he says, this judgement is intrinsically 'imbibed' in all their acts), in his conclusions he denies that proper judgement can be found in the interior senses of both humans and nonhuman animals. Regarding the interior senses of nonrational animals, he says that they are too imperfect and too material to be able to confer one item to another or proceed from one judgement to another. In his overall evaluation of the capacities of the interior sense's operations he claims that in order to explain animal behaviour and their sensory appetite judgement in the proper sense is not necessary. Their sensory appetite can well be moved by judgement in the broad sense. In general, judgements by composition and division are far from necessary to motivate the sensory appetite. It is also not necessary to assume this kind of

---

interior sense 'interprets' the confusedly apprehended incidental sensible, such as substance. It cannot be said that the substance of the per se sensibles is distinctly cognized by the interior sense because of the availability of the unsensed species that represent them to the power. As Suárez makes it clear, there are *no* unsensed species detected by the internal sense and the sensibles per accidens are not represented by modification of the species representing the proper sensibles either. They are only hidden intrinsic parts of the sensible accidental wholes.

[102]In this connection South (2001b, 135–6) pointedly says: 'After all, the incidental sensible, separate from the proper sensibles, would be another type of unsensed species, and we have already seen that Suárez has no use for such species. The result is that there is no proper sensed species of the substance itself. Instead, we properly sense only its accidents. Consequently, for Suárez, incidental perception is of little moment'.

[103]For Suárez's theory of the individuation of accidents, see Heider (2005).

judgement to explain the animals' 'prudent' behaviour based on their putative virtues or their ability to learn. Suárez also makes it clear that we do not have to posit a discursive capacity in nonhuman animals to explain their foresights and their *experimenta*. In all these cases it is enough to assume again judgement in the broad sense. He adds that all animals are driven purely by instincts. Since their cognition is rather confused and they apprehend only the *material* aspect of the good and do not cognize distinctly the good in its formal character (*ratio boni*), it cannot be said that they cognize more perfectly than humans since they apprehend through a simple act. Suárez thus rejects the argument claiming that a power that apprehends in a single act is more perfect than that which proceeds by a complex act. This would hold, if it were related to the same kind of cognition. Accordingly, their cognition would be more perfect, only if they apprehended the character of the good distinctly and formally. However, this is what they cannot do.[104]

Mutatis mutandis, the same negative conclusion holds also for the *human* interior sense: 'No human interior sense can apprehend or judge by composing and dividing, either, since such judgment in general exceeds the limits of a sensory power'.[105] Such judgment belongs only to the immaterial intellect where truth and falsity are to be located. However, immediately after this conclusion Suárez is quick to append: 'Not all approve of this conclusion [...]'[106] Aquinas and Cajetan, *inter alii*, are introduced as representatives of the discordant view. The general strategy of Suárez's argument against their views is based on relegating to the intellect all the above-mentioned acts they had attributed to the *cogitativa*. (1) He starts by rejecting the argument based on the Axiom of Continuity. In a rather dualistic way he says that it makes no sense to say that a material power, such as the cogitative power, gets perfected by being linked to the spiritual intellect. As a material power it can receive no power from it. (2) Judgements that proceed *componendo et dividendo* and any reasoning exceed the capacity of material powers. No matter how close the *cogitativa* stands to the intellect, it still remains 'intra latitudinem potentiae materialis'.[107] Moreover, if this act were conceded to this sense power, it would have to be granted to nonrational animals too. If the materiality of this power did not prevent it from having these operations, there would be no reason to deny it to nonrational animals, either. Even more, these operations should be granted to them since, while in humans they can be assigned to the intellect, in nonhumans they cannot. (3) If the *cogitativa* could function *componendo et dividendo*, and if it could conjoin a predicate with a subject, it could proceed discursively also by collating one proposition with another. If the former were affirmed, there would be no reason to

---

[104]See *CDA* disp. 5, q. 6, nn. 13–4, t. 2, 424–30.

[105]For this conclusion, see also Suárez, *TDA* book 3, ch. 6, n. 8, 639. Castellote's edition in *CDA* disp. 5, q. 6, n. 15, t. 2, 430 states imprecisely: 'Nullus etiam sensus interior hominis potest apprehendere vel iudicare componendo et dividendo. Unde huiusmodi iudicium in universum excedit limites potentiae *cognoscitivae*'. [my italics]. Of course, this kind of judgment does not exceed the possibility of the intellectual power.

[106]*CDA* disp. 5, q. 6, n. 15, t. 2, 430: 'Huic conclusioni non omnes consentiunt [...]'

[107]Ibid., n. 16, 432.

deny the latter. Moreover, if the *cogitativa* could infer a proposition from another proposition, it would have to be able to cognize the implications of this inference. Then it would have to cognize that the implications follow from the intrinsic composition of the terms. However, again, such operation exceeds the capacity of a material power since it would have to involve the cognition of universals. Moreover, if it could judge by composition and if it could exercise inferential operations, it would have to be capable of ordering means to ends and cognize the formal character of an end as such. But, as Suárez's *reductio* follows, if this were the case, apprehension of good and bad intentions would have to be conceded to the interior sense too. As operating discursively, the internal sense would have to be able to order means to ends. Concluding this whole *reductio*, this would lead to the attribution of some kind of *liberty* to the sensory appetite. By itself, the sensory appetite would have to be capable of receiving virtues and committing venial sins. Suárez is sure that all these consequences are utterly absurd and are to be regarded as devaluating the notion of the central human capacity, which is the will with its *liberum arbitrium*. Only an immaterial power, sc. the rational appetite or will, can be the subject of this free choice.[108]

In *CDA* disp. 11, q. 3 Suárez *ex professo* reprehends Cajetan for this claim about sensory liberty which leads to a conception according to which a subject houses a plurality of 'wills'. Again, he confirms that it holds for Cajetan that since the sensory appetite is rational by virtue of its participation in the will, it becomes somehow free and capable of receiving virtues and vices.[109] Referring not only to Aquinas but to 'omnes auctores et sancti', Suárez says that *all* sins can be found only in the will.[110] There can be no liberty outside the immaterial power of will. In keeping with *CDA* disp. 5, q. 6 he asserts that if there is neither discourse, nor deliberation, nor inquisition of means to their ends, nor universal cognition in the interior sense, there can be no liberty there, either. Attributing liberty to the *cogitativa* would entail another absurd state of affairs, which he views as a clear consequence of the Cajetanic position: If the operations of the sensory appetite were *ex propriis* free, and if there could be evil in this appetite without an interference of the will, one could find oneself in a schizophrenic situation in which one could at the same time commit a sin by means of the sensory appetite and do a good by virtue of the will. In order to avoid this consequence all moral virtues and vices in the proper sense must be subjected in the will.[111]

---

[108]Ibid., n. 17, 432–4.

[109]*CDA* disp. 11, q. 3, n. 3, t. 3, 362.

[110]He refers also to Aquinas's formulation from *STh.* I-IIae, q. 74, art. 1, c., 35: 'Cum autem proprium sit actuum moralium quod sint voluntarii, ut supra habitum est; sequitur quod voluntas, quae est principium actuum voluntariorum, sive bonorum sive malorum, quae sunt peccata, sint principium peccatorum. Et ideo sequitur quod peccatum sit in voluntate sicut in subiecto'. However, elsewhere Aquinas explicitly says that there is venial sin in sensuality (see *STh.* I-IIae, q. 74, art. 3, ad 3, 37; art. 4, ad 3, 38).

[111]For this, see Francisco Suárez, *De actibus qui vocantur passiones, tum etiam de habitibus, praesertim studiosis ac vitiosis*, disp. 3, s. 7, n. 5, 493: 'Omnes virtutes morales per se primo esse in

5.3.2 Memory and Reminiscence

Beside the *vis cogitativa* that detects intentions and operates with them, Aquinas considers also the power that retains and reactivates them. Following Avicenna,[112] Aquinas says that while the phantasy is a storehouse (*thesaurus*) of the *species sensata* formed by the common sense, the memory is a treasury of the *species nonsensata* or intentions received from the cogitative power. Although both apprehend objects *in absentia* and operate with notions of objects, they are different. Unlike the phantasy, which perceives timeless objects, some of which are impossible, such as chimeras, the memorative power is essentially linked to past experience.[113] But how are Avicennian intentions associated with the aspect of pastness for Aquinas? In *STh.* I, q. 78, art. 4, one of the crucial texts for Suárez's treatise of the interior senses in general, Aquinas states that the character of pastness (*ratio praeteriti*) is to be ranked among the intentions that are not sensed by the external senses. He confirms this by saying that animals and humans often remember sensibles in terms of their individual convenience and inconvenience.[114] Accordingly, the memory can be described as a storehouse of sensibles which are always associated with an intention, whether that of pastness or those of usefulness and harmfulness. In some places Aquinas suggests that the conservation of the past is related to our past sensory apprehensions.[115] Unfortunately, with his emphasis on the external objects he does not explain how we can remember these acts.[116] The phantasy, on the contrary, will be a storehouse of *phantasmata* without these 'connotational attributes'. Importantly, in many places of his *corpus* Aquinas states

---

voluntate. Nam actus istarum virtutum sunt eliciti a voluntate: ergo generant in illa virtutis habitum'. Even though in the second conclusion of this section he admits that some habits of moral virtues facilitating virtuous acts are created in the sensory appetite (habits can also be produced in the interior sense–for this, see *DM* disp 44, s. 4, n. 7, t. 2, 670), in the third conclusion he leaves no doubt that these habits are not virtues in the proper sense but only in a certain respect (*secundum quid*) and through an extrinsic denomination (*per denominationem extrinsecam*). Ibid., nn. 9–11, 494–5. For Suárez's denials of the existence of pardonable sin in the sensory appetite by claiming that without will there is no sin at all, see Suárez, *De actibus qui voncantur passiones*, disp. 5, s. 5, 562–3.

[112] Avicenna, *Liber de anima seu Sextus de naturalibus I–II–III*, p. 1, ch. 5, 89–90.

[113] Sancti Thomae de Aquino, *Quaestiones disputatae de veritate*, q. 10, art. 2, c., 300: 'Responsio. Dicendum, quod memoria secundum communem usum loquentium pro notitia praeteritorum accipitur. Cognoscere autem praeteritum ut est praeteritum, est illius cuius est cognoscere nunc ut nunc: hoc autem est sensus'.

[114] Thomas de Aquino, *STh.* I, q. 78, art. 4, c., 256: 'Ad conservandum autem eas, vis memorativa, quae est thesaurus quidam huiusmodi intentionum. Cuius signum est, quod principium memorandi fit in animalibus ex aliqua huiusmodi intentione, puta quod est nocivum vel conveniens. Et ipsa ratio praeteriti, quam attendit memoria, inter huiusmodi intentiones computatur'. For this connection, see also Pasnau (2002a, 280–1).

[115] Aquinas, *De memoria et reminiscencia*, ch. 1, 106: 'Semper enim, cum anima memoratur, pronunciat se prius audiuisse aliquid uel sensisse uel intellexisse.' See also ibid., ch. 3, 115: 'Si autem anima conuertatur ad ipsum [phantasma] in quantum est ymago eius quod prius uidimus aut intelleximus, hoc pertinet ad actum memorandi'.

[116] For this critique, see Perler (2020b, 260–8).

that the memory in the full sense, which is concerned with the past as such and the flow of time, can be located *only* in the sensitive part of the soul. Unlike the intellective part whose objects are timeless quiddities, the sensitive part does not abstract from singular conditions that are fundamental to memorative cognition.[117]

In line with the discursive nature of the *cogitativa* which, according to Aquinas, apprehends intentions intuitively, the human memory as its abstractive counterpart acquires also a new habilitating quality through which it gets elevated above the operations we usually associate with the memory of nonrational animals. These operations are called acts of reminiscence or recollection. The capacity to reminisce is essential for learning. As Aquinas says:

> As for the power of remembering, man has not only only memory, like the other animals, in the immediate recording of past things, but also reminiscence in inquiring quasi-syllogistically into the memory of the past by means of individual intentions.[118]

For Aquinas reminiscence is neither a first reception of a cognoscible object nor its mere re-take. It accrues to memory as a motion and *via* to an act of remembering. The endpoint of a discursive motion of recollecting is a memory. Reminiscence is a scrutiny towards a recollection which is sought. This motion proceeds inferentially from something that is familiar to something that is yet unknown. It is a process that leads to the extraction of something from the 'depth' of the memory that has already been forgotten for some time. Aquinas shows that there are various ways in which this 'pulling out of a memory' can proceed. It can proceed on the basis of a similarity between the known and the forgotten (someone remembering Socrates recollects also Plato since the two are similar in wisdom), of a contrariety (one remembering Hector recollects Achilles), or of a proximity (by remembering a father one recalls also his son). Due to its epistemological elevation through participation in the higher intellectual capacity the operation of reminiscence is similar to syllogizing. As in a syllogism, in which one proceeds from the premises to the conclusion, so in recollecting we proceed by way of searching from what we know, as if from a principle, with an intention to extract something that has been forgotten from the memory. Importantly, as Aquinas makes it clear, reminiscence is not an act of the intellective part but of the sensitive part.[119]

Following Aristotle, Aquinas shows that the human memory with the cooperation of the cogitative power is important also in the process of concept formation and, consequently, for the knowledge of first principles.[120] In the last chapter of the second book of *Posterior Analytics* and in the first chapter of the first book of the *Metaphysics* Aristotle formulates the famous sequence of cognitive steps leading

[117]Thomas de Aquino, *STh.* I, q. 79, art. 6, c., 270: 'Si vero de ratione memoriae sit quod eius obiectum sit praeteritum, ut praeteritum; memoria in parte intellectiva non erit, sed sensitiva tantum, quae est apprehensiva particularium. Praeteritum enim, ut praeteritum, cum significet esse sub determinato tempore, ad conditionem particularis pertinet'.

[118]Thomas de Aquino, *STh.* I, q. 78, art. 4, c., 256: 'Ex parte autem memorativae, non solum habet memoriam, sicut cetera animalia, in subita recordatione praeteritorum; sed etiam reminiscentiam, quasi syllogistice inquirendo praeteritorum memoriam, secundum individuales intentiones'. For the translation, see Freddoso in Aquinas (2010, 77).

[119]Thomas de Aquino, *De memoria et reminiscencia*, tr. 2, ch. 5, 121.

[120]For an emphasis on this capacity in Aquinas, see Lisska, *Aquinas's Theory of Perception*, 265–8.

from sensation via repeated memories to experience (*experimentum*) conceived as an accumulated practical knowledge about individuals of a certain class:

> Thus sense-perception gives rise to memory, as we hold; and repeated memories of the same thing give rise to experience; because the memories, though numerically many, constitute a single experience. And experience, that is the *universal* when established as a whole in the soul – the *One* that corresponds to the Many, the *unity* that is identically present in them all – provides the starting point of art and science [. . .] although it is the particular that we perceive, the act of perception involves the *universal*, for example, 'man', not 'a man, Callias'.[121] It is from memory that men acquire experience because the numerous memories of the same thing eventually produce the effect of a single experience. [. . .] To have a judgement that when Callias was suffering from this or that disease this or that benefited him, and similarly with Socrates and various other individuals, is a matter of experience. [my italics][122]

In his commentary on the *Metaphysics* Aquinas notes that experience requires some reasoning or collation of particulars received in the memory. This collation is exercised by the particular reason, alias the cogitative power.[123] Experience takes the one from the many that are retained in the memory. In his commentary on the passage from *Posterior Analytics*, Aquinas stresses that it is the sense, sc. the cogitative power, with the memory that somehow (*quodammodo*) attains the universal or what is in many singulars. This is characteristic of *experimentum*. He adds that the cogitative power apprehends Callias not only insofar as he is Callias, but also insofar as he is a particular *human being*.[124] If it did not apprehend a universal nature in a singular, it would be impossible to cause universal cognition from sense apprehension. The gap between these faculties and their objects would be too big. The first universal principles are cognized inductively by the senses. In the first step it is the sense that establishes the universal in the soul.[125]

Suárez agrees that in its memorative act the *phantasia* retains sensible species and (if attentive) elicits acts representing objects in absentia. He is clear that one and the same power is both retentive and productive of memorative acts. But since he dismisses non-sensed species as a distinct principle operative in the cogitative power and the memory, he cannot take the memory to be a distinct storehouse of

---

[121] Aristotle, *Posterior Analytics* II.19, 100a5–100b2, 257–9.

[122] Aristotle, *Metaphysics* I.1, 980b30–981a3, 5.

[123] Thomas de Aquino, *Sententia libri Metaphysicae*, book 1, l. 1, n. 15: 'Experimentum enim est ex collatione plurium singularium in memoria receptorum. Huiusmodi autem collatio est homini propria, et pertinet ad vim cogitativam, quae ratio particularis dicitur: quae est collativa intentionum individualium, sicut ratio universalis intentionum universalium'.

[124] Thomas de Aquino, *Expositio Posteriorum*, book 2, l. 20, n. 14: 'Qualiter autem hoc unum accipi possit, manifestat consequenter. Manifestum est enim quod singulare sentitur *proprie* et per se, *sed tamen sensus est quodammodo etiam ipsius universalis*. Cognoscit enim Calliam non solum in quantum est Callias, sed etiam in quantum est hic homo, et similiter Socratem in quantum est hic homo'. [my italics].

[125] Ibid.: 'Si autem ita esset quod sensus apprehenderet solum id quod est particularitatis, et nullo modo cum hoc apprehenderet universalem naturam in particulari, non esset possibile quod ex apprehensione sensus causaretur in nobis cognitio universalis [. . .] Quia igitur universalium cognitionem accipimus ex singularibus, concludit manifestum esse quod necesse est prima universalia principia cognoscere per inductionem. Sic enim, scilicet per viam inductionis, sensus facit universale intus in anima, in quantum considerantur omnia singularia'.

intentions. More in line with Scotus and other late medieval authors, his conception of the memory can be called a *personal* or *episodic* memory.[126] What does this personal or episodic memory mean? As a definition in contemporary cognitive psychology has it, it '[. . .] is the memory of autobiographical events (times, places, associated emotions, and other contextual who, what, when, where, why knowledge) that can be explicitly stated or conjured. It is the collection of past personal experiences that occurred at a particular time and place'.[127] Memory marked in this personal way is essentially linked to the remembering self, her subjective sense of time and 'autonoetic' consciousness.[128] Suárez's concept of memory meets all these attributes. As stated in 3.7 and 5.1, although Suárez (unlike Scotus) does not distinguish between the species representing remote objects, namely external sensibles, and the species standing for proximate objects, that is, the external senses's acts, the memory is for him related to the past because it is essentially related to the *acts* of the external senses through which the sensibles have been apprehended. The character of pastness is thus fundamentally secured because the species of the internal sense retained in the memory are 'autonoetically' modified by the acts of the external senses through which the sensibles have been perceived. Due to this modification I can remember Peter because I remember having seen Peter.[129]

Suárez significantly departs from Aquinas in his claim that the memory in the proper sense is not a sensory memory but an *intellectual* memory. Only the latter can conserve species in the human post-mortem state, which is an important capacity in the diachronic identity between the embodied and the separated soul. In his juxtaposition of sensory and intellectual memory, Suárez specifies the difference between these two as follows:

> A sense can cognize a past thing in an exercised act (considering a thing that is gone) and not pastness as such, in the same way as it perceives a present thing in an exercised act. Considering time differences is a function of the intellect.[130] A sense apprehends the past materially, the intellect also formally since it knows the character (*ratio*) of time.[131]

---

[126]For this notion of personal memory as a wide-spread view in late-medieval philosophy, see Knuuttila and Sihvola (2014, 217–8). In fact, Aristotle might have been the first who noticed this feature of memory and who considered it to be a true kind of memory. At the outset of *De memoria et reminiscencia*, ch. 1, 449b22–3, 291, he says that '[. . .] when a man is exercising his memory he always says in his mind that he has heard, or felt, or thought this before'. For this interpretation of Aristotle, see Sorabji (1972, x–xi). We have seen above that also Aquinas inclines to this view, which he, however, does not explain sufficiently.

[127]For this definition and properties of episodic memory, see Fleming (2019, 134).

[128]Ibid., 135.

[129]For this, see also *CDA* disp. 8, q. 2, n. 5, t. 2, 50. Speaking about the sense memory Suárez also concedes that the phantasma or the sensible species stored in the memory often get damaged to such an extent that a recollecting person fails to remember many or even all the circumstances associated with the place, time, etc., of a recalled object (ibid.).

[130]*CDA* disp. 6, q. 5, n. 9, t. 2, 526: '[. . .] sensus potest cognoscere rem praeteritam in actu exercito, [considerando scilicet rem quae praeteriit], non praeteritionem ut sic; sicut etiam percipit rem praesentem [in actu exercito]. Considerare vero temporum differentiam illus est intellectus'.

[131]*CDA* disp. 9, q. 10, n. 5, t. 3, 266: '[. . .] sensus tantum cognoscit praeteritum materialiter, intellectus vero etiam formaliter, nam cognoscit ipsam rationem temporis; ergo'.

The theoretical space which Suárez achieves by affirming the existence of intellectual memory as the memory *proper* that can 'isolate' or prescind the form of time from its material subject is also the reason why he indicates that discursive operations attributed by Aquinas to sense memory alias reminiscence are acts of intellectual memory. Suárez is aware of Aristotle's ambiguous position in his *De memoria et reminiscencia* where on the one hand he claims that (discursive) reminiscence is a sense power pertaining to a sense faculty,[132] while on the other hand (at the beginning of the same treatise) he asserts that '[...] the quick-witted and those who learn easily are better at recollecting'.[133] In fact, the latter formulation seems to attribute recollection to the intellect after all. Despite his verbally conciliatory approach Suárez leaves no doubt which position is right: '[...] reminiscence comes to be from the intellect, at least as far as discourse and compositions are concerned [...]'[134] The interior sense is not present in the acts of reminiscence actively, that is, as the proximate subject, but only collaterally.[135] Since *in statu isto* the intellect can exercise its operations only in parallel with the operations of the interior sense, when the intellect recollects, the phantasy must in a parallel way attend to these operations of recollection as well. We have seen that when one cognitive power operates, the other one is active too. However, its tracking (*indagatio*) does not run through its proper discourse but only by means of a 'mechanical' succession of apprehensions that are essentially dependent on the prior and operative intellectual discourse. The operations of reminiscence can be subjected only in the intellect.[136]

The same relegation to the intellect can be observed in the case of the cognitive operations constitutive of experience. In his reply to the question regarding the formation of *experimentum* Suárez sets out by distinguishing two kinds of experience. The first is experience in a broad sense (*late sumpta*), the second is experience in the proper sense (*proprie sumpta*). The first kind of experience can be caused already by the external senses when, for example, touch experiences that fire generates heat. In *DM*, while commenting on the introductory passages of Aristotle's *Metaphysics*, Suárez notes that this kind of experience can arise even from a single perception of an individual, such as a hand's perception of a fire's heat. Nonrational animals also have this kind of experience since they have simple memories of singulars. Clearly, such experience based on a single case is often fallacious. This is why Aristotle introduces the second kind of experience called by Suárez *experimentum* in the proper sense. This kind of experience is based on the cognition

---

[132]Aristotle, *On Memory and Recollection* ch. 2, 453a22–3, 311: '[...] so the man who is employed in recollecting and search sets in motion a bodily part in which the affection resides'.

[133]Ibid., ch. 1, 449b7–8, 289.

[134]*CDA* disp. 8, q. 1, n. 12, t. 3, 24: '[...] reminiscentiam potissime fieri ab intellectu, saltem ad discursum et compositiones [...]'

[135]For this parallel and harmonizing connection of these two powers based on rootedness in the same soul, see *CDA* disp. 9, q. 7, nn. 7–8, t. 3, 206–8.

[136]Ibid., 24–6.

of an experimental effect that is acquired on the basis of a number of memories. However, such experience also requires the collation of singulars. It involves the comparison that this medicine cured Peter who suffered from this disease and that the *same* medicine healed also Paul who fell sick with the *same* illness. Without such correlation of, for example, medicines and diseases *experimentum* often turns out to be very unreliable. Suárez concludes that although experience in the proper sense, which is based on judicative knowledge, is initiated by the senses, it is perfected and realized only in and by the *intellect*. He adds two other reasons for this conclusion. First, in experimental propositions such as 'This medicine *heals'* and 'This fire *generates heat'* the predicates are universal concepts. However, the senses as material powers do not apprehend universals because they do not penetrate the quiddity of a thing. In order to cognize universals, one must 'penetrate' a thing to reach its quiddity. Such 'intus-legere' can be done only by the immaterial intellect which prescinds the universal nature from an individual difference. The senses cognize things only according to their external accidents, though. Second, as noted by Aristotle, experience in the proper sense, which is based on accumulated practical experience about individuals of a given class, generates sciences and arts in the intellect. But the intellect does not assent to a universal proposition based on the sensory cognition of singulars. It assents to it only by virtue of its own apprehension.[137]

[137] *CDA* disp. 5, q. 6, n. 18, t. 2, 434–6; *DM* disp. 1, s. 6, n. 23, t. 1, 59–60.

Chapter 6
Conclusion

6.1 Review of Suárez's Philosophy of Perception

The representative features of Suárez's philosophy of perception as investigated in Chaps. 2, 3, 4 and 5 of this book can be summarized as follows:

1. *Reification of the hylomorphic principles and emphasis on the soul's efficient causality.* Commenting on Aristotle's *On the Soul*, Suárez starts from two well-known definitions of the soul. While one, namely the soul as the first actuality of a natural and organic body potentially possessing life, expresses the soul's *formal* causality, the other, viz. the soul as the internal principle of all vital operations or that whereby we live, perceive and think, states the soul's *efficient* causality. A typical doctrinal feature of Suárez's approach, when compared to Aquinas, is his *reification* of the hylomorphic principles. Prime matter and substantial form make two incomplete substances or incomplete beings with their incomplete essences and existences, in the cases of prime matter and the rational soul complemented by partial subsistences called in the latter instance a 'semi-person'. Albeit a rational soul actualizing an organically structured body constitutes with it a substantial unity per se (the soul cannot be the *forma assistens* but only the *forma informans*), its ontological and epistemic status becomes more robust than in the theories of those who do not reify these principles. The soul turns out to be an independent agent *sui generis* whose causal efficacy and causal influence (despite being mediated by really distinct powers) in acts is *proximate*, and not merely remote. The soul is not entirely shrouded by the powers that as such can be moved only by an external mover which is already in act. In a way, it tends to be conceived more as a self-mover. Analogically, reified prime matter, which does not get its being from substantial form but has its own entitative act, becomes ontologically robust enough to become a subject, or the material cause of accidents, before the advent of a substantial form. With inherent accidents it constitutes an organic body that potentially has the substantial life mentioned by

D. Heider, *Aristotelian Subjectivism: Francisco Suárez's Philosophy of Perception*, Studies in the History of Philosophy of Mind 28,
https://doi.org/10.1007/978-3-030-67341-3_6

Aristotle in the first definition of the soul. From this point of view, it is not surprising that in *CDA*, disp. 7 Suárez devotes so much space to the issue of the organs of the particular sensory modalities.

With respect to perceptual and cognitive theory in general, the soul's substantial proximate causal influence in operations, including those of the sense powers, is manifested above all by *substantial attention*. This attention is substantial since it, in a way, precedes the power's operations. The soul of a blind man can intentionally stretch out its attention to visible objects even though no attention (intention) can be 'fulfilled' on his side. Without attending, without an attentive mind, no cognitive act can be produced, no matter how many stimuli the perceiver receives. Without attention no intentionality of perceptual acts is possible. Suárez also characterizes the proximate causal influence of the soul in acts manifested by this attention as 'actus vitalis'. This theory of the soul's substantial attention directly accounts for several significant features of Suárez's philosophy of perception, such as concomitant perceptual self-awareness and non-complex perceptual judgment. Although vital cognitive acts are characterized as absolute entities (*res*), sc. qualities, which seems to make them freely transferable and implantable by God to the powers of other cognitive subjects, Suárez rejects this possibility as contradictory to the status of vital acts. The soul's causal influence in vital acts cannot be substituted even by God. Peter cannot see a sensible object through a perceptual act transitively implanted to his sight by God. Even more, God cannot implant an act of beatific vision in Peter's intellect. While the concepts of vital and immanent act are not identical, the two constitute a crucial conceptual tandem in Suárez's cognitive theory. The interrelation between vitality and immanency substantially determines Suárez's active account of perception. A perceptual act cannot be caused from outside. It cannot be brought about by an external agent's transitive operation, whether coming from God or the sensible species. Indeed, not even a power can transitively cause an effect in a distinct power in the same supposit.

The emphasis on the soul's efficient causality is far from being a 'Deus-ex-machina' of Suárez's philosophical psychology. This emphasis is foreshadowed by his theory of the natural resulting of really distinct powers from the soul. Albeit this flowing is not an instance of vital operating, it is the soul, not an external progenitor, such as God or parent(s) that is the efficient cause of the resultant powers. The soul's causality in respect to the powers is also exclusive. The lower powers, or powers of the lower part of the soul, do not effectively emanate from the higher powers, or from powers of the higher part of the soul. All equally flow directly from the soul. It cannot be said that the internal senses as a lower power are ontologically derived from the intellect as from a higher power. Accordingly, it cannot be said that the interior sense *participates in* the operations of the higher power and through this participation is 'elevated' in its operation above its material level. All powers effectively come from the soul which is a crucial coordinator in the activities of the distinct faculties.

2. *The Unicity View; the indivisibility of the rational soul and of the souls of perfect nonrational animals; their holenmeric character*. Suárez is a staunch advocate of

the Unicity View. Accordingly, there is and can be only a single substantial form in a whole substantial composite. He rejects all versions of the Plurality View according to which material substances are composed of two or more really distinct substantial forms, whether these are only souls or also inanimate forms, such as the (Scotistic) form of corporeity. Adopting a theory of a plurality of forms would be detrimental to the substantial unity of material composites and the central position of the soul. The main recurrent argument for the existence of the single coordinating soul is based on abductive reasoning from observable introspective phenomena. This argument occurs in various *loci* of *CDA* and in other treatises too. In order to explain the phenomena related to the mutual interaction and inferences between the distinct powers' acts there must be a single soul co-ordinating all these activities which (with its limited attentional reservoir) accounts also for their interferences. If Peter's attentional capacity is exhausted and consumed by a long-lasting intellectual work, there is little left for his sensomotoric vigilance or the optimal execution of vegetative acts.[1]

Endorsing the Unicity View, however, does not imply espousing the view according to which the lower souls (the vegetative and sentient souls) are included *eminently* in the higher (rational) soul. Implicitly referring to the Scotistic *distinctio formalis* Suárez says that the lower souls are contained in the rational soul *formally*. The main reason for rejecting eminent inclusion, consistent with the overall thrust of Suárez's approach, is that eminent inclusion of the lower part of the soul in the higher soul would entail an objectionable 'elevation' of the operations of the lower power. In the case of the internal senses this would lead to their inappropriate 'lift' through participation in the intellect's operations. They would then be able to exercise a number of proto-intellectual operations exceeding the level of material sensorial operations.

The same abductive argument is applied (and evaluated as *the* best argument) also to his reasoning for the *indivisibility* of both the rational and the sentient soul of perfect nonrational animals. If the soul were extended and divisible and if one part of the soul were in a foot and another in the brain, the part in the foot could not apprehend and 'obey' what is commanded by the other part in the brain. The indivisibility of the soul also has to do with its holenmeric character. Although in distinct bodily parts and organs the souls can exercise only some specific operations, which is primarily due to the souls' different modes of informing the body, the soul exists *as a whole* in all animated parts of the body. It is not too far-fetched to say that the indivisibility and holenmeric character of the souls, in a way, ushers in Suárez's

[1]This idea about the limited attentional reservoir of the soul is extensively utilized by Suárez in his treatment on spiritual extasis and contemplation which causes the suspension of the external senses's operations. For this see Francisco Suárez (1859), *Opera omnia*, vol. 14, *Opus de virtute et statu religionis*, treatise 4: *De oratione, devotione, et horis canonicis*, book 2: *De oratione mentali ac devotione*, chpts. 14–7, 186–97. No doubt, this fascinating topic waits for a future research.

view according to which the organ of touch lies in all animated parts of the body and prepares the way for the manifest interoceptual tendencies in his account of touch.

3. *Reification of Powers*. The view of the soul's powers conceived as distinct *res* is correlative to the theory of the proximately efficacious soul in the acts. Or, more precisely, the existence of the single soul coordinating the activities of the distinct powers can be conceived as a systematic consequence of Suárez's reification of the powers conceived as distinct modules within which the particular acts (and habits) are produced according to the principles governing efficient causality. Without the existence of the single soul in which all the powers are rooted no coordination of the powers' operations would be possible. All these acts are in mutual consonance indirectly through the mediation of this single soul. The powers as categorial qualities are *res* produced by the soul's special efficient causal action. They are not effects of an external progenitor's activity. Although, nomologically speaking, these powers-*res* flow necessarily from the soul and there can never be a 'naked' soul, at the supernatural level (metaphysically speaking) their emanation can be blocked by divine intervention. Accordingly, there must be two distinct causal actions terminating in two distinct effects that are two distinct *res* together forming an ontological aggregate. One is an action of the progenitor terminating in the production of the soul, the other is an action of the soul terminating in the production of powers.

In Chap. 3, devoted to a general analysis of the sense powers, whether methodologically abstracting from both the external and the internal senses or only from the external senses, seven features of Suárez's approach can be extracted from the above-mentioned analyses.

1. *Necessary existence of the sensible species and their 'vestigial' character*. Suárez's philosophy of perception is grounded in the assimilative view of cognition. All kinds of human cognition in all 'epistemic situations' up to the cognition of the separated soul require the existence of intentional likeness, with the only exception of the beatific vision (and even in this kind of cognition the existence of intelligible species *is* possible). Contrary to late medieval and post-Suarezian Jesuit critique of the theory of the sensible species, Suárez regards the sensible species as necessary for determining the sense powers and union with the sensible objects. The multiplication of these similitudes or representations of sensibles is a necessary feature in the first phase of the complex three-stage perceptual process. This complex cognitive process indicates that perception is not just about the reception of species in the powers. Although sensible species are distinct in kind from the objects they represent, they are material, divisible and their representing is only *virtual*. Formal and distinct representation occurs only at the level of perceptual acts. Employing a biological metaphor, the species is compared to a seed in which no figure, structure and colour of (say) a tree can be discerned. Its representation is obscure. It can become distinct only when the seed-species germinates, namely through the power's 'articulation'. The sensible species is a subtle material entity pertaining to the accident of quality. Its 'spirituality' does

not imply literal immateriality but only a material subtlety. As such the species is only a trace of the represented object. Unlike Cajetan, Suárez does not think that it is possible to distinguish two distinct aspects in the species, namely the entitative and the representational aspect. One cannot say that only the latter is relevant for perception in the way that its reception amounts to perception. Both aspects of the species are identical. As being identical, the sensible species can only *inhere* in the power. No intentional literal identification through a reception of an allegedly distinct representational aspect of the species is possible. The species can constitute only an accidental aggregate with its power. Being of vestigial character the species of the external senses is produced only by the intentional action of the sensible objects. No agent sense in the form of a separated intelligence, God or medium is necessary and possible to spiritualize these species. The naturalistic (in this specific sense) account of their origin is the only correct explanation of their production.

2. *Perceptual act as a distinct unit (quality).* The imperfect and obscure character of the sensible species implies that a power's reception of the species cannot be identified with a perceptual act. A perceptual act cannot be reduced to the vital reception of the species in the power. It is a non-eliminable and irreducible item in Suárez's metaphysics of the mind. Perceptual and cognitive acts make up robust entities that cannot be explained away by means of identifying them with modes, whether of the soul or of the powers. An act of perception is a special quality-*res* received in the power and intentionally re-presenting an object. This quality is really distinct from the power insofar as this is informed by the sensible species.
3. *Moderately active theory of perception.* The vestigial character of the species and the robust concept of perceptual act show that after the reception of the species other co-causes of the production of perceptual acts have to be involved. In line with his rejection of transitive operating in the realm of cognition (the only transitive operation is that of the object's emission of the species toward the external senses) and in line with the Augustinian axiom about the impossibility of ascendant causality according to which a lower entity (e.g. a nonvital or material one) cannot cause a more perfect (vital or immaterial) item, the presence of the material sensible species in the overall mechanism of the causes of the perceptual process seems to be a compromising factor after all. The species can neither be the only cause of the perceptual act, nor can it be the only or total instrumental cause. It can be only a partial instrumental cause which together with the second instrumental co-cause, sc. with the power, constitutes the total instrumental cause in the production of the perceptual act. Symptomatically, this total instrumental (accidental) cause composed of the species and the vital power can be operative only if directed by the principal (substantival) cause, sc. the soul. Only the soul together with this total instrumental cause can produce a perceptual act (the second stage of the perceptual process), which is then received in the same power (the third stage of the perceptual process).
4. *Productive character of all kinds of cognition including the external senses; non-representational notions of 'idolum' and objective being.* Suárez's analysis of the terminus of a cognitive act can be regarded as a consistent application of his

metaphysics of efficient causality to the problem of the cognitive endpoint, or the expressed species. The distinction between the cognitive act as a way to its terminus and the cognitive act as a quality is conceived by virtue of the distinction between the ontological concepts of *res* and *modus*. Suárez's view of the expressed species and many other topics related to his perceptual theory show how determining the ontology of modes is for Suárez's philosophy of perception. Accordingly, already *CDA* ushers in a great importance of the theory of modes which becomes evident in *DM*. While the terminus, namely the perceptual act *qua* quality, is a *res*, the perceptual act in becoming is a *mode* that 'parasitizes' on this *res*. In line with his understanding of the criterion of the modal distinction, while a mode cannot exist without its thing, a thing can exist without this or that mode. The modal conception of the category of action, the central predicament for Suárez's view of the efficient cause, is an ontological underpinning which grounds that a given endpoint can be produced by a distinct action – a particular mode makes a *res* into a separable item. Typically, such production of the expressed species is an inherent feature in all kinds of cognition including the external senses (even in the beatific vision, which is an instance of vital operation too, the mental word is and must be produced). The terminus is produced in both abstractive and intuitive cognition. Beside the above-mentioned assimilative feature of cognition, apprehension is also characterized by this inevitable expressive and articulatory trait.

The necessary production of the *idolum* in the external senses and thus a certain possible approximation to representationalism at the level of experiential cognition of the external senses which is supposed to be in touch with the extramental reality lead Suárez to his long reasoning for a view that can be called the doctrine of a non-representational concept of the *idolum*. The *idolum*, which can be understood as a quasi-formal concept at the level of the senses, is not that in which (*id in quo*) the sensibles are apprehended. As for all early Jesuits, it is only that through which (*id quo*) they are cognized. However, not only the character of the quasi-formal concept in perception but also Suárez's characterization of objective being (*esse objective*) by means of spiritual and intentional termination, supplemented by his eliminative statement about the non-production of other entities other than acts *qua* qualities, shows that his view of the quasi-objective concept in perception can be read reductively, that is, as the perceptual content of perceptual acts *qua* qualities. This objective being, or this being cognized, does not necessarily constitute a *tertium quid* with a diminished being between the perceptual acts and the extramental sensibles, but it corresponds to the aspect under which a thing is presented and appears to the perceiver.

5. *Scope of attention and Early Theory of Selection*. The theory of the substantial attention of the soul, the presence of the powers in the overall make-up of the instrumental cause of perceptual acts and the statement about the efficient causality of the virtually representing species point to Suárez's emphasis on the notion of attention. A cognitive power is not fully determined by the species in such a way that it can divide its attention only in respect to the parts of the

cognizable whole represented by the species. Selective attention, not being an all-or-nothing phenomenon, can be divided and distributed to different objects so that one and the same power can simultaneously produce two distinct acts representing two distinct sensible objects which are perceived in the manner of many objects. Such split perception in the manner of many objects involves perception through distinct sensible likenesses of objects that even need not be related to each other. Sight can produce two simultaneous visions of black and white colour through the reception of two distinct (contrary) species.

6. *Same-order perceptual self-awareness and non-complex perceptual judgment 'imbibed' in simple apprehension.* Since all operations of the external senses are vital acts, as said, all the sensorial acts they produce are necessarily accompanied by same-order self-awareness or, more precisely, by an awareness of one's own acts through which the external sensibles are apprehended. Albeit the intentional acts of the external senses tend 'teleologically' to their proper objects, that is, to extramental sensible qualities, we are aware not only of the external sensibles but simultaneously also of these acts. Our attention can be conceived as divided already at this level between the extramental sensibles and the acts. The concomitant consciousness of our acts cannot be doubted and makes our internal experience reliable. This kind of awareness is essential also for memorative acts. It would not be possible to have memorative acts of the past, if these acts were not somehow perceived while exercised.

   Since the simple apprehensions of the external senses are vital acts, they must be also judicial acts. By seeing white and black colour sight *ipso facto* judges that white colour is not black colour. Within the domain of its own proper sensible objects it can generate the non-complex perceptual judgments assimilated in the simple apprehensions of the external senses. Only due to these judgments can we speak about non-complex truth and falsity in the case of the external senses' acts. And only due to this judgment can the external senses' apprehension be assessed as inchoatively true or false or as giving an occasion to the intellect's true or false complex judgment in the proper sense, which proceeds through composing and dividing a predicate and a subject.

7. *Epistemological (sceptical) issue: natural and supernatural context.* The formulation of the sceptical *status quaestionis*, according to which we often perceive objects differently from the way they are, or perceive something that at the moment of perception does not exist at all, is divided into two main contexts. One is the context of the external senses' natural illusory perceptions, the other is a thought experiment related to a hypothetical divine deceptive intervention. In both contexts the crucial factor of the explanation of these two kinds of deception is the object, conceived as the motive principle of the cognitive act and as a termination of the perceptual act. Starting from the classical Aristotelian triple division of sensible objects into proper, common and incidental sensible objects, and leaving aside incidental sensibles in which error occurs more often than in other cases, Suárez formulates the properly Aristotelian claim that at the natural level an error of sense is more likely in the common sensibles than in the proper

sensibles. On the other hand, he also admits that if the sensibles are considered 'in particulari', that is, according to their kinds and sub-kinds, an error is possible with respect to both the common and the proper sensibles. We can go wrong not only about the size or distance of objects but also about their colours.

Admitting the efficient causality of species, the possibility of a divine substitution of all the secondary natural causes and of a spiritual intentional termination of all cognitive acts including those of the external senses according to which an existent extramental object is not necessary, it is not surprising to find Suárez conceding to the metaphysical possibility of the external senses' abstractive cognition of nonexistent (or absent) sensible objects. By inserting the sensible species of objects that do not represent objects existing in the presence of Peter God can cause an abstractive cognition of nonexistent objects in Peter's eyes. However, not to succumb to the temptation of scepticism which became later the commonplace in early modern philosophy, Suárez is quick to say that Peter will have an epistemic defence through which he will be able to discern the abstractive (and thus deceptive) character of his visual act. Either he will intellectually introspect this act more thoroughly, or his intellect will consider the coherence of this act and its judgment with other judgments based on normal (veridical) perceptual acts.

Moreover, not only that God can insert a sensible species of a nonexistent sensible object in Peter's eyes. He can also create an absolute quality materially identical with a vital act of sensation in Peter's sight without his soul's causal concurrence. This is a metaphysically possible state of affairs. As said, the action through which this quality-*res* is produced is a separable and variable mode. There is not only a single way of how Peter's quality corresponding to a vital act of sensation could have been caused. It could have been created otherwise, sc. by divine interposition. However, contrary to the impressed species, Suárez denies that such implementation would entail the production of Peter's vision. He introduces a crucial distinction between an act conceived as a mere quality and an act framed as an operation of seeing, intellection etc., in the proper sense. He makes it clear that all perfect vital acts, that is, cognitive and appetitive acts *qua* vital operations, are necessarily related to and caused by the internal principle. They cannot be caused transitively by a distinct subject, even the divine one. Importantly, the mere reception of an act in a power is not enough to explain the phenomenon of 'ownness' that is inherent to vital acts. If Peter's seeing were not intrinsically caused through his soul, his vision with its subjective phenomenology would remain unexplained.

In Chap. 4 I focused on Suárez's scientific views (especially physiology, anatomy and optics) which are, nevertheless, always considered in a philosophical context. These views are often associated with what I have called the first stage of the perceptual process. This stage involves the reception of species in the power through the medium of the organ. This reception assumes the emission of species from sensibles and their propagation *in medio*. Such multiplication implies the existence of the sensible qualities represented by the species. All the questions related to the ontological status of the sensible qualities, their subject, medium, the kinds of media of the propagation of the sensibles and their species, and the anatomophysiological questions about the sense organs were topics of Chap. 4.

Before approaching these topics, I started by presenting Suárez's view of the individuation and ordering of the senses. Following the world-based criterion of individuation, supplemented by other back-up criteria, such as the criterion based on the diversity of organs and especially that based on the manner of the powers' immutation by the sensibles, Suárez distinguishes five (classical) external senses. These five senses are first of all ordered according to their spirituality and perfection (sight, hearing, smell, taste and touch, in descending order), which is also the *ordo doctrinae* followed by Suárez in *CDA*.

In Suárez's treatment of the particular external senses, the following five doctrinal traits can be underlined:

1. *Non-reductionist account of affective qualities*. Contrary to the mainstream of early modern classical philosophy, for Suárez the proper sensibles of all external senses pertain to the category of affective qualities. They can be reduced neither to substances nor to the common sensibles. Every particular sense modality is specified by the existence of its own proper sensible. Even if these sensibles did not exist, the powers would be transcendentally related to them. Moreover, the ontological status of the proper sensibles cannot be explicated relationally. Although colours need to be illuminated to become proximately visible, they are fully colours also at night, that is, independently of percipients. Light (*lumen*) is also a real entity, sc. a quality inhering in a transparent medium that cannot be reduced to a mere manifestation of colours or to a sensible species of luminous objects. In his recurrent argument for the extramental existence of affective qualities Suárez often employs a distinction between 'to be *formally X'* and 'to be *subjectively X'*. Although odours require the existence of a smoky exhalation or vapour in which they exist, they are not formally identical with it. Although sound requires the existence of air ripples since it can exist only in them, it is not identical with them. The same holds also for sapours in their relation to tactile qualities.
2. *Non-eliminability of the sensible species in all external senses*. In all sensory modalities pertinent sensible species must be posited. Only the reception of species is relevant for the further cognitive process. Perhaps also a theological reason according to which resurrected bodies will sense but will not 'suffer' natural changes in their organs leads Suárez's view of the possibility of disjoining the propagation of the sensible qualities and the species in all the external senses. The following kinds of arguments for the conclusion about the necessity of the species and the dissociability of the multiplication of the species and the propagation of the sensible qualities are iterative: (1) Localizability of sounds/voices and odours – the objects must emit species *ab initio* in order to be perceptually localizable; (2) perception of distant objects – subtler entities, such as species, can attain more remote places than coarser ones, such as smoky evaporation or air undulation; (3) the impossibility of sensible qualities propagation in some media in which perception can be observed – fish can hear deep underwater; (4) the

imperceptibility of qualities inherent in the organs themselves – we tactually perceive external qualities and not qualities inherent in our bodies.

Speaking about vision, apart from the reception of species in an attentive power (intromission account) no other egression of visual rays or spirits is necessary (against extramission theory). Although Suárez embraces one important aspect of extramission theory, namely active substantial attention, which (as it was in Olivi) can be explicated as a virtual stretch to the visibles, his account cannot be interpreted in the materialist manner of extramission theory. It is absurd to 'localize' perception to the object or somewhere to a point of the medium at which the visual rays or spirits arrive. What is commonly considered to be the strong side of extramission theory, namely the perception of size and distance, is explained by employing the ontology of modes. The sensible species representing colours and light are modified through this or that size and this or that distance of the object seen.

3. *Dispensability of medium.* Following Aristotle, Suárez endorses the existence of medium as an important transmitter of qualities and their likenesses. A visual and auditive experience in a vacuum would be impossible. In the case of colours, the existence of a (transparent) medium is necessary. If a coloured object were not illuminated, no vision of it would be possible. However, and this is a non-Aristotelian feature of Suárez's philosophy of perception, in all other cases the contiguity of the sensible qualities with the organ is *not* a decisive impediment for producing the pertinent perceptual acts. Although intentional and natural action in the distal senses of sight, hearing and smell can be disassociated so that only species representing qualities can attain the organs, the propagation of qualities up to the organs does not prevent the powers from perceiving. In the case of touch the medium's dispensability can be seen clearly. The organ of touch is not the heart but the external animated parts of the body, such as skin, flesh and nerves. The flesh is not an internal medium but the adequate organ of touch. Medium ceases to be a necessary spiritualizing device 'elevating' the species from natural being to intentional being and making them spiritual so that a mere reception of them would be sufficient to produce perceptual acts. Sensible qualities together with species representing them can be contiguous to the organs and yet perceptual acts can be elicited.
4. *The problem of interoception; dolorogenic quality as the object of touch; Suárez's ambivalent position.* Influenced not only by the broadly conceived Aristotelian tradition but also by the medical tradition, Suárez seems to be caught in an ambivalent position. On the one hand, he endorses the Aristotelian position according to which our external sensation is essentially exteroceptual. Each sense has its own external sensible quality constituting its proper object. Accordingly, bodily pain cannot be a tactile object but only an emotion or an act of the sensory appetite that includes other conceptual (subjective) signs such as active bitterness and active avoidance of the harmful object. In his explicit list of the objects of touch in both *CDA* and his *GC* no interoceptual quality is mentioned. However, in line with the medical tradition he admits a special dolorogenic quality as part of the total object of touch that cannot be reduced to other tactile qualities. While

this latter view seems to be, on the one hand, more consistent with his emphasis on the indivisibility and the holenmerical character of the soul which through the sense of touch registers all events occurring in the body, on the other hand it goes against his explicit statement about the imperceptibility of qualities inherent in the percipient's body.

5. *Material 'correlation' of organs and objects*. The organs and through them the powers must be mutable by objects and their species. An organ must be accommodated to the respective power since the power is its 'final cause'. From this point of view objects require consonance with powers, that is, they must be able to affect them. If a part of the object were a part of the organ, the organ and the power would be prevented from such affection. According to this 'logic of mutability', the organ of sight, sc. the pupil, must be deprived of colours. It must be translucent, soft and watery. The inner ear, the organ of hearing, must be free of the motion of air. The organ of touch must be set in a medial disposition to be able to detect the first qualities. The tongue, the organ of taste, must be deprived of sapours. This logic of proportionality between objects and organs gives Suárez a main tool for his rejection of the view advocated by some representatives of the medical tradition according to which the organs of the senses should be placed either in the optic chiasma (sight) or in the frontal part of the brain (smell).

In Chap. 5 I aimed to show how our perceptual experience is unified. Our senses are endowed with the capacity to discriminate between various kinds of sensibles. Since our perception, followed by the appetitive and the locomotive power, is an important instrument for self-preservation, it must be largely 'practical'. We perceive the sensible qualities and objects in terms of their convenience or inconvenience. We perceive the sensible objects not only as being present but also as absent. These aspects constituting a part of our perceptual experience should be 'reflected' in a crucial representative device, namely the sensible species of the internal senses. Suárez's theory of the origin of the *species phantastica* is regulated by this idea of perceptual unity. The sensible species of the interior sense are more perfect than those of the external senses since, contrary to the latter, they can also be principles of abstractive cognition. Following his mindset about intra-power encapsulation, Suárez says that rather than directly caused by an external sense through its operation, it is more probable that this species is produced by the soul through the interior sense. An act of an external sense serves at most as an exemplary cause for the soul's production of the interior sense's species. In this case the soul through the internal sense can be called 'the agent sense'. However, it can be so only with an important qualification that mirrors Suárez's distrust regarding transitive causality in his metaphysics of cognition. The soul through the interior sense will be both the agent and the passive sense. A real distinction between these two senses is neither necessary nor intelligible.

The practical 'intentions' of convenience and inconvenience or amity and enmity are not distinct objects that would have to be represented by sensible species entirely distinct from the likenesses representing the sensible aspects of the objects. These 'non-sensed' aspects are included either as concomitant features in the operations of the external senses or, which is more probable, as being added through the soul by means of the interior sense. In both cases these intentions are conceived as *modifications* of the sensible species representing sensible objects.

Apart from the specific origin of the sensible species of the interior senses, two main features of Suárez's theory of the internal senses can be singled out from Chap. 5:

1. *Reduction of the number of internal senses (Unicity View).* Since Avicenna three main criteria of the individuation or specification of the internal senses had been adopted by numerous scholastics. One is based on the existence of distinct objects, namely those that are perceived by the external senses and those that are not (intentions); the second criterion was based on the distinction between reception (intuitive cognition) and retention (abstractive cognition); the third was based on the distinction between active and passive powers. For Suárez, all these criteria are insufficient for positing a multiplicity of internal powers. Internal senses are not to be multiplied if one power can do more than another power. Since a power generating abstractive cognition can produce an act of intuitive cognition, it is not necessary to distinguish two interior senses, one of which apprehends intuitively (the common sense and the estimative or cogitative power) and another cognizing abstractively (the phantasy and the memory). Since sensible qualities and practical intentions do not constitute objects that are so distinct that their cognition should be attributed to two distinct powers, no real distinction between the common sense and the estimative power on the one hand and between the phantasy and the memory on the other hand is necessary either. The argument for a plurality of internal senses based on the distinction between the plural brain cavities with their distinct material dispositions (the moist part as good for reception and the dry part as suitable for retention) is outdated since only a single cavity, namely the back one called cerebellum, is the adequate organ of internal sense apprehension. The upshot of these reductions is the existence of a single internal sense, called phantasy by Suárez, which is in charge of a variety of functions and operations. It is only possible to speak about a plurality of interior senses if one keeps in mind that this distinction is purely conceptual, that is, based on the *fundamenta* of the distinct acts exercised by this power.
2. *Reduction of the cognitive scope of the interior senses.* The reduction in the number of internal senses is accompanied by a reduction in the number of internal cognitive functions. Suárez does not keep all the cognitive functions of all the internal senses advocated by supporters of the Plurality View. Concerning perceptual self-awareness and the perception of the common sensibles, the common sense, or, more precisely, the interior sense *qua* the common sense, does not bring an additional interpretative value to the external senses' apprehension. The common sensibles are represented through modes of the species representing the proper sensibles. As such they are represented also by species of the external sense. This is why the interior sense apprehends them in the same way. The internal sense does not bring a distinct cognition of incidental sensible objects

either. This sense is not able to go 'under' the external accidents conceived as *res* and to reveal a hidden substance. A distinct cognition with an additional interpretative value can be realized only by the intellect, which is the only power with the capacity of 'intus-legere'. Only the intellect can discursively detect a persistent subject underneath the accidents-*res*. *Mutatis mutandis* the same holds for the cognition of the soul, the powers, the habits and the intentional species themselves.

Following the fundamental claim about the non-elevation of the interior sense through participation in the intellect, Suárez also refuses to attribute proto-intellectual operations to the internal sense. It is not true that the interior sense *qua* the cogitative power engages the individual intentions representing substances in complex judicial, inferential and discursive operations. Since the primary objects of the intellect are material and sensible singulars, there is no need to posit a power, such as the cogitative faculty, that is in charge of these functions. Compared to Aquinas for whom the intellect is only concerned with material quiddities, in this regard Suárez, in a way, blurs the distinction between the intellect and the senses and conceives the former, to a certain degree, as a perceptual power too. In line with his refusal to attribute these operations to the *vis cogitativa* he consistently refuses to attribute a kind of freedom to the sensory appetite. He strictly denies the theory of two or more wills housed in a single person. Only the intellect and the rational appetite, sc. the will, are responsible for these operations. Suárez assigns the cognitive and volitional functions Aquinas had attributed to the cogitative power and the sensory appetite only to the intellect and the will. Thus, unlike Aquinas, he attributes no virtues and no venial sin to the sensory appetite.

Mutatis mutandis the same reduction holds also for the sensory memory, endowed by Aquinas with the proto-intellectual operation of recollection. Sensory memory is not only about spontaneous memories or associations. In the case of human memorative acts it is about seeking syllogistically from the known to unknown. Since for Suárez memory is above all the intellectual memory that can exist even after death, he concedes such recollecting only to the intellect. Accordingly, an *experimentum* in the proper sense (the cognition of an experiential effect acquired based on a plurality of memories detecting 'one in the many') can be allocated only to the intellect.

6.2 Elements of Modal Ontology in Suárez's Theory of Perception

As I hope to have shown, Suárez's account of sense perception is a highly systematized body of propositions grounded in anthropological, metaphysical, scientific and theological principles, of which the last are above all related to the voluntarist axiom that God can produce a *res* in separation from any other really distinct *res*. Like his metaphysics, his philosophy of perception captivates by a 'holenmeric' ability to see the whole in the whole and in every part of the whole. It evinces Suárez's great capacity to see principles through their conclusions. In fact, this is in line with the overall style of *CDA*, which is a systematical treatise rather than a commentary. Like many other Jesuits of the first generations who aimed to defend an 'opinio communis', Suárez's primary goal was not to introduce original and provocative ideas. Instead, his approach can be characterized as a creative and highly systematical development of the past philosophical and theological traditions. Having said this, however, I believe that there *is* a systematical element in Suárez's theory that constitutes a doctrinal innovation. This theoretical element is his ontology of modes. Although his theory of modes, as any other metaphysical topic, is at length elaborated at various places of his *DM*,[2] especially in disp. 7 titled 'De variis distinctionum generibus' and in disp. 32–53 on the categories, it is not difficult to trace its presence in *CDA* too. Despite the fact that this theory is not *ex professo* introduced in the early text, there are many references to the concept of mode throughout *CDA*, including the segment concerning his perceptual theory. His use of the notion of mode and of the correlative term of *res* shows that Suárez often conceives mode in this text in *the* proper sense, and in the same way as in the later *DM*: modes are real and positive features, which in comparison with *res* are classified as lesser entities. They must be introduced to supplement the imperfect character of finite beings. Due to creatures' imperfection, sc. their dependence, limitation, mutability and compositionality, a necessary metaphysical complement, modally distinct from the thing, must be assumed.[3] A crucial sign of this modal distinction, which is a non-mutual or asymmetrical separability,[4] can be observed in *CDA* too. In consonance with this criterion, it holds that while a subject-*res* can be separated from this or that numerical mode – I intentionally speak about a particular mode to include also those kinds of modes of which some particular instance is necessary for a thing – by divine power at least, a mode cannot be detached from its subject-*res* even by divine power. Accordingly, modes in the proper sense are negatively contrasted in a twofold way, namely as distinct from *res* and as distinct from mere *rationes*.

As most scholars agree, Suárez's ontology of modes is a new accomplishment.[5] The novelty of this theory is related not only to Suárez's metaphysics of categories

[2]For the examples of modes, see *DM* disp. 7, s. 1, n. 18, t. 1, 256.

[3]*Ibid.*, s. 1, n. 19, 256.

[4]*Ibid.*, s. 2, n. 6, 263.

[5]For this claim, see Pasnau (2011a, 253). His ontology of modes brings something new especially if compared with Fonseca's theory of modes, an important historical precedent of Suárez's theory. Menn (1997, esp. 242–50) explains that '[. . .] Suárez goes beyond Fonseca, and far beyond earlier thinkers, in the way he develops the theory of modes' (*ibid.*, 242). Although referring to Fonseca's tripartite division of modes with the last genus identified with the mode 'properly and by a special title' as an imperfect entity which can exist only as attaching to and modifying its subject (for this division, see Fonseca, *In Metaphysicorum Aristotelis Stagiritae Libros*, vol. 2: *In quintum librum Metaphysicorum Aristotelis*, ch. 6, q. 6, s. 2, 400–1), Suárez rejects almost all of his examples of modes from these categories. Whiteness, indicated by Fonseca as an example of a mode of the first

but it applies also to his philosophical psychology in general and perceptual (cognitive) theory in particular.[6] Although such historiographical evaluation would require a detailed comparative analysis of Suárez's theory with the theories of sense perception of other Jesuit and non-Jesuit scholastic contemporaries (an important goal for further research), on the basis of my current knowledge I dare say that there is no parallel systematic and wide-ranging application of the notion of modes in their cognitive theories.

These applications can be briefly summarized as follows:

1. Form (soul), matter (body), power, habit, sensible species and perceptual (cognitive) act *qua* quality are conceived as *res* that are actually related to their complements either through a substantial mode of union or through an accidental mode of inherence.
2. While total subsistence (personality) is grasped as a mode of the individual (rational) nature, the partial subsistences of the rational soul and prime matter are conceptualized as partial modes.
3. While perceptual acts *qua* qualities are framed as *res*, cognitive acts *qua* becoming (*fieri* or *via*) through which the former are produced are taken as separable and variable modes.
4. The non-sensed intentions, such as those of enmity or amity, are represented as the variable *modi* of the (indeterminate) sensed species standing for the qualities-*res* that are sensed by the external senses, while in a wolf's perception of another wolf this modification is that of amity, in the case of a sheep's apprehension it is that of enmity.
5. The common sensibles, including distance, are represented through modes of the sensible species standing for the proper sensible qualities-*res*, although it holds that the sensible species representing a proper sensible such as colour is necessarily modified by a mode representing continuous quantity (size), this or that

genus, is a *res*; a determination contracting the notion of being to the highest genera, qualified by Fonseca as a mode of the second genus, is only rationally distinct from it; a determination contracting *ens* to its 'being necessary' or 'being contingent', introduced by Fonseca as an example of the third genus, is only rationally distinct from it too. For an exposition of Fonseca's division and Suárez's critique, see *DM* disp. 7, s. 1, n. 19, t. 1, 257. Unlike Fonseca, other Jesuit contemporaries did not even recognize this modal distinction and considered the division of distinction into real and conceptual to be entirely exhaustive. See, e.g. Francisco de Toledo, *Commentaria in universam Aristotelis logicam, De praedicamentis in communi*, q. 3, 193. For this reference, see also Pasnau (2011a, 256, note 14).

[6] This statement about Suárez's innovation does not mean that he is an originator of the ontology of modes as such. For example, a sophisticated theory of the so-called *modi intrinseci* (among others, those of finiteness and infiniteness determining being as such), conceived by the Jesuit as mere *rationes* only conceptually distinct from a determined item such as being, with its wide application in metaphysics, was an integral part of Scotistic metaphysics from the early fourteenth century onwards. For the historical context of the Scotist doctrine of intrinsic modes and its distinction to the *modi extrinseci* (widespread in later Jesuit philosophy) with a special focus on Mastrius, see Andersen (2016, 405–26).

(numerically) particular mode referring to this or that common sensible is a variable item.

6. The external senses' operations are modified by a concomitant self-awareness; their accompanying self-awareness can be conceptualized through the notion of *modus*, or rather it can be squarely reduced to the mode which is identified with the perceptual act as the *via* to its endpoint, of which the main and proximate principle is the soul; again, it seems that this mode with respect to this or that degree of our vacillating attention is a variable item.
7. Accordingly, the interior sense's awareness of the lower external senses' acts is conceived through the *modi* of the former's species conceived as *res*.

6.3 Historical Significance of Suárez's Theory of Perception: The Emergence of Aristotelian Subjectivism

No matter how systematic and well arranged Suárez's theory of perception is, it is a historical fact that the philosophical courses including that on Aristotle's *De anima* with its theory of perception, delivered by the Jesuit in the first half of the 1570s in Segovia, the only place in his whole academic career where he taught 'pure philosophy', were not favourably received by his Jesuit superiors. Our author was charged with novelties at both the methodological and the doctrinal level. He was threatened with removal from his *cátedra* and his academical career was in danger. The Jesuit Provincial of Castille Juan Suárez (no kinship to Suárez), who received many letters of complaints about his novel teaching, had to deal with a new situation. He had to have the texts of Suárez's lectures examined by several experts. Unfortunately, their expertise did not end up well for the young philosopher. The Provincial invited him and asked for an explanation. Fortunately, Suárez was able to refute all their objections that had been raised against him and reassured the Provincial about his philosophical and theological orthodoxy. As Scorraille has pointed out, we do not know which doctrines were the targets of their criticism.[7] Nevertheless, based on the analysis in this book, I find it most likely that as far as the doctrinal level is concerned these critiques, which no doubt came from philosophers and theologians that were well versed in Aristotle and Thomism,[8] were focused on what I call the 'Aristotelian subjectivism' of Suárez's philosophical psychology in general and his theory of sense perception in particular.

As is notorious among scholars of intellectual history, Renaissance Aristotelians were far from constituting a single and monolithic school, as suggested by the

[7]Scorraille (1911, vol. 1, 139–42).

[8]Presumably, these evaluators were literally guided by St. Ignatius's recommendations (Sanctus Ignatius de Loyola, 1938, *Monumenta ignatiana, Constitutiones Societatis Jesu*, 3:150–1): 'In Theologia legetur vetus et novum Testamentum, et doctrina scholastica divi Thomae [...] In Logica, et Philosophia naturali, et morali, et Metaphysica, doctrina Aristotelis sequenda est'.

rhetoric of early modern anti-Aristotelians.[9] Renaissance Aristotelianism is much rather an umbrella term encompassing a wide spectrum of diverse approaches and interpretations of the Stagirite. This diversity came about mainly by two interconnected factors, one of which is internal to the Aristotelian philosophy of this period, whereas the second is external. As for the first, for many sixteenth-century authors including Suárez it gradually became evident that the *corpus Aristotelicum* does *not* constitute a coherent whole (see 4.5.4 regarding the tactile organ, 3.7 concerning the common sense's role in perceptual consciousness, 5.2 related to the number of internal senses, as well as many other issues discussed in this book). This textual incoherence as such would surely suffice for competing exegeses of the ancient philosopher's texts to emerge. The second factor bears upon the new cultural conditions. Many pages have been written about the influence of humanism on peripatetic and scholastic philosophy. The humanists' return *ad fontes*, stimulating new Greek and Latin editions of Aristotle's complete works, including the Venice edition of Aristotle's complete works with Averroes's comprehensive commentary from 1550–52, translations of the Stagirite's texts to vernacular languages, such as Italian, Spanish, English, German and French, and the recovery and diffusion of Greek commentaries on Aristotle (Simplicius, Philoponus, Themistius, Alexander of Aphrodisias and others),[10] significantly contributed to the deep doctrinal and methodological diversification within Aristotelian philosophy in the time span from the fifteenth to the early seventeenth century. Another fundamental historical factor was of course the advent of the Reformation, and the emergence of the radicalized secular and naturalist Aristotelianism that was cultivated especially in Padua.[11] In line with this new cultural-ideological climate, it is not surprising to see that the various forms of Renaissance Aristotelianism have been classified exclusively according to historiographical and purely denominational criteria. This approach has led many historians of ideas to speak about the secular, the Catholic, the Protestant, the Alexandrist, the Averroist, the Neoplatonic, the Thomistic, the Scotistic and the Ockhamist strands of Renaissance Peripateticism.[12] Although I agree that Suárez's eclectic Aristotelianism (eclectic here taken in a non-pejorative sense) was influenced by many of these historical transformations of Aristotle, there is, in my opinion, no more poignant characterization of Suárez's philosophy of perception than 'Aristotelian subjectivism', that is, Aristotelianism modified by an emphasis on the subject. However, this characteristic needs some qualification.

---

[9]For a wide range of different forms of Renaissance Aristotelianism, see Copenhaver and Schmitt (1992, esp. 60–76); Schmitt (1973, 1983) and Salatowsky (2006, 20–8); *ibid.*, 277: 'Der Renaissance-Aristotelismus war kein monolitischer Block [...]'

[10]For these editions and translations, see Schmitt (1983, 121–5).

[11]For this secular and naturalist Aristotelianism, see Kristeller (1964, 74).

[12]In the second half of the seventeenth century the Jesuit Honoré Fabri (1608–1688) specifies Aristotelianism even 'numerically'. He says that while scholastics have their own Aristotle depending on this or that school leader, Fabri has his own (Blum, 2012, 214). Sortally speaking, Blum considers Fabri's Aristotelianism to be 'one of Cartesian imprint' (Blum, 2012, 209).

While, as we have seen above in 3.7 and elsewhere, there is anything but unanimity among the scholars on Aristotle's and Aquinas's views of self-awareness, selfhood, ownness, subjectivity, self-reference, first-person agency and cognitive activity/passivity issue,[13] there is and, I think, there can be no such contention about Suárez's position. The soul is aware or, to put it in the first person, I am implicitly and indirectly aware of all occurrent cognitive (and appetitive) operations. Moreover, I am aware that these acts are *mine*. This immediate (imperfect, though) experiential awareness of the self that amounts to the underlying subject of all my mental states epitomizes a pivotal trait of Suárez's *scientia de anima* which in its thematic focus is centred on man.[14] By experiencing a perceptual act related to an extramental sensible object I somehow experience myself in this act too. Since sense perception is active and since it has to necessarily attend to sensibles I apprehend not only what is extramental but also my activity and myself. Despite an anthropological hylomorphic account Suárez, due to his specific reified version of this theory, can much better than other Aristotelians conceive this perception as being 'coloured' by the 'ownness'. He can better refer sensory acts to the subject that can be exposed in the first person as a self since the soul called a semi-person comes to be for Suárez an agent of its kind. This semi-person generates a proximate efficient causality in all its vital operations. Due to its substantial attention the soul happens to be also 'an agent sense' which produces the crucial principle of the internal sense apprehension, namely its sensible species. At the same time, the soul is the main principle that ensures the unity of perceptual experience. It is a substantial form that co-ordinates all activities that take place in the really distinct modules or powers of the soul and the composite. True, as most of scholastics, Suárez has little to say about phenomenal consciousness. However, his metaphysics of the soul, active concept of cognition, views related to the bodily self-awareness including pain perception and his view of objective being based on the aspectual and perspectival perception of sensibles, among others, can be considered as opening a *theoretical space* for a further investigation of human subjectivity. As is well known, this space or this not yet filled 'slot' was to be later completed by early modern non-scholastic philosophers and even more and later by phenomenologists. Suárez's sensibility for individuality and subjectivity palpable in his theory of sense perception makes him a modern author rather than a Thomist Aristotelian.[15]

---

[13]As regards a variety of interpretations of Aquinas's self-knowledge, see Marenbon (2019, 268–73).

[14]By now it should be clear that the adjectives 'subjectivism' and 'subjective' employed in my evaluation do *not* signify the view according to which sensible objects, such as colours and sounds, have (only) mind-dependent (or in the contemporary sense: subjective) reality. For Suárez these proper sensibles exist *independently* of our knowledge. Suárez's subjective transformation of Aristotle systematically supplements his object-oriented philosophy with the no less important viewpoint related to the knowledge of one's own subjective (and bodily) state. For Suárez this knowledge constitutes the second and intergal part of our general awareness.

[15]In this respect, Sven Knebel's view (formulated in line with the historiographical 'Continuity Thesis') of Suárez's virtual reflection and its relation to Descartes's *cogito* (for which the virtual

Beside the influence of Augustine's and medieval Augustinian theories – the latter's influence being largely indirect – and also of the Jesuit spirituality exemplified by St. Ignatius's *Spiritual Exercises*,[16] Suárez's subjectivist twist seems to be motivated by his reception of medical tradition too. Leaving aside his critical remarks against this tradition concerning above all its naturalism, Suárez not only often accepts the latest physiological and anatomical findings but is not hesitant to criticize Aristotle's old-fashioned scientific views either. In the context of the topic of bodily self-awareness and interoception he clearly draws his inspiration from this medical lore, even though, as we have seen, he really struggles with it within the traditional framework given by exteroceptual Aristotelianism.

As I have suggested in Chap. 1 of this study, every philosophy of perception is expected to answer not only the enquiry into how percipients become aware of their own mental states (the subjective aspect) but also how these states are related to the objects of the extramental world (the objective aspect). Philosophy of perception is supposed to explain the intentionality of our cognitive operations too. From what has been said in this book, it may be concluded that Suárez's philosophy of perception is not unilaterally subject-centred, and this is what makes it at the same time an '*Aristotelian* subjectivism'. Although the active external senses also produce the internal representations called expressed species Suárez seeks to expound their nature as anti-representationalist and object-centred. Contrary to what can be found in medical tradition, Suárez shows that the sensibles' causal influence upon the organs and powers cannot be 'naturalized', that is, it cannot be reduced to a mere natural (physical) action or affection. As we would say today, the issue of intentionality cannot be studied solely by natural sciences. Although he 'naturalizes' the impressed sensibles species since he refuses the literal spiritualist interpretation of these items, Suárez makes it clear that there is a 'specific difference' between the intentional and the natural actions and their effects. They are distinct since there can be one without another, namely an intentional action *without* a natural action. Moreover, these species are not directly known. They constitute only a partial instrumental cause that co-produces a perceptual operation. Their systematic

reflection seems to be essential) is telling: According to standard historiography, 'the Cogito must unconditionally mark a historical turn. But it does not' ('Deswegen muß das Cogito unbedingt historisch einen Wendepunkt markieren. Das tut es aber nicht', Knebel 2016, 87). In connection with Antoine Arnauld's scholastic interpretation of Descartes's *cogito* based on the notion of the virtual reflection, Schmal (2020, 732) concludes in a similar way: 'If the proposal that Arnauld managed to bring out the complexity of this account on the basis of his scholastic expertise [one based on the notion of the virtual reflection; D.H.] is acceptable, a final conclusion suggests itself. The novelty of the Cartesian approach did not consist so much in a groundbreaking new theory as to how the nature of the *cogitatio* is to be construed, as in a new role assigned to it in the "project of pure enquiry"'.

[16]St. Ignatius's *Spiritual exercises* with its turn to subjectivity and with the emphasis on the the practitioner's control over his or her spiritual growth could have well motivated Suárez's subjective twist which he gave to his Aristotelianism too. For St. Ignatius of Loyola's *Spiritual Exercises* and their relation to modern introspectional subjectivity, see Sluhovsky (2013). Of course, the analysis of this motivation would require a different study from the one of this volume.

significance has a direct counterpart in extramental reality too. Contrary to the early modern philosophical mainstream, affective sensible qualities which are the proper sensible objects of the external senses are the *real* and *mind-independent* qualities. They are objects to which our external senses are essentially related. They are not produced as late as in the subject and they are not reducible to the common sensibles. Suárez's philosophy of perception successfully finds a new and dynamic balance between criteria related to the subjective and the objective aspects, which is the main task to be expected from every philosophy of perception.

Bibliography

Primary Sources

Antonius a Matre Dei. (1628). *In octo libros Physicorum Aristotelis*. Paris: Apud M. Sonnium.

Aquinas, T. (1886). *Commentaria in libros Aristotelis De caelo et mundo. De generatione et corruptione et Meteorologicorum*. In *Opera omnia* (Vol. 3). Rome: Ed. Leonina.

Aquinas, T. (1888). *Pars Prima Summae Theologiae. A quaestione I ad quaestionem XLIX*. In *Opera omnia* (Vol. 4). Rome: Ed. Leonina.

Aquinas, T. (1889). *Pars Prima Summae Theologiae. A quaestione L ad quaestionem CXIX*. In *Opera omnia* (Vol. 5). Rome: Ed. Leonina.

Aquinas, T. (1891). *Prima Secundae Summae Theologiae. A quaestione I ad quaestionem LXX*. In *Opera omnia* (Vol. 6). Rome: Ed. Leonina.

Aquinas, T. (1892). *Prima Secundae Summae Theologiae. A quaestione LXXI ad quaestionem CXIV*. In *Opera omnia* (Vol. 7). Rome: Ed. Leonina.

Aquinas, T. (1918). *Summa contra Gentiles. Liber 1–2*. In *Opera omnia* (Vol. 13). Rome: Ed. Leonina.

Aquinas, T. (1926). *Summa contra Gentiles. Liber 3*. In *Opera omnia* (Vol. 14). Rome: Ed. Leonina.

Aquinas, T. (1970). *Quaestiones disputatae de veritate*. Volumen II. Fasc. 1. qq. 8–12. In *Opera omnia* (Vol. 22, 2/1). Rome: Ed. Leonina.

Aquinas, T. (1984). *Sentencia libri De anima*. In *Opera omnia* (Vol. 45, 1). Rome/Paris: Ed. Leonina/Vrin.

Aquinas, T. (1985). *Sentencia libri De sensu et sensato. De memoria et reminiscencia*. In *Opera omnia* (Vol. 45, 2). Rome/Paris: Ed. Leonina/Vrin.

Aquinas, T. (2010). *Treatise on Human Nature. The Complete Text (Summa Theologiae I, Questions 75–102)* (A. J. Freddoso, Trans.). South Bend, IN: St. Augustine's Press.

Aquinas, T. (2012). *The Power of God* (R. J. Regan, Trans.). Oxford, UK: Oxford University Press.

Aquinas, T. (2018). *New English Translation of St. Thomas Aquinas's Summa Theologiae (Summa Theologica)* (A.J. Freddoso, Trans.). URL = https://www3.nd.edu/~afreddos/summa-translation/TOC.htm

Aquinas, T. *Quodlibeta*. URL = http://www.corpusthomisticum.org/q03.html

Aquinas, T. *Quaestio disputata de spiritualibus creaturis*. URL = https://www.corpusthomisticum.org/qds.html

D. Heider, *Aristotelian Subjectivism: Francisco Suárez's Philosophy of Perception*, Studies in the History of Philosophy of Mind 28,
https://doi.org/10.1007/978-3-030-67341-3

Aquinas, T. *Quaestiones disputatae de potentia. A quaestione VIII ad quaestionem X*. URL = http://www.corpusthomisticum.org/qdp8.html
Aquinas, T. *Scriptum super Sententiis. Liber III a distinctione XXIII ad distinctionem XXV*. URL = http://www.corpusthomisticum.org/snp3023.html
Aquinas, T. *Sententia libri Metaphysicae*. URL = https://www.corpusthomisticum.org/cmp00.html
Aquinas, T. *Expositio libri Posteriorum Analyticorum*. URL = https://www.corpusthomisticum.org/cpa1.html
Aristotle. (1927). *The Works of Aristotle* (W. D. Ross, Ed.). Oxford, UK: Oxford University Press.
Aristotle. (1929). *Physics*. Books 1–4 (P. H. Wicksteed, F. M. Cornford, Trans.). Cambridge, MA/London: Harvard University Press.
Aristotle. (1934). *Physics*. Books 5–8 (P. H. Wicksteed, F. M. Cornford, Trans.). Cambridge, MA/London: Harvard University Press.
Aristotle. (1952). *Meteorologica* (H. D. P. Lee, Trans.). Cambridge, MA/London: Harvard University Press.
Aristotle. (1960). *Posterior Analytics. Topica* (E. S. Forster, Trans.). Cambridge, MA/London: Harvard University Press.
Aristotle. (1961a). *Parts of Animals. Movement of Animals. Progression of Animals* (E. S. Forster, Trans.). Cambridge, MA/London: Harvard University Press.
Aristotle. (1961b). *Generation of Animals* (A. L. Peck, Trans.). Cambridge, MA/London: Harvard University Press.
Aristotle. (1965a). *Problems* II, XXII–XXXVIII. *Rhetorica ad Alexandrum*. (W. S. Hett, H. Rackham, Trans.). Cambridge, MA/London: Harvard University Press.
Aristotle. (1965b). *History of Animals*. Books I–III (A. L. Peck, Trans.). Cambridge, MA/London: Harvard University Press.
Aristotle. (1970). *History of Animals*. Books IV–VI (A. L. Peck, Trans.). Cambridge, MA/London: Harvard University Press.
Aristotle. (1984). *The Complete Works of Aristotle* (J. Barnes, Ed.). The Revised Oxford Translation, Vol. 2, Bollingen Series 71.2. Princeton, NJ: Princeton University Press.
Aristotle. (1991). *History of Animals*. Books VII–X (D. M. Balme, Trans.). Cambridge, MA/London: Harvard University Press.
Aristotle. (1993). *Metaphysics*. Books 1–9 (H. Tredennick, Trans.). Cambridge, MA/London: Harvard University Press.
Aristotle. (2000). *On the Soul. Parva Naturalia. On Breath* (W. S. Hett, Trans.). Cambridge, MA/London: Harvard University Press.
Aristotle. (2002). *Categories. On Interpretation. Prior Analytics* (H. P. Cooke, H. Tredennick, Trans.). Cambridge, MA/London: Harvard University Press.
Augustine. (1978). *De moribus ecclesiae catholicae*. In J. K. Coyle (Ed.), *Augustine's "De moribus ecclesiae catholicae." A Study of the Work, its Composition and its Sources*. Fribourg, Switzerland: The University Press.
Augustine. (1991). *The Trinity* (De trinitate), 2nd ed. (E. Hill, Trans.). The Works of St Augustine I/5. Hyde Park, NY: New City Press.
Augustine. (1950). *The Greatness of the Soul (De quantitate animae). The Teacher (De magistro)* (J. M. Colleran, Trans.). New York/Mahwah, NJ: The Newman Press.
Augustine. De Genesi ad Litteram libri duodecim. In *S. Aurelii Augustini Opera omnia*. Patrologiae latinae elenchus (Vol. 34). URL = http://www.augustinus.it/latino/
Aureolus, P. (1952). *Scriptum super primum Sententiarum*. Vol. I: Prologue – Distinction I (E. M. Buytaert, Ed.). St. Bonaventure, NY: The Franciscan Institute.
Averroes. (1953). *Commentarium magnum in Aristotelis De anima libros* (F. S. Crawford, Ed.). Cambridge, MA: The Mediaeval Academy of America.
Aversa, R. (1650). *Philosophia metaphysicam physicamque complectens in duos tomos distribute* (Vol. 2). Bologna, Italy: Ex Typographia Iacobi Montii.
Báñez, D. (1614). *Scholastica commentaria in primam partem Angelici Doctoris S. Thomae: a Quaestione LXV vsque ad CXIX & vultimam* (Vol. 2). Douai, France: Apud Petrum Borremannum.

Boethius, S. (1847). *Liber de Persona et Duabus Naturis Contra Eutychen et Nestorium* (H. F. Stewart, Ed.). *Patrologia latina* (1337–1354) (J. P. Migne, Ed., Vol. 64). Paris.

Bayle, P. (1740). *Dictionnaire historique et critique* (4 Vols, 5th ed.). Amsterdam/Leyde/La Haye/ Utrecht, The Netherlands: P. Brunel.

Cajetan, T. de Vio. (1939). *Scripta philosophica, Commentaria in De anima Aristotelis*. Rome: Institutum "Angelicum".

Chatton, G. (1989). *Reportatio et Lectura super Sententias*. Collatio ad librum primum et prologus (J. C. Wey, Ed.). Toronto, Canada: Pontifical Institute of Mediaeval Studies.

Collegium Conimbricensis. (2006). *Commentarii in tres libros de Anima*. Hildesheim, Germany: Olms Verlag (reprint of Cologne edition: Impensis Lazari Zetzneri, 1609).

D'Abano, P. (1985). *Conciliator controversiarum quae inter philosophos et medicos versantur* (E. Riondato, L. Olivieri, Eds.). Padua, Italy: Editrice antenore (reprint of Venice edition: Apud Iuntas, 1565).

Dandini, G. (1611). *De corpore animato lib. VII*. Paris: C. Chappelet.

Descartes, René. (1902). *Discours de la méthode & Essais. Oeuvres de Descartes* (Ch. Adam, P. Tannery, Eds., Vol. 6). Paris: Léopold Cerf.

Dionysios the Areopagite. (1920). *On the Divine Names and the Mystical Theology* (C. E. Rolt, Trans.). Grand Rapids, MI: Christian Classics Ethereal Library.

Durandi de Sancto Porciano. (2012). *Scriptum super IV libros Sententiarum. Buch II, dd. 1–5* (F. Retucci, Ed.). Leuven, Belgium/Paris/Walpole, MA: Peeters.

Eustachius a S. Paulo. (1609). *Tertia pars Summae Philosophicae, Quae est Physica*, Tomus posterior. Paris: C. Chastellain.

Fonseca, P. da. (1615). *In quintum librum Metaphysicorum Aristotelis* (Vol. 2). Cologne, Germany: Lazarus Zetzner.

Galen, C. (1528). *De usu partium corporis humani: magna cura ad exemplaris Graeci veritatem castigatum, universo hominum generi apprime necessarium*. Paris: Ex officina Simonis Colinaei.

Galen, C. (2009). In B. Gundert (Ed.), *De symptomatum differentiis*. Berlin, Germany: Walter de Gruyter.

Hurtado de Mendoza, P. (1624). *Universa philosophia, De anima*. Lyon, France: Ludovici Prost.

Ioannes a S. Thoma. (2008a). *Cursus philosophicus Thomisticus*. Vol. II: Naturalis philosophiae pars I et III. Hildesheim, Germany: Olms Verlag (reprint of Rome edition, 1929).

Ioannes a S. Thoma. (2008b). *Cursus philosophicus Thomisticus*. Vol. III: Naturalis philosophiae pars IV et Indices. Hildesheim, Germany: Olms Verlag (reprint of Rome edition, 1929).

James, W. (1890). *The Principles of Psychology* (Vol. 1). Boston, MA: Henry Holt and Co..

Janduno, I. de. (1552). *Super libros Aristotelis De Anima quaestiones subtilissimae*. Venice, Italy: Apud Hieronymum Scotum.

Mastrius, B. (1727). *Philosophiae ad mentem Scoti cursus integer* (Vol. 4), *In libros metaphysicorum*, Pars prior. Venice, Italy: Ed. Pezzana.

Mastrius, B., Belluto, B. (1727). *Philosophiae ad mentem Scoti cursus integer* (Vol. 3), *Libros de anima, Libros de generatione & corruptione, Libros de coelo & metheoris*. Venice, Italy: Ed. Pezzana.

Mayronis, F. de. (1520). *In primum Sententiarum (Conflatus)*. Venice.

Nifo, A. (1517). De sensu agente. In *In librum Destructio Destructionum Averrois commentationes*, 124–129. Venice.

Ockham, G. de. (1967). *Scriptum in librum primum Sententiarum (Ordinatio)*, Prologus et Distinctio prima. Opera theologica 1 (G. Gál & S. F. Brown, Eds.). St. Bonaventure, NY: St. Bonaventure University.

Ockham, G. de. (1980). *Quodlibeta septem*. Opera theologica 9 (J. C. Wey, Ed.). St. Bonaventure, NY: St. Bonaventure University.

Ockham, G. de. (1982). *Quaestiones in librum tertium Sententiarum (Reportatio)*. Opera Theologica 6 (F. E. Kelley, G. I. Etzkorn, Eds.). St. Bonaventure, NY: St. Bonaventure University.

Olivi, Petrus Iohannis (1924). *Quaestiones in secundum librum Sententiarum* (Vol. 2, qq. 49–71, B. Jansen S.I., Ed.). Ad claras aquas (Quarachi): Ex typographia Collegii S. Bonaventurae.

Olivi, Petrus Iohannis (1926). *Quaestiones in secundum librum Sententiarum* (Vol. 3, qq. 72–118, B. Jansen, S.I. Ed.). Ad claras aquas (Quarachi): Ex typographia Collegii S. Bonaventurae.

Oviedo, F. de. (1663). *Cursus philosophiae, Vol. 2. Metaphysica. De anima.* Lyons, France: L. Borde, L. Arnaud, P. Borde, G. Barbier.

Philoponus, J. (2005). *On Aristotle's On the Soul 2.1–6* (W. Charlton, Trans.). Ithaca, NY: Cornell University Press.

Plato. (1997). *Complete Works* (J. M. Cooper, Ed.). Indianapolis, IN/Cambridge, UK: Hackett Publishing Company.

Proclus. (1963). *The Elements of Theology* (E. R. Dodds, Trans.). Oxford, UK: The Oxford Clarendon Press.

Sanctus Ignatius de Loyola. (1938). *Monumenta Ignatiana ex autographis vel ex antiquioribus exemplis collecta*, Series tertia: *Constitutiones Societatis Iesu*, Tomus tertius: Textus latinus. Rome: Borgo S. Spirito.

Scotus, J. D. (1894a). Reportata Parisiensia. Liber primus, Dist. 1–48. Liber secundus, Dist. 1–11. In *Opera omnia* (Vol. 22). Paris: L. Vivès.

Scotus, J. D. (1894b). Reportata Parisiensia, Liber secundus, Dist. 12–44. Liber tertius, Dist. 1–35. Liber quartus, Dist. 1–6. In *Opera omnia* (Vol. 23). Paris: L. Vivès.

Scotus, J. D. (1895). Quaestiones Quodlibetales. A quaestione prima usque ad tertiam decimam. In *Opera omnia* (Vol. 25). Paris: L. Vivès.

Scotus, J. D. (1954). Ordinatio I, Dist. 3. In *Opera omnia* (Vol. 3). Civitas Vaticana, Italy: Typis Vaticanis.

Scotus, J. D. (1963). Ordinatio I, Dist. 26–48. In *Opera omnia* (Vol. 6). Civitas Vaticana, Italy: Typis Vaticanis.

Scotus, J. D. (1973). Ordinatio II, Dist. 1–3. In *Opera omnia* (Vol. 7). Civitas Vaticana, Italy: Typis Vaticanis.

Scotus, J. D. (2001). Ordinatio II, Dist. 4–44. In *Opera omnia* (Vol. 8). Civitas Vaticana, Italy: Typis Vaticanis.

Scotus, J. D. (2010). Ordinatio IV, Dist. 8–13. In *Opera omnia* (Vol. 12). Civitas Vaticana, Italy: Typis Vaticanis.

Simplicius. (1544). *Commentaria in quatuor libros de coelo Aristotelis*. Venice, Italy: Hieronymus Scotus.

Soncinas, P. (1579). *Quaestiones metaphysicales acutissimae*. Lyons, France: Carolus Pesnot.

Suárez, F. (1856a). *Opera omnia (Vol. 1): De deo uno et trino*. Paris: L. Vivès.

Suárez, F. (1856b). *Opera omnia (Vol. 2): De angelis*. Paris: L. Vivès.

Suárez, F. (1856c). *Opera omnia (Vol. 3): De opere sex dierum. Tractatus De anima*. Paris: L. Vivès.

Suárez, F. (1856d). *Opera omnia (Vol. 4): De fine hominis. De voluntario et involuntario. De bonitate et malitia humanorum actuum. De actibus qui vocantur passiones tum etiam de habitibus, praesertim studiosis, ac vitiosis. De vitiis et peccatis*. Paris: L. Vivès.

Suárez, F. (1859). *Opera omnia (Vol. 14): Opus de virtute et statu religionis*. Paris: L. Vivès.

Suárez, F. (1861a). *Opera omnia (Vol. 25): Disputationes metaphysicae. Vol. 1*. Paris: L. Vivès.

Suárez, F. (1861b). *Opera omnia (Vol. 26): Disputationes metaphysicae. Vol. 2*. Paris: L. Vivès.

Suárez, F. (1978, 1981, 1991). *Commentaria una cum quaestionibus in libros Aristotelis De anima* (3 Vols, S. Castellote, Ed.). Madrid: Sociedad de estudios y publicaciones.

Suárez, F. (1994). *On Efficient Causality. Metaphysical Disputations 17, 18, and 19* (A. J. Freddoso, Trans.). New Haven, CT: Yale University Press.

Suárez, F. (2000). *On the Formal Cause of Substance. Metaphysical Disputation XV* (J. Kronen, J. Reedy, Trans.). Milwaukee, WI: Marquette University Press.

Suárez, F. (2002). *On Creation, Conservation, and Concurrence. Metaphysical Disputations 20–22* (A. J. Freddoso, Trans.). South Bend, IN: St. Augustine Press.

Suárez, F. (2012). *Selections from De Anima. On the Nature of the Soul in General–On the Immateriality and Immortality of the Rational Soul* (J. Kronen, J. Reedy, Trans.). Munich, Germany: Philosophia.

Suárez, F. (2014). *De actibus qui vocantur passiones. Sobre os atos que se chaman paixões. De los actos que se llaman pasiones*. Porto, Portugal: Edições Afrontamento.

Suárez, F. (2017). *Thesaurus doctrinae circa libros Aristotelis De Generatione et corruptione* (S. Castellote, Ed.). URL = http://www.catedraldevalencia.es/castellote/degetc2.pdf

Suárez, F. (forthcoming 2021). *Expositio et Quaestiones in quinque priores primae secundae divi Thomae Aquinatis quaestiones materiam de beatitudine continentes* (J. A. Senent De Frutos, Ed.).

Themistius. (1533). *Paraphrasis Themistii peripatetici acutissimi, In Aristotelis Posteriora, & Physica*. Basel, Switzerland: Apud Ioannes Valderum.

Toledo, F. de. (1985). *Opera omnia philosophica. Introductio in universam Aristotelis logicam. Commentaria in universam Aristotelis logicam. Commentaria una cum quaestionibus in tres libros Aristotelis de Anima*. Hildesheim, Gremany: Georg Olms Verlag (reprint of Cologne edition: Sumptibus Hermanni Mylii, 1615–6).

Vallés, F. (1582). *Controversiarum medicarum et philosophicarum libri decem*. Frankfurt am Main, Germany: Andreus Wechelus.

Vesalius, A. (1543). *De humani corporis fabrica*. Basel, Switzerland: Ex officina Ioannis Oporini.

Secondary Sources

Abercrombie, N. (1938). *Saint Augustine and French Classical Thought*. New York: Russell & Russell.

Abrams, M. H. (1953). *The Mirror and the Lamp: Romantic Theory and the Critical Tradition*. Oxford, UK: Oxford University Press.

Aho, T. (2009). The Status of Psychology as Understood by Sixteenth-Century Scholastics. In S. Heinämaa, M. Reuter (Eds.), *Psychology and Philosophy. Inquiries into the Soul from Late Scholasticism to Contemporary Thought* (pp. 47–66). Dordrecht: Springer.

Åkerlund, E. (2015). Material Causality–Dissolving a Paradox: The Actuality of Prime Matter in Suárez. In J. L. Fink (Ed.), *Suárez on Aristotelian Causality* (pp. 43–64). Leiden, The Netherlands/Boston, MA: Brill.

Alejandro, J. M. (1948). *La gnoseología del Doctor Eximio y la acusación nominalista*. Santander, Spain: Universidad Pontificia Comillas.

Amaral, P. (1987–1988). Harmony in Descartes and the Medical Philosophers. *Philosophy Research Archives, 13*, 499–556.

Andersen, C. A. (2016). *Metaphysik im Barockscotismus. Untersuchungen zum Metaphysikwerk des Bartholomaeus Mastrius. Mit Dokumentation der Metaphysik in der scotistischen Tradition ca. 1620–1750*. Amsterdam, The Netherlands/Philadelphia, PA: John Benjamins Publishing Company.

Anfray, J.-P. (2017). Suárez sur la causalité de l'agent et la notion de cause libre. In J.-B. Guillon (Ed.), *Le libre arbitre: Perspectives contemporaines* (pp. 29–56). Paris: Collège de France.

Armstrong, D. M. (1981). What Is Consciousness? In D. Armstrong (Ed.), *The Nature of Mind and Other Essays* (pp. 55–67). Ithaca, NY: Cornell University Press.

Aureolus, P. (1956). *Scriptum super primum Sententiarum*. Vol. II: Distinctions II-VIII (E. M. Buytaert, Ed.). St. Bonaventure, NY: The Franciscan Institute.

Avicenna Latinus (1972). *Liber De Anima seu Sextus De Naturalibus*, I-II-III (S. van Riet, Ed.). Louvain, Belgium: E. Peeters, Leiden, The Netherlands: Brill.

Aydede, M. (2019). Pain. In *The Stanford Encyclopedia of Philosophy* (Spring 2019 Edition), E. N. Zalta (Ed.), URL = <https://plato.stanford.edu/archives/spr2019/entries/pain/>.

Bakker, P. J. J. M. (2012). Petrus Trapolinus on the Nature and Place of Psychology. In P. J. J. Bakker, S. W. de Boer, C. Leijenhorst (Eds.), *Psychology and the Other Disciplines. A Case of Cross-Disciplinary Interaction (1250–1750)* (pp. 11–59). Leiden, The Netherlands/ Boston, MA: Brill.

Black, D. L. (2000). Imagination and Estimation: Arabic Paradigms and Western Transformations. *Topoi, 19*, 59–75.

Blum, P. R. (2012). *Studies on Early Modern Aristotelianism*. Leiden, The Netherlands/Boston, MA: Brill.

Brower-Toland, S. (2014). Ockham on the Scope and Limits of Consciousness. *Vivarium, 52(3–4)*, 197–219.

Buchenau, S., Lo Presti, R. (2017). *Human & Animal Cognition in Early Modern Philosophy & Medicine*. Pittsburgh, PA: University of Pittsburgh Press.

Burillo, S. F. (1992). Introduccion a la teoria del conocimiento de Francisco Suárez. *Pensamiento. Revista de Investigación e Información filosófica, 48(190)*, 211–230.

Burnett, C., Fend, M., Gouk, P. (1991). *The Second Sense. Studies in Hearing and Musical Judgment from Antiquity to the Seventeenth Century*. London: Warburg Institute, University of London.

Burnett, C. (1991). Sound and its Perception in the Middle Ages. In C. Burnett, M. Fend, P. Gouk (Eds.), *The Second Sense. Studies in Hearing and Musical Judgment from Antiquity to the Seventeenth Century* (pp. 43–69). London: Warburg Institute, University of London.

Burnyeat, M. (2001). Aquinas on 'Spiritual Change in Perception'. In D. Perler (Ed.), *Ancient and Medieval Theories of Intentionality* (pp. 129–153). Leiden, The Netherlands/Boston, MA: Brill.

Carraud, V. (2002). *Causa sive ratio. La raison de la cause, de Suarez à Leibniz*. Paris: Presses Universitaires de France–PUF.

Casati, R., Dokic, J., Di Bona, E. (2020). Sounds. In *The Stanford Encyclopedia of Philosophy* (Winter 2020 Edition), E. N. Zalta (Ed.), URL = <https://plato.stanford.edu/archives/win2020/entries/sounds/>.

Casini, L. (2006). *Cognitive and Moral Psychology in Renaissance Philosophy. A Study of Juan Luis Vives' 'De anima et vita'*. Uppsala, Sweden: Universitetstryukeriet.

Castellote, S. (1962). *Die Anthropologie des Suarez. Beiträge zur spanischen Anthropologie des XVI. und XVII. Jahrhunderts*. Freiburg, Germany/Breisgau, Germany: Alber Verlag Freiburg.

Chabada, M. (2005). *Cognitio intuitiva et abstractiva: Die ontologischen Implikationen der Erkenntnislehre des Johannes Duns Scotus mit Gegenüberstellung zu Aristoteles und I. Kant*. Mönchengladbach, Germany: B. Kühlen.

Copenhaver, B. P., Schmitt, C. B. (1992). *Renaissance Philosophy*. A History of Western Philosophy: 3. Oxford, UK/New York: Oxford University Press.

Cory, T. S. (2014). *Aquinas on Human Self-Knowledge*. Cambridge, UK: Cambridge University Press.

Cory, T. S. (2015). Attention, Intentionality, and Mind-Reading in Aquinas's 'De Malo 16.8'. In M. V. Dougherty (Ed.), *Aquinas's Disputed Questions on Evil* (pp. 164–191). Cambridge, UK: Cambridge University Press.

Cory, T. S. (2021). Consciousness. In R. Cross and J. T. Paasch (Eds.), *The Routledge Companion to Medieval Philosophy* (pp. 249–262). New York: Routledge.

Cronin, T. S. (1966a). *Objective Being in Descartes and in Suarez*. Rome: Gregorian University Press.

Cronin, T. S. (1966b). Objective Reality of Ideas in Human Thought: Descartes and Suarez. In V. F. Daues, M. R. Holloway, L. Sweeney (Eds.), *Wisdom in Depth. Essays in Honor of Henri Renard, S.J.* (pp. 68–79). Milwaukee, WI: Bruce Publishing Company.

Cross, R. (2009). The Mental Word in Duns Scotus and Some of His Contemporaries. In T. Shimizu, C. Burnett (Eds.), *The Word in Medieval Logic, Theology and Psychology*. (Acts of the XIIIth International Colloquium of the Société Internationale pour l'Étude de la Philosophie Médiévale, Kyoto, 27 September–1 October 2005) (pp. 291–332). Turnhout, Belgium: Brepols.

Cross, R. (2014). *Duns Scotus's Theory of Cognition*. Oxford, UK: Oxford University Press.

Courtine, J.-F. (1990). *Suarez et le système de la métaphysique*. Paris: Épiméthé, Presses Universitaires de France.

De Boer, S. W. (2017). The (Human) Soul. In H. Lagerlund, B. Hill (Eds.), *The Routledge Companion to Sixteenth Century Philosophy* (pp. 411–435). New York/London: Routledge.

De Haan, D. D. (2014). Perception and the Vis Cogitativa: A Thomistic Analysis of Aspectual, Actional, and Affectional Percepts. *American Catholic Philosophical Quarterly, 88(3)*, 397–437.

De Vignemont, F., Massin, O. (2015). Touch. In M. Matthen (Ed.), *The Oxford Handbook of Philosophy of Perception* (pp. 294–313). Oxford: Oxford University Press.

Deely, J. (1994). *New Beginnings. Early Modern Philosophy and Postmodern Thought*. Toronto, Canada: University of Toronto Press.

Des Chene, D. (1996). *Physiologia. Natural Philosophy in Late Aristotelian and Cartesian Thought*. Ithaca, NY/London: Cornell University Press.

Des Chene, D. (2000). *Life's Form. Late Aristotelian Conceptions of the Soul*. Ithaca, NY/London: Cornell University Press.

Des Chene, D. (2012). Suárez on Propinquity and the Efficient Cause. In B. Hill, H. Lagerlund (Eds.), *The Philosophy of Francisco Suárez* (pp. 89–100). Oxford, UK: Oxford University Press.

Doyle, J. P. (1967). Suarez on the Reality of Possibles. *The Modern Schoolman, 65*, 29–48.

Edwards, M. (2012). Suárez in Late Scholastic Context: Anatomy, Psychology, and Authority. In B. Hill, H. Lagerlund (Eds.), *The Philosophy of Francisco Suárez* (pp. 25–37). Oxford, UK: Oxford University Press.

Ehret, C. (2017). The Flow of Powers: Emanation in the Psychology of Avicenna, Albert the Great, and Aquinas. In R. Pasnau (Ed.), *Oxford Studies in Medieval Philosophy* (Vol. 5, pp. 87–121). Oxford, UK: Oxford University Press.

Ellis, J. (1990). The Trouble with Fragrance. *Phronesis, 35(3)*, 290–302.

Elorduy, E. (1963). La accion de resultancia en Suárez. *Anales de la Cátedra Francisco Suárez, 3(1/2)*, 45–71.

Ewbank, M. B. (1987). The Route to Substance in Suarez's *Disputationes Metaphysicae*. In D. O. Dahlstrom (Ed.), *The Metaphysics of Substance. Proceedings of the American Catholic Philosophical Association* (pp. 98–111). Washington, DC: American Catholic Philosophical Association.

Fabro, C. (1938). Knowledge and Perception in Aristotelico-Thomistic Psychology. *The New Scholasticism, 12*, 337–365.

Farges, A. (1921). Deux déviations de la théorie thomiste sur l'action transitive. *Revue néo-scolastique de philosophie, 23(90)*, 163–190.

Fish, W. (2010). *Philosophy of Perception. A Contemporary Introduction*. New York/London: Routledge.

Fleming, S. (2019). *Cognitive Psychology*. Waltham Abbey Essex, UK: Ed-Tech Press.

Forigari, L. (1968). Chain of Being. In P. P. Wiener (Ed.), *Dictionary of the History of Ideas* (Vol. 1, pp. 325–335). New York: Charles Scribner's Son.

Fragenberg, T. (1991). Auditus visu praestantior: Comparison of Hearing and Vision in Charles de Bovelles's Liber de sensibus. In C. Burnett, M. Fend, P. Gouk (Eds.), *The Second Sense. Studies in Hearing and Musical Judgment from Antiquity to the Seventeenth Century* (pp. 71–94). London: The Warburg Institute, University of London.

Freeland, C. (1995). Aristotle on the Sense of Touch. In M. C. Nussbaum, A. O. Rorty (Eds.), *Essay on Aristotle's "De Anima"* (pp. 227–248). Oxford, UK: Oxford University Press.

García Cuadrado, J. A. (2005). La armonía de facultades en la gnoseología de Francisco Suárez. *Pensamiento. Revista de Investigación e Información filosófica, 71(267)*, 587–615.

García Cuadrado, J. A. (2017a). Los tratados médicos renacentistas en la gnoseología de Francisco Suárez. *Scientia et Fides, 5(1)*, 37–59.

García Cuadrado, J. A. (2017b). Suarez on Genius and Melancholy. *Vestnik SPbSU. Philosophy and Conflict Studies, 33*(2), 182–190.

Giglioni, G. (2019). Medical Approaches to the Mind in the Late Middle Ages and the Renaissance. In S. Schmid (Ed.), *Philosophy of Mind in the Late Middle Ages and Renaissance*. In R. Copenhaver, C. Shields (Eds.), *The History of the Philosophy of Mind* (Vol. 3, pp. 41–62). London/New York: Routledge.

González Ayesta, C. (2012). Sobre el "Escotismo" de Suárez: El caso de la causalidad de la especie. In A. Echavarría, J. F. Franck (Eds.), *La causalidad en la filosofía moderna: De Suárez al Kant precrítico* (pp. 9–16). Cuadernos de Anuario Filosófico, Serie Universitaria 246. Pamplona: Servicio de Publicaciones de la Universidad de Navarra.

Gracia, J. J. E. (1982). Introduction. In *Suárez on Individuation. Metaphysical Disputation V: Individual Unity and its Principle* (pp. 1–27). Milwaukee, WI: Marquette University Press.

Gracia, J. J. E. (1991). Suárez's Conception of Metaphysics: A Step in the Direction of Mentalism? *The American Catholic Philosophical Quarterly, 65*(3), 287–309.

Grahek, N. (1991). Objective and Subjective Aspects of Pain. *Philosophical Psychology, 4*(2), 249–265.

Grendler, P. F. (2019). Philosophy in Jesuit Schools and Universities. In C. Casalini (Ed.), *Jesuit Philosophy on the Eve of Modernity* (pp. 13–33). Leiden, The Netherlands/Boston, MA: Brill.

Grice, H. P. (1962). Some Remarks about the Senses. In R. J. Butler (Ed.), *Analytical Philosophy, first series* (pp. 35–54). Oxford, UK: Basil Blackwell.

Hardcastle, V. G. (2015). Perception of Pain. In M. Matthen (Ed.), *The Oxford Handbook of Philosophy of Perception* (pp. 530–541). Oxford: Oxford University Press.

Harvey, E. R. (1975). *The Inward Wits. Psychological Theory in the Middle Ages and the Renaissance*. London: The Warburg Institute, University of London.

Hasse, D. N. (2000). *Avicenna's De anima in the Latin West*. In *The Formation of a Peripatetic Philosophy of the Soul 1160–1300*. London/Turin, Italy: The Warburg Institute–Nino Aragno Editore.

Hasse, D. N. (2014). The Soul's Faculties. In R. Pasnau, C. van Dyke (Eds.), *The Cambridge History of Medieval Philosophy* (Vol. 1, pp. 305–319). Cambridge, UK: Cambridge University Press.

Hatfield, G. (1998). Attention in Early Scientific Psychology. In R. D. Wright (Ed.), *Visual Attention* (pp. 3–25). New York/Oxford, UK: Oxford University Press.

Hattab, H. (2009). *Descartes on Forms and Mechanicisms*. Cambridge, UK: Cambridge University Press.

Hattab, H. (2011). Suárez and Descartes: A Priori Arguments against Substantial Forms and the Decline of the Formal Cause. *Studia Neoaristotelica, 8*(2), 143–162.

Hattab, H. (2017). The Metaphysics of Substantial Forms. In H. Lagerlund, B. Hill (Eds.), *The Routledge Companion to Sixteenth-Century Philosophy* (pp. 436–457). New York/London: Routledge.

Heider, D. (2004). F. Suárez and J. Senftleben: Reifikace první látky. In *Slánské hovory: Španělsko*, 37–42. Slaný.

Heider, D. (2005). Suárez über das Individuationsprinzip der Akzidenten in dem Licht des thomistischen Lösung. *Acta Commeniana et Historica, 19*, 71–90.

Heider, D. (2008). Suárez on Material Substance. Reification of Intrinsic Principles and the Unity of Material Composites. *Organon F, 15*(4), 423–438.

Heider, D. (2009). The Nature of Suárez's Metaphysics. *Disputationes Metaphysicae* and Their Main Systematic Strains. *Studia Neoaristotelica, 6*(1), 99–110.

Heider, D. (2011a). *Suárez a jeho metafyzika. Od pojmu jsoucna přes transcendentální jednotu k druhům transcendentální jednoty*. Praha: Filosofia.

Heider, D. (2011b). Reifikace hylemorfických principů, substanciální jednota člověka a nesmrtelnost lidské duše ve filosofii F. Suáreze. In T. Nejeschleba, V. Němec, M. Recinová

(Eds.), *Pojetí člověka v dějinách a současnosti filosofie* (pp. 139–152). Brno, Czech Republic: CDK.

Heider, D. (2013). Význam transsubstanciačního pojetí eucharistie pro Suárezovu ontologii spojité kvality. In P. Ambros (Ed.), *Tajemství eucharistie: Pascha, oběť, iniciace a spiritualita, 2013* (pp. 99–121). Olomouc, Czech Republic: Refugium.

Heider, D. (2014). *Universals in Second Scholasticism. A Comparative Study with Focus on the Theories of Francisco Suárez S.J. (1548–1617), João Poinsot O.P. (1589–1644) and Bartolomeo Mastri da Meldola O.F.M. Conv. (1602–1673)/Bonaventura Belluto O.F.M. Conv. (1600–1676)* (Bochumer Studien zur Philosophie 54). Amsterdam, The Netherlands/Philadelphia, PA: John Benjamins Publishing Company.

Heider, D. (2015a). Suárez on the Metaphysics and Epistemology of Universals. In V. Salas, R. Fastiggi (Eds.), *A Companion to Francisco Suárez* (pp. 164–191). Leiden, The Netherlands/Boston, MA: Brill.

Heider, D. (2015b). Idea řádu a metodologie Suárezovy psychologie. In D. Heider, L. Novák, J. Samohýl (Eds.), Pluralita tradic: od antiky po novověk (pp. 103–117). In *Studia Neoaristotelica: Supplementum II*.

Heider, D. (2016a). Aquinas on Sensitive Appetitive Powers and Suárez's Reductionism in *De anima*. In I. H. Zorroza (Ed.), *Las pasiones y las virtudes en la época de "El Greco"* (pp. 129–140). Pamplona, Spain: Cuadernos de pensamiento Español, Universidad de Navarra.

Heider, D. (2016b). Introduction. In D. Heider (Ed.), *Cognitive Psychology in Early Jesuit Scholasticism* (pp. 1–11). Neunkirchen-Seelscheid, Germany: Editiones scholasticae.

Heider, D. (2016c). The Notitia Intuitiva and Notitia Abstractiva of External Senses in Second Scholasticism: Suárez, Poinsot and Francisco de Oviedo. *Vivarium, 54(2–3)*, 173–203.

Heider, D. (2016d). Suárez on the Lower External Senses. *Filosofický časopis,* Special Issue: Explorations in Late Scholasticism (P. Dvořák, T. Machula, Eds.), 81–102.

Heider, D. (2017a). The Internal Sense(s) in Early Jesuit Scholasticism. *Filosofický časopis*, Special issue: Perception in Scholastics and Their Interlocutors (D. Heider, L. Lička, M. Otisk, Eds.), 133–148.

Heider, D. (2017b). Suárez on Visual Theory. *Scientia et Fides*, *5(1)*: The Rational Psychology of Francisco Suárez (J. Á. García Cuadrado, Ed.), 61–84.

Heider, D. (2017c). Suárez on Sound and Hearing. *Anuario Filosófico*, *50(2)*: Escolástica y Mundo Moderno (V. S. Santacruz, I. Zorroza, Eds.), 323–344.

Heider, D. (2017d). Suárez on the Functional Scope of the Imaginative Power. *Cauriensia. Revista anual de ciencias eclesiásticas, 12*, 135–152.

Heider, D. (2018). Suárez on Pain and Touch. *Pensamiento. Revista de Investigación e Información filosófica, 74*(279 Extra): Francisco Suárez (1548–1617): ¿un mero antecedente de la modernidad o una variante posible? (I. S. del Río, J. A. Senent de Frutos, Eds.), 75–90.

Heider, D. (2019a). Psychology and the Theory of Knowledge. In C. Casalini (Ed.), *Jesuit Philosophy on the Eve of Modernity* (pp. 115–134). Leiden, The Netherlands/Boston, MA: Brill.

Heider, D. (2019b). Late Scholastic Debates about External and Internal Senses: In the Direction of Francisco Suárez (1548–1617). In S. Schmid (Ed.), *Philosophy of Mind in the Late Middle Ages and Renaissance*. In R. Copenhaver, C. Shields (Eds.), *The History of the Philosophy of Mind* (Vol. 3, pp. 165–184). London/New York: Routledge.

Heider, D. (2019c). Francisco Suárez's Metaphysics of Cognitive Acts. In R. A. Maryks, J. A. Senent de Frutos (Eds.), *Francisco Suárez (1548–1617). Jesuits and the Complexities of Modernity* (pp. 23–45). Leiden, The Netherlands/Boston, MA: Brill.

Heider, D. (2020). ¿Fue Suárez un representacionalista de la percepción? In M. Santiago de Carvalho, M. Lázaro Pulido, S. Guidi (Eds.), *Francisco Suárez: Metaphysics, Politics and Ethics* (pp. 263–284). Coimbra, Portugal: Coimbra University Press.

Heider, D., Machula, T. (2019). Sentidos internos, appetito sensitivo y virtudes cardinales en Cayetano y Suárez. *Estudios Filosóficos, 68*, 258–270.

Hernandéz, M. C. (1948). La intencionalidad en la Filosofía de Francisco Suárez. *Revista de Filosofía, 7*, 767–791.
Hilbert, D. R. (2015). Vision. In M. Matthen (Ed.), *The Oxford Handbook of Philosophy of Perception* (pp. 257–273). Oxford, UK: Oxford University Press.
Kaczmarek, L. (1990). Vitalis immutatio. Erkundungen zur erkenntnispsychologischen Terminologie der Spätscholastik. In A. Heinekamp, W. Lenzen, M. Schneider (Eds.), *Mathesis rationis: Festschrift für Heinrich Schepers* (pp. 189–206). Münster, Germany: Nodus.
Kemp, S., Fletcher, G. J. O. (1993). The Medieval Theory of the Inner Senses. *The American Journal of Psychology, 106(4)*, 559–576.
King, P. (2002). Late Scholastic Theories of the Passions. Controversies in the Thomist Tradition. In H. Lagerlund, M. Yrjönsuuri (Eds.), *Emotions and Choice from Boethius to Descartes* (pp. 229–258). Dordrecht, The Netherlands: Kluwer Academic Publisher.
King, P. (2007). Rethinking Representation in the Middle Ages. In H. Lagerlund (Ed.), *Representation and Objects of Thought in Medieval Philosophy* (pp. 83–102). Aldreshot, UK: Ashgate.
Klubertanz, G. P. (1952). *The Discursive Power. Sources and Doctrines of the Vis Cogitativa According to St. Thomas Aquinas*. St. Louis, MO: The Modern Schoolman.
Knebel, S. K. (2009). Aureol and the Ambiguities of the Distinction of Reason. In S. F. Brown, T. Dewender, T. Kobusch (Eds.), *Philosophical Debates at Paris in the Early Fourteenth Century* (pp. 325–337). Leiden, The Netherlands/Boston, MA: Brill.
Knebel, S. K. (2014). Durand, Quirós, Consciousness. In A. Speer, F. Retucci, T. Jeschke, G. Guldentops (Eds.), *Durand of Saint-Pourçain and His Sentences Commentary: Historical, Philosophical, and Theological Issues* (Recherches de Theologie et Philosophie médiévales–Bibliotheca 9) (pp. 343–384). Leuven, Belgium: Peeters.
Knebel, S. K. (2016). Das Cogito und die Krise des Schulbegriffs der Reflexion. *Freiburger Zeitschrift für Philosophie und Theologie, 63(1)*, 57–88.
Knight, D. M. (1957). Suárez's Approach to Substantial Form. *The Modern Schoolman, 34*, 147–191.
Knuuttila, S. (2008). Aristotle's Theory of Perception and Medieval Aristotelianism. In S. Knuuttila, P. Kärkkäinen (Eds.), *Theories of Perception in Medieval and Early Modern Philosophy* (Studies in the History of Philosophy of Mind) (Vol. 6, pp. 1–22). Dordrecht, The Netherlands: Springer.
Knuuttila, S. (2011). Pain. In H. Lagerlund (Ed.), *Encyclopedia of Medieval Philosophy. Philosophy Between 500 and 1500* (Vol. 2: M–Z) (pp. 909–910). Dordrecht, The Netherlands: Springer.
Knuuttila, S. (2014). The Connexions Between Vital Acts in Suárez's Psychology. In L. Novák (Ed.), *Suárez's Metaphysics in its Historical and Systematic Context* (pp. 259–274). Berlin, Germany/Boston, MA: De Gruyter.
Knuuttila, S. (2015). Suárez's Psychology. In V. Salas, R. Fastiggi (Eds.), *A Companion to Francisco Suárez* (pp. 192–220). Leiden, The Netherlands/Boston, MA: Brill.
Knuuttila, S., Sihvola, J. (Eds.). (2014). *Sourcebook for the History of the Philosophy of Mind. Philosophical Psychology from Plato to Kant* (Studies in the History of Philosophy of Mind) (Vol. 12). Dordrecht, The Netherlands: Springer.
Kriegel, U. (2007). The Same-Order Monitoring Theory of Consciousness. Second Version. *Synthesis Philosophica, 22*(2), 361–384.
Kristeller, P. O. (1964). *Eight Philosophers of the Italian Renaissance*. Stanford, CA: Stanford University Press.
Lagerlund, H. (Ed.). (2010). *Rethinking the History of Scepticism: The Missing Medieval Background*. Leiden, The Netherlands/Boston, MA: Brill.
Langston, D. C. (1993). Scotus's Doctrine of Intuitive Cognition. *Synthese, 96*, 3–24.
Lázaro Pulido, M., Zorroza, I. H. (2018). Memoria peninsular de un legado para el siglo XXI: Francisco Suárez (1617–2017). *Pensamiento. Revista de Investigación e Información filosófica, 74*(279 Extra): Francisco Suárez (1548–1617): ¿un mero antecedente de la modernidad o una variante posible? (I. S. del Río, J. A. Senent de Frutos, Eds.), 299–318.

Leijenhorst, C. (2007). Cajetan and Suarez on Agent Sense: Metaphysics and Epistemology in Late Aristotelian Thought. In H. Lagerlund (Ed.), *Forming the Mind. Essays on the Internal Senses and the Mind/Body Problem from Avicenna to the Medical Enlightenment* (Studies in the History of Philosophy of Mind) (Vol. 5, pp. 237–262). Dordrecht, The Netherlands: Springer.

Leijenhorst, C. (2014). Active Perception from Nicholas of Cusa to Thomas Hobbes. In J. F. Silva, M. Yrjönsuuri (Eds.), *Active Perception in the History of Philosophy* (Studies in the History of Philosophy of Mind) (Vol. 14, pp. 167–185). Dordrecht, The Netherlands: Springer.

Leijenhorst, C. (2012). Suárez on Self-Awareness. In H. Lagerlund, B. Hill (Eds.), *The Philosophy of Francisco Suárez* (pp. 137–153). Oxford, UK: Oxford University Press.

Leinsle, U. G. (2006). *Dilinganae Disputationes. Der Lehrinhalt der gedruckten Disputationen an der Philosophischen Fakultät der Universität Dillingen 1555–1648*. Regensburg, Germany: Schnell & Steiner.

Leinsle, U. G. (2016). Ontology and Experience – Rodrigo de Arriaga and Christoph Haunold on the 'Species Sensibiles'. *Filosofický časopis*, Special Issue: Explorations in Late Scholasticism (P. Dvořák, T. Machula, Eds.), 103–118.

Lička, L. (2016). Perception and Objective Being: Peter Auriol on Perceptual Acts and Their Objects. *American Catholic Philosophical Quarterly, 90(1)*, 49–76.

Lička, L. (2017). Attention, Perceptual Content, and Mirrors: Two Medieval Models of Active Perception in Peter Olivi and Peter Auriol. *Filosofický časopis*, Special Issue: Perception in Scholastics and Their Interlocutors (D. Heider, L. Lička, M. Otisk, Eds.), 101–119.

Lička, L. (2019). What Is in the Mirror? The Metaphysics of Mirror Images in Albert the Great and Peter Auriol. In B. Glenney, J. F. Silva (Eds.), *The Senses and the History of Philosophy* (pp. 131–148). London: Routledge.

Lička, L. (2020). The Visual Process: Immediate or Successive? Approaches to the Extramission Postulate in 13th-Century Theories of Vision. In E. Băltuță (Ed.), *Medieval Perceptual Puzzles: Theories of Sense-Perception in the 13th and 14th Centuries* (pp. 73–110). Leiden, The Netherlands/Boston, MA: Brill.

Lindberg, D. C. (1976). *Theories of Vision from Al-Kindi to Kepler*. Chicago/London: The University of Chicago Press.

Lindberg, D. C. (1978). The Science of Optics. In D. C. Lindberg (Ed.), *Science in the Middle Ages* (pp. 338–368). Chicago: University of Chicago Press.

Lindberg, D. C. (1998). *Roger Bacon's Philosophy of Nature*. A Criticial Edition, with English Translation, Introduction, and Notes, of *De multiplicatione specierum* and *De speculis comburentibus*. South Bend, IN: St. Augustine's Press.

Lisska, A. J. (2016). *Aquinas's Theory of Perception: An Analytic Reconstruction*. Oxford, UK: Oxford University Press.

Lo Presti, R. (2017). (Dis)embodied Thinking & the Scale of Beings: Pietro Pomponazzi & Agostino Nifo on the "Psychic" Processes in Men & Animals. In S. Buchenau, R. Lo Presti (Eds.), *Human & Animal Cognition in Early Modern Philosophy & Medicine* (pp. 37–54). Pittsburgh, PA: University of Pittsburgh Press.

Lohr, C. H. (1976). Jesuit Aristotelianism and Sixteenth-Century Metaphysics. In H. G. Fletcher III, M. B. Schulte (Eds.), *Paradosis: Studies in Memory of Edwin A. Quain*, 203–220. New York: Fordham University Press.

López Prieto, L. J. (2006). Suárez, crocevia nella filosofia tra medioevo e modernità. *Alpha Omega, 9*, 3–38.

Lovejoy, A. O. (1936). *The Great Chain of Being: A Study of the History of an Idea*. Cambridge, MA: Harvard University Press.

Ludwig, J. (1929). *Das akausale Zusammenwirken (sympathia) der Seelenvermögen in der Erkenntnislehre des Suárez*. München, Germany: Karl Ludwig-Verlag.

Mahoney, E. P. (2000). Pseudo-Dionysius's Conception of Metaphysical Hierarchy and Its Influence on Medieval Philosophy. In T. Boiadjiev, G. Kapriev, A. Speer (Eds.), *Die Dionysius-Rezeption im Mittelalter* (pp. 429–475). Turnhout, Belgium: Brepols.

Mantovani, M. (2020). *Visio per sillogismum*: Sensation and Cognition in the 13th Century Theories of Vision. In E. Băltuţă (Ed.), *Medieval Perceptual Puzzles. Theories of Sense Perception in the 13th and 14th Centuries* (pp. 111–152). Leiden, The Netherlands/Boston, MA: Brill.

Marenbon, J. (2019). The Self. In M. Cameron (Ed.), *Philosophy of Mind in the Early and High Middle Ages*. In R. Copenhaver, C. Shields (Eds.), *The History of the Philosophy of Mind* (Vol. 2, pp. 257–278). London/New York: Routledge.

Mattens, F. (2009). Perception, Body, and the Sense of Touch: Phenomenology and Philosophy of Mind. *Husserl Studien, 25*(2), 97–120.

Matthen, M. (2015). The Individuation of the Senses. In M. Matthen (Ed.), *The Oxford Handbook of Philosophy of Perception* (pp. 567–586). Oxford, UK: Oxford University Press.

Maurer, A. A. (1990). Francis of Meyronnes's Defense of Direct Realism. In A. A. Maurer (Ed.), *Being and Knowing: Studies in Thomas Aquinas in Later Medieval Philosophers* (pp. 311–331). Toronto Canada: Pontifical Institute of Medieval Studies

Maier, A. (1967). Das Problem der "species in medio" und die neue Naturphilosophie des 14. Jahrhunderts. In A. Maier, *Ausgehendes Mittelalter. Gesammelte Aufsätze zur Geistesgeschichte des 14. Jahrhunderts* (pp. 419–451). Rome: Edizioni di storia e letteratura.

Mendía, B. (1948). Influencia de los maestros franciscanos en la psicología del conocimiento intelectual de Suárez. *Verdad y Vida, 6*, 421–453.

Menn, S. (1997). Suárez, Nominalism, and Modes. In K. White (Ed.), *Hispanic Philosophy in the Age of Discovery* (pp. 226–256). Washington, DC: The Catholic University of America Press.

Mole, C. (2017). Attention. In *The Stanford Encyclopedia of Philosophy* (Fall 2017 Edition), E. N. Zalta (Ed.), URL = <https://plato.stanford.edu/archives/fall2017/entries/attention/>.

Montagnes, B. (1968). L'axiome de continuité chez Saint Thomas. *Revue des Sciences Philosophiques et Théologiques, 52*, 201–221.

Müller, H. J. (1968). *Die Lehre vom verbum mentis in der spanischen Scholastik: Untersuchungen zur historischen Entwicklung und zum Verständnis dieser Lehre bei Toletus, den Conimbricensern und Suarez*. Münster.

Novotný, D. D. (2013). *Ens Rationis from Suarez to Caramuel: A Study in Scholasticism of the Baroque Era*. New York: Fordham University Press.

Novotný, D. D. (forthcoming 2021). The Systematic Framework of the *Disputationes metaphysicae*. *Review of Metaphysics*.

O'Callaghan, C. (2007). *Sounds*. Oxford, UK: Oxford University Press.

O'Keefe, T. (2000). Epicurus (341–271 B.C.E.). *Internet Encyclopedia of Philosophy*, URL = https://iep.utm.edu/

Park, K. (1988). The Organic Soul. In C. B. Schmitt, Q. Skinner, E. Kessler, J. Kraye (Eds.), *The Cambridge History of Renaissance Philosophy* (pp. 464–484). Cambridge, UK: Cambridge University Press.

Pasnau, R. (1997). *Theories of Cognition in the Later Middle Ages*. Cambridge, UK: Cambridge University Press.

Pasnau, R. (1999a). Olivi on Human Freedom. In A. Boureau, S. Piron (Eds.), *Pierre de Jean Olivi (1248–1298). Pensee scolastique, dissidence spirituelle et societe* (Actes du colloque de Narbonne) (pp. 15–25). Paris: Vrin.

Pasnau, R. (1999b). What Is Sound? *The Philosophical Quarterly, 49*(*196*), 310–324.

Pasnau, R. (2000). Sensible Qualities: The Case of Sound. *Journal of the History of Philosophy, 38*(*1*), 27–40.

Pasnau, R. (2002a). *Thomas Aquinas on Human Nature. A Philosophical Study of Summa Theologiae Ia 75–89*. Cambridge, UK: Cambridge University Press.

Pasnau, R. (2002b). *The Cambridge Translations of Medieval Philosophical Texts*, Vol. 3: *Mind and Knowledge*. Cambridge, UK: Cambridge University Press.

Pasnau, R. (2004). Form, Substance, and Mechanism. *The Philosophical Review, 113*(*1*), 31–88.

Pasnau, R. (2011a). *Metaphysical Themes 1274–1671*. Oxford, UK: Oxford University Press.

Pasnau, R. (2011b). Scholastic Qualities, Primary and Secondary. In L. Nolan (Ed.), *Primary and Secondary Qualities. The Historical and Ongoing Debate* (pp. 41–61). Oxford, UK: Oxford University Press.

Pasnau, R. (2017). *After Certainty. A History of Our Epistemic Ideals and Illusions*. Oxford, UK: Oxford University Press.

Pattin, A. (1988). *Pour l'historie du sens agent. La controverse entre Barthélemy de Bruges et Jean de Jandun. Ses antécédents et son évolution*. Leuven, Belgium: Leuven University Press.

Penner, S. (2020). *Bibliography of Works on Francisco Suárez, 1850–present*. Compiled by Jacob Schmutz and updated by Sydney Penner. URL = http://www.sydneypenner.ca/bib.shtml.

Perler, D. (2008a). Seeing and Judging: Ockham and Wodeham on Sensory Cognition. In S. Knuuttila, P. Kärkkäinen (Eds.), *Theories of Perception in Medieval and Early Modern Philosophy* (Studies in the History of Philosophy of Mind) (Vol. 6, pp. 151–169). Dordrecht, The Netherlands: Springer.

Perler, D. (2008b). Thought Experiments: The Methodological Function of Angels in Late Medieval Epistemology. In I. Iribarren, M. Lenz (Eds.), *Angels in Medieval Philosophical Inquiry* (pp. 143–153). Aldershot, UK: Ashgate.

Perler, D. (2004). *Theorien der Intentionalität im Mittelalter*. Frankfurt am Main, Germany: Vittorio Klostermann.

Perler, D. (2006). *Zweifel und Gewissheit: skeptische Debatten im Mittelalter*. Frankfurt am Main, Germany: Vittorio Klostermann.

Perler, D. (2009). Does God Deceive Us? Skeptical Hypotheses in Late Medieval Epistemology. In H. Lagerlund (Ed.), *Rethinking the History of Skepticism: The Missing Medieval Background* (pp. 171–192). Leiden, The Netherlands/Boston, MA: Brill.

Perler, D. (2011). What are Intentional Objects? A Controversy Among Early Scotists. In D. Perler (Ed.), *Ancient and Medieval Theories of Intentionality* (pp. 203–226). Leiden, The Netherlands/Boston, MA: Brill.

Perler, D. (2013). What are the Faculties of the Soul? Descartes and his Scholastic Background. In J. Marenbon (Ed.), *Continuity and Innovation in Medieval and Modern Philosophy. Knowledge, Mind, and Language* (pp. 9–38). Oxford, UK: Oxford University Press.

Perler, D. (2014a). Suárez on Consciousness. *Vivarium, 52(3–4)*, 261–286.

Perler, D. (2014b). Skepticism. In R. Pasnau (Ed.), *The Cambridge History of Medieval Philosophy* (pp. 384–396). Cambridge, UK: Cambridge University Press.

Perler, D. (2015). Faculties in Medieval Philosophy. In D. Perler (Ed.), *The Faculties: a History* (pp. 97–139). Oxford, UK/New York: Oxford University Press.

Perler, D. (2018). Suárez on the Metaphysics of Habits. In N. Faucher, M. Roques (Eds.), *The Ontology, Psychology and Axiology of Habits (Habitus) in Medieval Philosophy* (Historical-Analytical Studies on Nature, Mind and Action) (Vol. 7, pp. 365–384). Dordrecht, The Netherlands: Springer.

Perler, D. (2019). Mind and Method. In S. Schmid (Ed.), *Philosophy of Mind in the Late Middle Ages and Renaissance*. In R. Copenhaver, C. Shields (Eds.), *The History of the Philosophy of Mind* (Vol. 3, pp. 22–40). London/New York: Routledge.

Perler, D. (2020a). Suárez on the Unity of Material Substances. *Vivarium, 58(3)*, 143–167.

Perler, D. (2020b). *Eine Person sein. Philosophische Debatten im Spätmittelalter*. Philosophische Abhandlungen, Band 119. Frankfurt am Main: Vittorio Klostermann.

Perler, D., Schierbaum, S. (2014). *Selbstbezug und Selbstwissen: Texte zu einer mittelalterlichen Debatte*. Frankfurt am Main, Germany: Vittorio Klostermann.

Pini, G. (2011). Can God Create My Thoughts? Scotus's Case Against the Causal Account of Intentionality. *Journal of the History of Philosophy, 49/1*, 39–63.

Pink, T. (2018). Agents, Objects and Their Powers in Suarez and Hobbes. *Philosophical Explorations, 21(1)*, 3–24.

Piñon, M. (1975). The Will as Principal Agent and Primary Cause in Its Particular Psychological Sphere: The Thomistic and Suarezian View. *Philippiniana sacra, 10*(*29*), 330–348.

Poncela, G. Á. (Ed.). (2015). *La escuela de Salamanca. Filosofía y Humanismo ante el mundo moderno*. Madrid, Spain: Editorial Verbum.

Prieto López, L. J. (2016). Ockham, Suárez y Descartes: transición de las noéticas tardomedievales a la filosofia moderna. *Anales del Seminario de Historia de la Filosofía, 33*(*1*), 33–57.

Rabade Romeo, S. (2003). La metafisica de Suárez: subjetivización y dinamismo. *Cuadernos salmantimos de filosofía, 30*, 145–155.

Richardson, K. (2015). Formal Causality: Giving Being by Constituting and Completing. In J. L. Fink (Ed.), *Suárez on Aristotelian Causality* (pp. 65–84). Leiden, The Netherlands/Boston, MA: Brill.

Rinaldi, T. (1998). *Francisco Suárez. Cognitio singularis materialis: De anima* (Vestigia: Studi e strumenti di storiografia filosofica). Bari, Italy: Levante Editori.

Rode, C. (2015). *Zugänge zum Selbst: Innere Erfahrung in Spätmittelalter und Früher Neuzeit*. Münster, Germany: Aschendorff Verlag.

Ronchi, V. (1970). *The Nature of Light: An Historical Survey* (V. Barocas, Trans.). London: Heinemann.

Ross, P. (2015). Primary and Secondary Qualities. In M. Matthen (Ed.), *The Oxford Handbook of Philosophy of Perception* (pp. 405–421). Oxford, UK: Oxford University Press.

Rösseler, H. (1922). Die Entstehung der sinnlichen Wahrnehmung und der Verstandeserkenntnis nach Suarez. *Philosophisches Jahrbuch der Görres-Gesellschaft, 35*(*3*), 185–198.

Rozemond, M. (2012). Unity in the Multiplicity of Suárez's Soul. In H. Lagerlund, B. Hill (Eds.), *The Philosophy of Francisco Suárez* (pp. 154–172). Oxford, UK: Oxford University Press.

Rozemond, M. (2015). Holenmerism (Holenmerianism). In L. Nolan (Ed.), *The Cambridge Descartes Lexicon* (pp. 366–368). Cambridge, UK: Cambridge University Press.

Salatowsky, S. (2006). *De anima. Die Rezeption der aristotelischen Psychologie im 16. und 17. Jahrhundert*. Amsterdam, The Netherlands/Philadelphia, PA: John Benjamins Publishing Company.

Sander, C. (2017). For Christ's Sake. Pious Notion of the Human and Animal Body in Early Jesuit Philosophy and Theology. In S. Buchenau, R. Lo Presti (Eds.), *Human & Animal Cognition in Early Modern Philosophy & Medicine* (pp. 55–73). Pittsburgh, PA: University of Pittsburgh Press.

Santiago De Carvalho, M. (2019). Álvares, Baltasar. In *The 'Cursus Conimbricensis'* (coord. by Mário Santiago de Carvalho), URL: http://www.conimbricenses.org/encyclopedia/baltasar-alvarez/.

Schmal, D. (2020). Virtual Reflection: Antoine Arnauld on Descartes' Concept of *Conscientia*. *British Journal for the History of Philosophy, 28*(*4*), 714–734.

Schmaltz, T. M. (2008). *Descartes on Causation*. Oxford, UK: Oxford University Press.

Schmid, S. (2015a). Efficient Causality: The Metaphysics of Production. In J. L. Fink (Ed.), *Suárez on Aristotelian Causality* (pp. 84–121). Leiden, The Netherlands/Boston, MA: Brill.

Schmid, S. (2015b). Faculties in Early Modern Philosophy. In D. Perler (Ed.), *The Faculties: A History* (pp. 150–197). Oxford, UK: Oxford University Press.

Schmitt, C. B. (1973). Towards a Reassessment of Renaissance Aristotelianism. *History of Science, 11*, 159–193.

Schmitt, C. B. (1983). *Aristotle and the Renaissance*. Cambridge, MA/London: Harvard University Press.

Scorraille, R. de. (1911). *François Suarez de la Compagnie de Jésus* (2 Vols.). Paris: P. Lethielleux.

Shields, C. (2012). The Reality of Substantial Form: Suárez. Metaphysical Disputation XV. In D. Schwartz (Ed.), *Interpreting Suárez. Critical Essays* (pp. 39–61). Cambridge, UK: Cambridge University Press.

Shields, C. (2013). The Unity of the Soul in Francisco Suárez. In R. L. Friedman, J.-M. Counet (Eds.), *Medieval Perspectives on Aristotle's De anima* (pp. 359–382). Louvain, Belgium: Éditions de l'institut supérieur de philosophie & Peeters.

Shields, C. (2014). Virtual Presence: Psychic Mereology in Francisco Suárez. In K. Corcilius, D. Perler (Eds.), *Partitioning the Soul. Debates from Plato to Leibniz* (pp. 199–218). Berlin, Germany: De Gruyter.

Siewert, Ch. (2017). Consciousness and Intentionality. In *The Stanford Encyclopedia of Philosophy* (Spring 2017 Edition), E. N. Zalta (Ed.), URL = https://plato.stanford.edu/archives/spr2017/entries/consciousness-intentionality/.

Siegel, R. E. (1970). *Galen on Sense Perception*. Basel, Switzerland/New York: S. Karger.

Silva, J. F., Yrjönsuuri, M. (Eds.). (2014). *Active Perception in the History of Philosophy. From Plato to Modern Philosophy* (Studies in the History of Philosophy of Mind) (Vol. 14). Dordrecht, The Netherlands: Springer.

Silva, J. F., Thörnqvist, T. (2019). Introduction: Assimilation and Representation in Medieval Theories of Cognition. *Vivarium, 57(3–4)*, 223–243.

Silva, J. F. (2014a). Augustine on Active Perception. In J. F. Silva, M. Yrjönsuuri (Eds.), *Active Perception in the History of Philosophy. From Plato to Modern Philosophy* (Studies in the History of Philosophy of Mind) (Vol. 14, pp. 79–98). Dordrecht, The Netherlands: Springer.

Silva, J. F. (2014b). Medieval Theories of Active Perception: An Overview. In J. F. Silva, M. Yrjönsuuri (Eds.), *Active Perception in the History of Philosophy. From Plato to Modern Philosophy* (Studies in the History of Philosophy of Mind) (Vol. 14, pp. 117–146). Dordrecht, The Netherlands: Springer.

Silva, J. F. (2017). Perceptual Judgment in Late Medieval Philosophy. *Filosofický časopis*, Special Issue: Perception in Scholastics and Their Interlocutors (D. Heider, L. Lička, M. Otisk, Eds.), 29–60.

Silva, J. F. (2018). Intentionality in Medieval Augustinianism. *Phänomenologische Forschungen 2018–2*: Modes of Intentionality. Phenomenological and Medieval Perspectives (J. Müller, M. Summa, Eds.), 25–44.

Silva, J. F. (2020). *Stop Making Sense(s): Some Late Medieval and Very Late Medieval Views on Faculty Psychology*. In S. N. Mousavian, J. L. Fink (Eds.), *The Internal Senses in the Aristotelian Tradition* (Studies in the History of Philosophy of Mind) (Vol. 22, pp. 45–67). Dordrecht, The Netherlands: Springer.

Simmons, A. (1999). Jesuit Aristotelian Education: The De anima Commentaries. In J. W. O'Malley et al. (Eds.), *The Jesuits: Culture, Learning and the Arts, 1540–1773* (pp. 522–537). Toronto, Canada: Toronto University Press.

Sluhovsky, M. (2013). St. Ignatius of Loyola's Spiritual Exercises and Their Contribution to Modern Introspective Subjectivity. *The Catholic Historical Review, 99(4)*, 649–674.

Smith, A. M. (2001a). *Alhacen's Theory of Visual Perception*. A Critical Edition, with English Translation and Commentary, of the First Three Books of Alhacen's De Aspectibus, the Medieval Latin Version of Ibn al Haytham's Kitāb al-Manāẓir. Vol. 1: Introduction and Latin Text. Philadelphia, PA: American Philosophical Society.

Smith, A. M. (2001b). *Alhacen's Theory of Visual Perception*. A Critical Edition, with English Translation and Commentary, of the First Three Books of Alhacen's De Aspectibus, the Medieval Latin Version of Ibn al Haytham's Kitāb al-manāẓir. Vol. 2: English Translation. Philadelphia, PA: American Philosophical Society.

Smith, B. C. (2015). The Chemical Senses. In M. Matthen (Ed.), *The Oxford Handbook of Philosophy of Perception* (pp. 314–352). Oxford, UK: Oxford University Press.

Smith, C. U. M., Frixione, E., Finger, S., Clower, W. (2012). Animal Spirit in an Age of Faith. In C. U. M. Smith, E. Frixione, S. Finger, W. Clower (Eds.), *The Animal Spirit Doctrine and the Origins of Neurophysiology* (pp. 71–94). Oxford, UK: Oxford University Press.

Söchting, J. (2019). Encarnación y subsistentica en las Disputationes metafísicas de Francisco Suárez: Algunas cuestiones en torno a los fundamentos de la modernidad. In R. A. Maryks, J. A. Senent-De Frutos (Eds.), *Francisco Suárez (1548–1617): Jesuits and the Complexities of Modernity* (pp. 154–177). Leiden, The Netherlands/Boston, MA: Brill.

Solmsen, F. W. (1961). Greek Philosophy and the Discovery of the Nerves. *Museum Helveticum, 18(4)*, 169–197.

Sorabji, R. (1971). Aristotle on Demarcating the Five Senses. *The Philosophical Review, 80(1)*, 55–79.

Sorabji, R. (1972). *Aristotle on Memory*. London: Gerald Duckworth.

South, J. B. (2001a). Suárez and the Problem of External Sensation. *Medieval Philosophy and Theology, 10*(2), 217–240.

South, J. B. (2001b). Suárez on Imagination. *Vivarium, 39(1)*, 119–158.

South, J. B. (2002). Singular and Universal in Suárez's Account of Cognition. *The Review of Metaphysics, 55(4)*, 785–823.

South, J. B. (2012). Immortality, and the Soul's Dependence on the Body. In B. Hill, H. Lagerlund (Eds.), *The Philosophy of Francisco Suarez* (pp. 121–136). Oxford, UK: Oxford University Press.

South, J. B. (2017). Aspects of Intentionality in Two 16th Century Aristotelians. *Gregorianum, 4*, 725–741.

South, J. B. (2018). Some Themes in Suárez's Account of the Separated Soul. *Pensamiento. Revista de Investigación e Información filosófica, 74*(279 Extra): Francisco Suárez (1548–1617): ¿un mero antecedente de la modernidad o una variante posible? (I. S. del Río, J. A. Senent de Frutos, Eds.), 63–74.

Spruit, L. (1995). *Species Intelligibilis: From Perception to Knowledge*. Vol. 2: Renaissance Controversies, Later Scholasticism and the Elimination of the Intelligible Species in Modern Philosophy. Leiden, The Netherlands/Boston, MA: Brill.

Spruit, L. (2008). Renaissance Views of Active Perception. In S. Knuuttila, P. Kärkkäinen (Eds.), *Theories of Perception in Medieval and Early Modern Philosophy* (Studies in the History of Philosophy of Mind) (Vol. 6, pp. 203–224). Dordrecht, The Netherlands: Springer.

Szlezák, T. A. (1976). Unsterblichkeit und Trichotomie der Seele im zehnten Buch der Politeia. *Phronesis, 21(1)*, 31–58.

Tachau, K. H. (1988). *Vision and Certitude in the Age of Ockham. Optics, Epistemology and the Foundations of Semantics 1250–1345*. Leiden, The Netherlands/Boston, MA: Brill.

Tachau, K. H. (1993). What Senses and Intellect Do: Argument and Judgment in Late Medieval Theories of Knowledge. In K. Jakobi (Ed.), *Argumentationstheorie. Scholastische Forschungen zu den logischen und semantischen Regeln korrekten Folgerns* (pp. 653–670). Leiden, The Netherlands/Boston, MA: Brill.

Taylor, C. C. W. (1999). *The Atomists: Leucippus and Democritus. Fragments*. Toronto, Canada/Buffalo, NY/London: University of Toronto Press.

Tellkamp, J. A. (2006). Aquinas on Intentions in the Medium and in the Mind. *Proceedings of the American Catholic Philosophical Association, 80*, 275–289.

Tellkamp, J. A. (2012). Francisco Suárez on the Intentional Species. *Quaestio, 12*, 3–23.

Toivanen, J. (2013a). Perceptual Self-Awareness in Seneca, Augustine, and Olivi. *Journal of the History of Philosophy, 51(3)*, 355–382.

Toivanen, J. (2013b). *Perception and the Internal Senses: Peter of John Olivi on the Cognitive Functions of the Sensitive Soul*. Leiden, The Netherlands/Boston, MA: Brill.

Tropia, A. (2014). Scotus and Suárez on Sympathy: The Necessity of the "Connexio Potentiarum" in the Present State. In L. Novák (Ed.), *Suárez's Metaphysics in Its Historical and Systematic Context* (pp. 275–292). Berlin, Germany/Boston, MA: De Gruyter.

Tropia, A. (2020). Three Jesuit Accounts of Cognition: Differences and Common Ground in the De Anima Commentaries by Maldonado, Toledo and Dandini (1564–1610). In V. Decaix, A. M. Mora-Márquez (Eds.), *Active Cognition. Challenges to an Aristotelian Tradition* (Studies in the History of Philosophy of Mind) (Vol. 23, pp. 103–131). Dordrecht, The Netherlands: Springer.

Tuttle, J. (2016). Suárez's Non-Reductive Theory of Efficient Causation. In R. Pasnau (Ed.), *Oxford Studies in Medieval Philosophy* (pp. 125–158). Oxford, UK: Oxford University Press.

Van Gulick, R. (2018). Consciousness. In *The Stanford Encyclopedia of Philosophy* (Spring 2018 Edition), E. N. Zalta (Ed.), URL = <https://plato.stanford.edu/archives/spr2018/entries/consciousness/>.

Wells, N. J. (1993). Esse Cognitum and Suárez Revisited. *American Catholic Philosophical Quarterly, 67(3)*, 339–348.

Wittmann, M. (1987). *Vox atque sonus. Studien zur Rezeption der Aristotelischen Schrift "De anima" und ihre Bedeutung für die Musiktheorie*. Pfaffenweiler, Germany: Centaurus-Verlagsgesellschaft.

Wolf-Devine, C. (1993). *Descartes on Seeing. Epistemology and Visual Perception*. Carbondale, IL/Edwardsville, IL: Southern Illinois University Press.

Wolfson, H. A. (1935). The Internal Senses in Latin, Arabic, and Hebrew Philosophic Texts. *Harvard Theological Review, 38*(*2*), 69–133.

Woods, R. (1982). Adam Wodeham on Sensory Illusions with an Edition of 'Lectura Secunda,' Prologus, Quaestio 3. *Traditio, 38*, 213–252.

Woods, R. (1987). Intuitive Cognition and Divine Omnipotence: Ockham in Fourteenth Century Perspective. In A. Hudson, M. Wilks (Eds.), *From Ockham to Wyclif* (pp. 51–61). Oxford, UK: Blackwell.

Wu, W. (2014). *Attention*. London/New York: Routledge.

Wu, T. (2017). The Ontological Status of the Body in Aquinas's Hylomorphism. *Studia Neoaristotelica, 14*(*1*), 5–38.

Wuellner, B. (2011). *Dictionary of Scholastic Philosophy*. Heusenstamm, Germany: Editiones-scholasticae.

Yrjönsuuri, M. (2006). Types of Self-Awareness in Medieval Thought. In V. Hirvonen, T. J. Holopainen, M. Tuominen (Eds.), *Mind and Modality: Studies in the History of Philosophy in Honour of Simo Knuuttila* (pp. 153–169). Leiden, The Netherlands/Boston, MA: Brill.

Yrjönsuuri, M. (2008). Perceiving One's Own Body. In S. Knuuttila, P. Kärkkäinen (Eds.), *Theories of Perception in Medieval and Early Modern Philosophy* (Studies in the History of Philosophy of Mind) (Vol. 6, pp. 101–116). Dordrecht, The Netherlands: Springer.

Yrjönsuuri, M. (2014). Seeing Distance. In J. F. Silva, M. Yrjönsuuri (Eds.), *Active Perception in the History of Philosophy. From Plato to Modern Philosophy* (Studies in the History of Philosophy of Mind) (Vol. 14, pp. 187–206). Dordrecht, The Netherlands: Springer.

Author Index

D. Heider, *Aristotelian Subjectivism: Francisco Suárez's Philosophy of Perception*, Studies in the History of Philosophy of Mind 28,
https://doi.org/10.1007/978-3-030-67341-3

Subject Index

D. Heider, *Aristotelian Subjectivism: Francisco Suárez's Philosophy of Perception*, Studies in the History of Philosophy of Mind 28,
https://doi.org/10.1007/978-3-030-67341-3

The manufacturer's authorised representative in the EU is Springer Nature Customer Service Centre GmbH, Europaplatz 3, 69115 Heidelberg, Germany. If you have any concerns regarding our products, please contact ProductSafety@springernature.com

Printed and bound by CPI Group (UK) Ltd, Croydon, CR0 4YY
22/09/2026
02220945-0011